PETROLEUM GEOCHEMISTRY

PETROLEUM GEOCHEMISTRY

Dr. D. Satyanarayana
Visiting Professor, Delta Studies Institute &
Former Professor of Marine Chemistry
Andhra University, Visakhapatnam

2011
DAYA PUBLISHING HOUSE
Delhi - 110 035

© 2011, D. SATYANARAYANA
ISBN 9789351241409

Published by	:	**Daya Publishing House** **A Division of** **Astral International Pvt. Ltd.** **– ISO 9001:2008 Certified Company –** 4760-61/23, Ansari Road, Darya Ganj, New Delhi-110 002 Ph. 011-43549197, 23278134 E-mail: info@astralint.com Website: www.astralint.com
Laser Typesetting	:	**Classic Computer Services** Delhi - 110 035
Printed at	:	**Chawla Offset Printers** Delhi - 110 052

PRINTED IN INDIA

Acknowledgements

I am grateful to the following publishers, Authors/Editors of books for according permission to use their published illustrations.

☆ M/S Academic Press, R.C. Selley. Elements of Petroleum Geology (1998); J. Brooks (Ed). Organic Maturation Studies and Fossil Fuel Exploration (1981).

☆ M/S Applied Science Publishers, G.D. Hobson (Ed). Developments in Petroleum Geology, Vol. I (1977) and Vol. II (1980).

☆ M/S Burgess Publishing Company, Douglas Waples. Organic Geochemistry for exploration Geologists (1981).

☆ M/S Chapman & Hall, Steven A Tedesco. Surface Geochemistry in Petroleum Exploration (1995).

☆ M/S Directorate General of Hydrocarbons (DGH). India–Petroleum Exploration and production Activities–2005-2006.

☆ M/S Elsevier Publishers, E.C Donaldson (Ed). Enhanced oil Recovery I : Fundamentals and analysis (1985).

☆ M/S Indian Petroleum Publishers, S.K. Biswas et al., (Eds). Proceedings of 2nd Seminar on Petroliferous Basins of India. Vol. I and Vol. II (1993); Lakshman Singh. Oil and Gas Fields in India (2000).

☆ M/S International Human Resource Development Corporation (IHRD), Douglas, W. Waples. Geochemistry in Petroleum Exploration (1985).

☆ M/S Kluwer Academic Publishers, M.D. Max (Ed). Natural Gas Hydrates (2000).

☆ M/S Springer–Verlag, B.P Tissot and D.H. Welte. Petroleum Formation and Occurrence (1984).

☆ M/S. Unwin Hyman Publishers, F.K. North. Petroleum Geology, 2nd edition (1990).

☆ M/S Freeman and Company, J.M. Hunt. Petroleum Geochemistry and Geology, 2nd edition (1996); A.I. Levorsen. Geology of Petroleum, 2nd edition (1967).

D. Satyanarayana

Foreword

Petroleum Geochemistry is a branch of Organic Geochemistry that deals with the study of distribution, composition and constitution of petroleum, its constituents and its precursors in sedimentosphere at gross and molecular levels to define principles of occurrence and origin of petroleum in sedimentary basins, the ubiquitous natural habitats of economically attractive accumulations.

The vicissitudes of basin evolutionary history of each sedimentary basin are complex and no two points along dip or strike theoretically have identical depositional conditions. No technique of petroleum exploration is, therefore, equally sensitive and accurate in space and time. Petroleum exploration and exploitation is thus both knowledge and technology intensive and is essentially a multidisciplinary task where an inadequacy of one discipline or technique is covered by strengths of one or other discipline or technique or set of disciplines or techniques. While geology and geophysics are excellent in unraveling the types of rocks, their physical attributes, structural attitudes and their distribution and superposition in space and time in a basin, they are unable to deal directly with aspects of origin, migration and accumulation of petroleum.

One of the imperatives of the industry is to have knowledge on aspects of origin and occurrence of petroleum specific to each basin or part thereof. This is essential to ensure reduction of risk in petroleum exploration. Petroleum Geochemistry helps exploration and development effort in each phase, right from reconnaissance and stage of general assessment to the final stage of extraction of the last producible content of hydrocarbons in a locale of accumulation, by providing geochemical concepts and data sets on various aspects of origin, migration and accumulation of petroleum specific to each basin under exploration and development. Tremendous developments have taken place in acquisition of petroleum geochemical data since the nineteen sixties through availability of robust equipments and methods of chemical analysis. It is now possible to have practically real time geochemical data to be integrated timely for making an exploration or development decision.

Synergy amongst Petroleum Geology, Petroleum Geochemistry and Geophysics has become an accepted practice for laying down risk based priorities in a portfolio of exploration and development opportunities. The synergy also leads to working out manageable investments to initialize an E&P activity with cash flows that can be deployed for further exploration and development efforts while assuring market competitive returns during the entire cycle of any given exploration and development activity.

Dr. D. Satyanarayana of the Delta Studies Institute of Andhra University has done yeoman's service to Indian academia to bring out a maiden Indian text book on Petroleum Geochemistry. This book will be found useful to graduate and postgraduate students taking courses in petroleum geology, petroleum exploration and petroleum engineering. The book touches upon authentically all aspects of petroleum geochemistry that are relevant to exploration and development effort.

In India, Oil and Natural Gas Corporation commenced undertaking petroleum geochemical studies in 1957 and established formal laboratories in 1960. ONGC Limited has since 1963 best in class research facilities in Petroleum Geochemistry at Keshav Dev Malviya Institute of Petroleum Exploration (KDMIPE) along with a strong Basin Studies Group at Dehradun. ONGC Limited also has, since 1985, standard data acquisition laboratories in all operational areas. Oil India has modern Petroleum Geochemistry laboratories since 1990. National Geophysical Research Institute has a National Centre of Excellence in Surface Geochemical Prospecting backed by Oil Industries Development Board and Director General Hydrocarbons since 2001-02 which caters to the requirements of geochemical data acquisition of various private sector oil companies and DGH to help evaluate prospectively of various acreages and exploration blocks. All private sector and joint venture major upstream oil companies profusely integrate geochemical data in exploration decision making. This book of Dr. Satyanarayana is therefore has great practical value for the upstream petroleum industry in general and especially for India where large acreages still remain explored, where deep and ultra deep water exploration has just begun and where the ageing oil fields are being operated for maximizing recovery factors.

To students with enquiring and creative minds and aptitude for innovation, the book surely would inspire to undertake basic and applied research. Dr. Satyanarayana joins the illustrious eminent academicians of the world, who brought out books born of lecture notes meant to disseminate in depth knowledge of Petroleum Geochemistry for geoscientists actively engaged in upstream activities.

The book is going to be a prized possession of academia in geosciences and of all the scientists and engineers of the upstream petroleum industry who are desirous to contribute their mite towards sustaining and augmenting global energy security through long time availability of petroleum.

Kuldeep Chandra
Former Executive Director R&D ONGC Limited,
and
Chairman, Afro-Asian Association of Petroleum Geochemists

Dehradun

Preface

During the last three decades there has been a rapid progress in petroleum geochemistry. The aim of this book is to cover the advances in a comprehensive way providing a background for understanding the basic concepts and principles. It is designed to develop principles of petroleum geochemistry and emphasise its applications to hydrocarbon exploration. Several geochemical and analytical techniques are described along with their relative merits and limitations so that an appropriate technique can be selected in a particular exploration programme. Application of principles of petroleum geochemistry to common exploration programmes are lucidly brought out by introducing worked out examples and case studies.

The book is written primarily for postgraduate students of earth sciences and graduate students of engineering taking courses in petroleum science or engineering. It is also useful for those working in oil industry dealing with exploration and related fields. Each geochemical concept is explained in detail prior to its application. Adequate references are cited in the text. Exhaustive list of books, memoirs, papers published in journals and proceeding of World Congress, International Seminars and Symposia are included in the bibliography for further reading. The contents are arranged in the following sequence so as to interpret geochemical data of varied reliability and to unravel complex geochemical processes involved in petroleum exploration and production.

☆ Role of petroleum geochemistry in prospect identification, prioritisation and risk reduction in hydrocarbon exploration and production. Composition, properties and genetic classification of crude oil and natural gas.

☆ Occurrence of petroleum in sedimentary basins–Basin type, classification and hydrocarbon richness. Resource estimation and production potential of typical Indian petroliferous basins of proven commercial production.

☆ The origin of Petroleum-Inorganic (abiogenic) and organic (biogenic) theories and their relative merits and limitations. Polymerisation theory of petroleum formation and its implications.

☆ Geochemical processes involved in generation, migration and accumulation of hydrocarbons in sedimentary basins. Factors effecting them and optimum conditions for commercial production of petroleum.

☆ Application of geochemical methods for the study of organic maturation, evaluation of stratigraphic units and distribution of organo facies for delineation of hydrocarbon kitchens. Principles of source rock evaluation. Typical examples and case studies.

☆ Geochemical surface prospecting of hydrocarbons to identify anomalous areas that prioritise targets for future exploration. Relative merits and limitations of surface geochemical methods.

☆ Application of hydrogeochemical surveys in exploration. Genetic indicators for delineation of oil wells. Role of oil field waters in enhanced oil recovery operations, and *in situ* oil degradation in reservoirs. Implications of scale formation and corrosion in petroleum exploration and production operations.

☆ Role of biomarkers for characterization and evaluation of source organo facies and depositional environments. Application of high resolution geochemical techniques such as GC-MS, CF-IRMS, MRM-GC-MS for molecular level studies involving thermal cracking of light oil and condensates; oil-oil, oil-source rock correlations to understand migration pathways; petroleum system and paleodepositional environments at micro level.

☆ Geochemical modeling of hydrocarbon generation based on kinetics of kerogen degradation. Integration of the model with geological history of the basin to delineate prospective areas. Validity of the model and its application for evaluation of petroleum potential and determination of timing of its formation for comparison with the age of traps.

☆ Application of geochemical techniques to unconventional petroleum resources such as shale gas and oil shale; bituminous sands; basin centered (tight) gas sands; coal bed methane (CBM); gas hydrates. Study of their characteristics, depositional environments, resource potential, exploration and production strategies, and environmental concerns. Global and Indian unconventional petroleum resources.

I acknowledge with thanks the authorities of Andhra University for offering an honorary professorship in Delta Studies Institute (DSI) which prompted me to write this book. It evolved from a series of lectures delivered on petroleum geochemistry to the students of M.Tech. Petroleum Exploration and discussions with several distinguished visiting faculty members. I take this opportunity to thank Prof. D. Rajasekhara Reddy, Director (DSI) for offering facilities, and to Sri T. Karunakarudu and V. Jayasundar Reddy (Teaching Assistants) for their assistance in the institute. My thanks are also due to several visiting faculty of DSI and former scientists of ONGC– Sri P.V. Ramana Rao, GGM; Sri S.S. Yalamarty, GGM & Basin Manager; Sri P.V. Ramana, G.G.M & Director, IRS; Sri G.S. Chari, D.G.M; Sri B.V. Rao, D.G.M; Dr. P.V.L.P. Babu, D.G.M for their encouragement.

I am particularly grateful to Dr. Kuldeep Chandra, Former Executive Director, ONGC and President of Afro-Asian Association of Petroleum Geochemists, for offering several constructive suggestions and improving the quality of the text. I am very much thankful and obliged for his kind gesture of writing the Foreword.

My special thanks are due to Prof. V.R.R.M. Babu, Former Professor of Geology and Geoengineering; Prof. U. Muralikrishna, Former Professor of Chemistry of Andhra University; Sri K. Ananta Krishna, Former DGM, ONGC for critically going through the text and offering valuable suggestions. I am thankful to Prof. G.S.Roonwal, Former Professor of Geology, Delhi University; Dr. V.V. Sarma, Scientist G, Regional Centre of NIO, Visakhapatnam; Dr. I. Nageswara Rao, Research Scientist, School of Chemistry; Dr. P. Prabhakar, Chief Chemist, ONGC for their assistance.

My special appreciation goes to Sri V. Hari Prasad and Sri M. Santhosh Kumar, Technical Assistants of DSI, for diligently bringing out the text including Tables and Figures.

D. Satyanarayana

Visiting Professor, Delta Studies Institute &
Former Professor of Marine Chemistry
Andhra University, Visakhapatnam

Contents

Chapter 1
Introduction

Geochemistry has been defined as the basic science concerned with chemistry of the earth, as a whole and its component parts. More specifically, it deals with the distribution of chemical elements within the earth in both time and space, and provides fundamental insight of geological processes and composition of geological materials. The discipline of petroleum geochemistry devotes itself to comprehensive understanding of chemical composition and formation of petroleum, and its precursor source organics within the geological frame work of sedimentary basins. It deals with the application of chemical principles to the study of the origin, generation, migration, accumulation and alteration of petroleum, and the use of this knowledge in exploration and production of oil and natural gas. Petroleum geochemistry has now become a useful and increasingly applied aid to exploration programmes. The basic concepts of petroleum geochemistry and the wide scope of modern geochemical techniques are of immense help in exploration at different stages from the initial basin study to investigations leading to prediction of petroleum accumulation and production evaluation.

Petroleum geochemistry is a relatively young and rapidly developing discipline. It came into prominence since 1960's because of rapid developments in analytical instruments such as Gas Chromatography (GC), Mass Spectrometry (MS), combined GC-MS with computer systems. Modern pyrolysis (Rock-Eval) technique, improved and sensitive microprocessor based analytical instruments have wide range of applications in petroleum exploration. Sensitive analytical techniques such as high resolution geochemistry technology (HRGT) have helped in molecular level characterization of complex mixtures of compounds in oil and natural gas (found in petroleum) to understand the system from source to trap. This in turn, allowed a clear assessment of migration and accumulation pathways and reduce exploration risks (Jokhan Ram, 2008; Bhowmic, 2009).

Since the early 1900's chemists were involved in the study of physical, chemical and optical properties of coals for a good understanding and prediction of their behaviour as fuels. Methods used by them such as transmitted and reflected microscopy, maceral analysis, pyrolysis and dispersed sedimentary organic matter analysis of coals made them to realise that kerogen, the principal component of coal, has a great deal in common with bituminites which are akin to petroleum. In fact, kerogen is the starting material for the formation of both coal and bitumen under different time temperature controlled and metamorphic conditions. It is now widely realised that three important factors of sedimentary organic matter containing kerogen namely the quantity, quality and thermal maturity, determine to a large extent, the oil and gas generation potential of a source rock.

Surface geochemical methods involving identification of oil or gas microseepages (particularly methane through butane) over petroleum accumulations have historically offered useful information for exploring important oil producing regions of the world. These methods coupled with geological and geophysical surveys are still widely employed for surface prospecting of oil and gas fields. Geochemical prospecting has several advantages such as: (*i*) cost and time effective tool for hydrocarbon exploration, (*ii*) helpful in carving interesting areas for exploration in large frontier basins, and (*iii*) prioritising areas for further exploration activities and extending existing fields for further exploration in petroliferous basins. Recent techniques such as remote sensing with satellites added further impetus to the surface prospecting methods.

Detailed geochemical studies largely helped to improve our understanding on (*i*) the distribution of different types and classes of organic compounds in sedimentary rocks, (*ii*) organic diagenesis and maturation, (*iii*) source rock identification, and (*iv*) effect of time and temperature on petroleum generation. Development of quantitative models of migrational processes by which bitumen moves from the source rock to reservoir, integrated basin analysis and development of conceptual basin models are the most challenging tasks of a petroleum geochemist at present.

Studies on biomarkers primarily present in oils and source rocks such as isoprenoids, diamondiods, steranes, hopanes, triterpenoids, porphyrins and carbazoles are now finding extensive application in petroleum exploration. They are widely used in oil-oil and oil-source rock correlations, reconstruction of depositional environments, of source rocks of different lithologies and age thermal, maturity of organic matter (kerogen) and in the basin modeling studies.

From the late 19th century, there was a controversy on the origin of petroleum which was basically attributed to two origins: (*i*) inorganic (abiogenic), and (*ii*) organic (biogenic). The inorganic theory assumes that petroleum is of primordial (extra-terrestrial) origin and produced by the interaction of carbonaceous chondrite (a meteorite rock) with water at high temperature and pressure within the earth's interior. On the other hand, the organic theory is based on the accumulation of organic matter in the sediments and its subsequent maturation yielding petroleum. However, the overwhelming geochemical and geological evidence during the past few decades clearly supports the organic theory.

The criteria as widely accepted now for commercial accumulation of petroleum hydrocarbons is the occurrence of:

1. Organic rich and mature source rock.
2. Favourable micro biological and hydrochemical conditions.
3. Reservoir rock with optimum porosity and permeability.
4. Impermeable cap rock for effective sealing.
5. Efficient trapping mechanism, and
6. Favourable timing of trap formation and hydrocarbon migration.

The subject of petroleum geochemistry is highly inter-disciplinary. Successful application of geochemical techniques to petroleum exploration requires close collaboration with other disciplines such as petroleum geology, geophysics, palaeontology, drilling and production engineering, in addition to physics, chemistry, mathematics, statistics, computer science and remote sensing applications.

Chapter 2

Composition and Properties of Petroleum

The composition of petroleum in a reservoir is initially controlled by the type of organic matter on the source rock; however, migration processes (particularly adsorption in the source rock) tend to even out differences so that petroleum in reservoir rocks shows a smaller range of compositional changes than do source rock extracts (Tissot and Welte, 1984). However, major changes in oil composition may be induced by alteration and maturation (particularly the latter) in the reservoir. Thus petroleum in the reservoir is frequently not uniform in composition, either in space or in time. Indeed, two adjacent wells in a single reservoir may produce petroleum with markedly different characteristics. On a molecular basis, petroleum is a complex mixture of hydrocarbon compounds, essentially comprising carbon and hydrogen, in addition to small amounts of sulphur, nitrogen, oxygen and trace quantities of metals such as vanadium, nickel, iron, copper etc. With few exceptions, the elemental composition of petroleum varies over narrow limits (Table 2.1) irrespective of its origin and place of occurrence.

Petroleum mainly exists in the gaseous or liquid state in its natural reservoir conditions. Its principal forms are: (i) natural gas which does not condense at standard temperature and pressure (STP, temperature: 15.6° C, pressure: 760 mm.Hg) conditions; (ii) condensate which is in gaseous state in the reservoir but condenses when brought to the surface; and (iii) crude oil which exists in the liquid state. The border line between gas and condensate is butane with carbon number 4. It is the heaviest among gases (C_1 to C_4) and lightest among condensates (C_4 to C_8). Compounds with carbon number C_9 to C_{15} are in general, categorised as liquids. The smallest molecule in petroleum is methane (CH_4) with a molecular weight 16, and the largest molecules are asphaltenes with molecular weights varying in thousands. Between these two extremes, there are hundreds of compounds with simple to very complex structures.

Table 2.1: Elemental Composition of Petroleum

Sl.No.	Element	Composition (Per cent)
1.	Carbon	83–87
2.	Hydrogen	10–14
3.	Nitrogen	0.1–2.0
4.	Oxygen	0.05–1.5
5.	Sulphur	0.05–6.0
6.	Metals (V, Ni, Fe, Cu, etc.)	< 1000*

* ppm (parts per million).

Source: Speight, 2007.

Petroleum also commonly known as crude or rock oil is separated into various molecular sizes by fractional distillation (based on differences in the boiling points of different constituents). A typical oil refinery will yield products in the order (smallest to largest) gas, gasoline, kerosene, diesel fuel, heavy gas oil, lubricating oil and residuum (Figure 2.1).

Petroleum hydrocarbons can be conveniently divided into two categories –saturated and un-saturated hydrocarbons. The saturated hydrocarbons in turn are divided into two types (*i*) normal and iso-alkanes commonly known as branched chain alkanes. (*ii*) cyclo alkanes commonly known as naphthenes. The second category, namely unsaturated hydrocarbons are also broadly divided into two types (*i*) alkenes, commonly known as olefins, and (*ii*) arenes, commonly known as aromatics. These groups show wide differences in relative abundances and in their contribution to the various boiling point fractions (Figure 2.1). The straight chain and branched

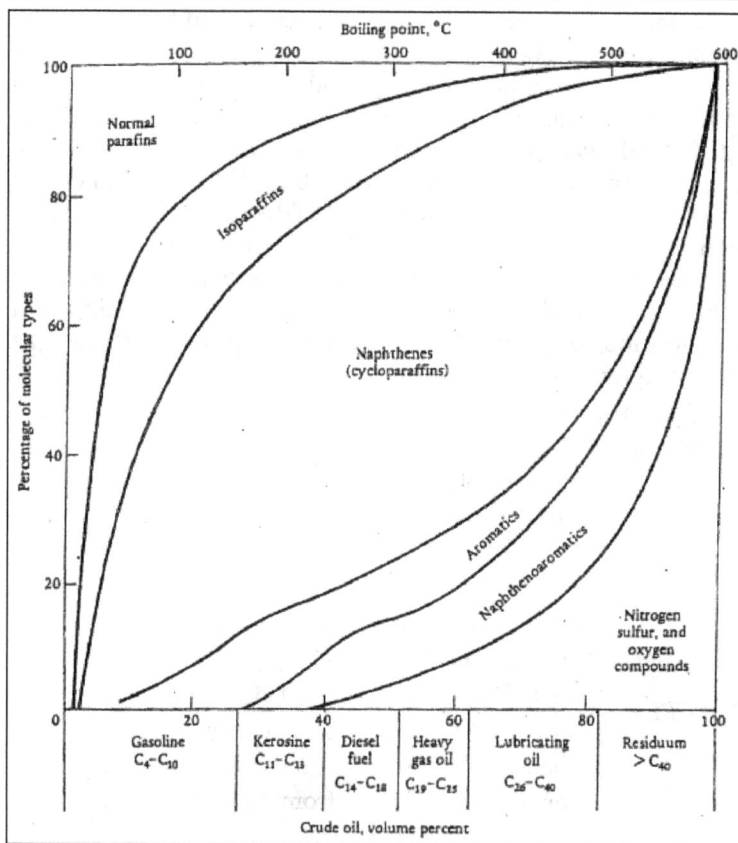

Figure 2.1: Chemical Composition of Crude Oil
(*Source*: Selley, 1998)

chain alkanes are relatively more abundant in the lower boiling point fractions. Naphtheno-aromatics and NSO (nitrogen, sulphur and oxygen containing) compounds dominate the higher boiling point fractions and the residuum. Naphthenes are significant over the whole range, except for the lightest and the heaviest fractions.

2.1 Saturated Hydrocarbons

2.1.1 Normal Alkanes

They are also commonly known as paraffins or aliphatic (saturated) hydrocarbons. They occur in most crude oils in the range

15-20 per cent ; in some cases upto 35 per cent and dominate the gasoline fraction (Figure 2.1). They form straight chain homologous series of compounds having the general formulae C_nH_{2n+2}, in which each member differs from the next member by a (CH_2) group. Table 2.2. shows the nomenclature, composition and abbreviated structure of each compound upto ten carbon numbers (C_{10}). Compounds less than C_4 (methane to butane) are gases at STP conditions. Compounds with C_5 (Pentane) to C_{15} (Penta decane) are liquids and compounds with greater than C_{15} ($C_{16,}$ n-hexa decane) are solids, commonly known as paraffin waxes. Relative abundance of individual compounds upto C_{35} are routinely reported using gas chromatography (GC) technique (10.6.0).

Distribution of n-alkanes in crude oils shows a slight predominance with odd number of carbon atoms over those with even numbers. The ratio of odd to even carbon numbers in a specific chain length (molecular weight) range is expressed by a term called carbon preference index (CPI). It is defined (Barker, 1985) as

$$CPI = \frac{1}{2}\left[\frac{\begin{array}{c}\text{Sum of odd - chain}\\ \text{length alkane}\\ \text{from } nC_{21} \text{ to } nC_{33}\end{array}}{\begin{array}{c}\text{Sum of even - chain}\\ \text{length alkanes}\\ \text{from } nC_{20} \text{ to } nC_{32}\end{array}} + \frac{\begin{array}{c}\text{Sum of odd - chain}\\ \text{length alkane}\\ \text{from } nC_{21} \text{ to } nC_{33}\end{array}}{\begin{array}{c}\text{Sum of even - chain}\\ \text{length alkanes}\\ \text{from } nC_{22} \text{ to } nC_{34}\end{array}}\right] \quad (2.1)$$

It is important to note that other ratios such as shown under are also in use (Bray and Evans, 1961).

$$CPI = \frac{1}{2}\left[\frac{\begin{array}{c}\text{Sum of odd - chain}\\ \text{length alkane}\\ \text{from } nC_{25} \text{ to } nC_{33}\end{array}}{\begin{array}{c}\text{Sum of even - chain}\\ \text{length alkanes}\\ \text{from } nC_{24} \text{ to } nC_{32}\end{array}} + \frac{\begin{array}{c}\text{Sum of odd - chain}\\ \text{length alkane}\\ \text{from } nC_{25} \text{ to } nC_{33}\end{array}}{\begin{array}{c}\text{Sum of even - chain}\\ \text{length alkanes}\\ \text{from } nC_{26} \text{ to } nC_{34}\end{array}}\right] \quad (2.2)$$

CPI values approach 1.0 in normal oils. It is used as an indicator of thermal maturity of source rock bitumen.

Table 2.2: Names and Abbreviations for *n*-Alkanes

Name		Abbreviations	
Methane	CH_4	CH_4	None
Ethane	C_2H_6	CH_3CH_3	None
Propane	C_3H_8	$CH_3CH_2CH_3$	∧
Butane	C_4H_{10}	$CH_3(CM_2)_2CH_3$	∧∨
Pentane	C_5H_{12}	$CH_3(CH_2)_3CH_3$	∧∧∧
Hexane	C_6H_{14}	$CH_3(CH_2)_4CH_3$	∧∧∨
Heptane	C_7H_{16}	$CH_3(CH_2)_5CH_3$	∧∧∧∧
Octane	C_8H_{18}	$CH_3(CH_2)_6CH_3$	∧∧∧∨
Nonane	C_9H_{20}	$CH_3(CH_2)_7CH_3$	∧∧∧∧∧
Decane	$C_{10}H_{22}$	$CH_3(CH_2)_8CH_3$	∧∧∧∧∨

Source: Waples, 1981.

2.1.2 Iso Alkanes

Iso-alkanes together with normal alkanes known as paraffin hydrocarbons constitute one of the major components of petroleum, present to an extent of 25 per cent in most crude oils. They are substances having the same chemical composition of n-alkanes but different molecular structure and different physical and chemical properties. The first member in the isoalkane series is isobutane (C_4H_{10}) having the same composition of n-butane but different structure and physical properties. Similarly there are three forms of isopentane, having the same molecular formula $(C_5 H_{12})$ of n-pentane but with different structures and properties (Figure 2.2). The number of isomers increase rapidly for higher members of isoalkanes. For example, there are five possible isomers of hexane $(C_6 H_{14})$ and 18 of octane $(C_8 H_{18})$ and so on. Like normal alkanes, the branched isoalkanes are not equally distributed among the various crude oil fractions but are more abundant in the light and middle point range (Figure 2.1).The boiling point of an isoalkane is lower than any n-alkane with the same structure and molecular weight (Figure 2.2). Thus the boiling point (B.P) of isobutane (–10.2°C) is lower than n-

n-butane: $CH_3(CH_2)_2CH_3$
B.P. −0.5°C

```
    H  H  H  H
    |  |  |  |
H—C—C—C—C—H
    |  |  |  |
    H  H  H  H
```

isobutane: $(CH_3)_2CHCH_3$
B.P. −10.2°C

```
    H  H  H
    |  |  |
H—C—C—C—H
    |  |  |
    H  |  H
       |
    H—C—H
       |
       H
```

n-pentane: $CH_3(CH_2)_3CH_3$
B.P. 36°C

```
    H  H  H  H  H
    |  |  |  |  |
H—C—C—C—C—C—H
    |  |  |  |  |
    H  H  H  H  H
```

isopentane: $CH(CH_3)_2 \cdot CH_2CH_3$
B.P. 28°C

```
    H  H  H  H
    |  |  |  |
H—C—C—C—C—H
    |  |  |  |
    H  |  H  H
       |
    H—C—H
       |
       H
```

tertiary pentane, neopentane, tetramethylmethane,
or 2.2-dimethylpropane: $C(CH_3)_4$
B.P. 9.45°C

```
            H
            |
    H   H—C—H   H
    |     |     |
H—C————C————C—H
    |     |     |
    H   H—C—H   H
            |
            H
```

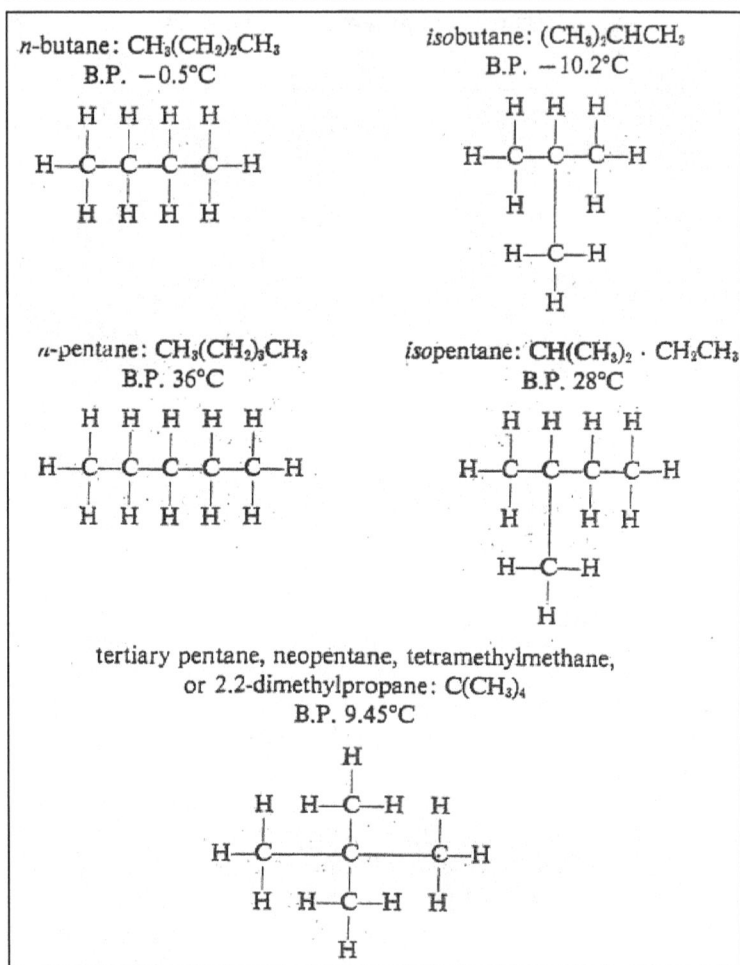

Figure 2.2: Structure of Butane, Pentane and their Isomers
***(Source*: Levorsen, 1967)**

butane (−0.5°C). Similarly the BP of isopentane (28°C) is lower than n-pentane (36°C). For the lower-molecular weight compounds all the isomers are known. Higher fractions are however, not well characterized. It has been established that the mono-substituted branched alkanes predominate over poly-substituted ones and that 2 or 3 positions are the preferred ones. A marked relationship exists between the relative amounts of individual branched hydrocarbons. Thus significant correlations were found between 2-and 3-methyl pentane, and between 2-and 3- methyl hexane. Similar correlations were also evident between iso-hexane and iso-heptane. The total amount of iso hexanes and iso heptanes vary significantly and systematically with geological age.

Isoprenoids are an important group of components in isoalkane series. They are having a characteristic branched chain (a methyl group on every fourth carbon atom such as 2,6,10 etc). These members which occur in crude oils and in ancient sediments, are believed to be derived from chlorophyll, and constitute a group of biomarkers useful in identification of the origin of oils. The commonest members among them (Table 2.3) are pristane (2,6,10,14 tetramethyl pentadecane) and phytane (2,6,10,14 tetramethyl hexadecane). The ratio of pristane to phytane in a crude oil or rock extract indicates the type of organic matter, depositional environment and the level of thermal maturity of source rocks. Crude oils dominated by normal and isoalkanes are known as paraffinic crudes. Typical examples of this type include Pennslvanian crude of USA, Southern Chile, Eastern Brazil, Southern Russia and some African crudes (North, 1990).

2.1.3 Cycloalkanes (Naphthenes)

Cycloalkanes commonly known as naphthenes are the major components of crude oils accounting for 30-60 per cent. This group has a general formula $C_n H_{2n}$ and occur as liquids under STP conditions. They are present in all fractions from C_5 and higher but show less variation between the various boiling point ranges than the normal and isoalkanes (Figure 2.1). The most abundant cycloalkane compounds are based on five and six membered rings. Notable among them are cyclopentane and cyclohexane (Figure 2.3). Methyl cyclohexane (C_7H_{14}) predominates over cyclohexane (C_6H_{12}) in all most all crudes, their ratio ranging from 0.9 to 4.8. Similarly methyl cyclopentane predominates over cyclopentane. Multi ringed nephthenes that are directly traceable to biologically produced molecules are receiving attention as biomarkers. They include 4-ring steranes and 5-ring hopanes (Figure 2.4). In each case there is a whole series of compounds formed by different side chains and varying spatial configurations. Some of the members are recognized as useful indicators of oil source, maturity and migration (10.1.0).

Table 2.3: Structures of Geochemically Important Isoprenoids

Structure	Name	Number of Carbon Atoms
	Phytane	20
	Pristane	19
	Norpristane	18
	—	17
	—	16
	Farnesane	15
	Squalane	30
	Lycopane	40
	Perhydro-β-carotene (β-carotene)	40

Source: Waples, 1981.

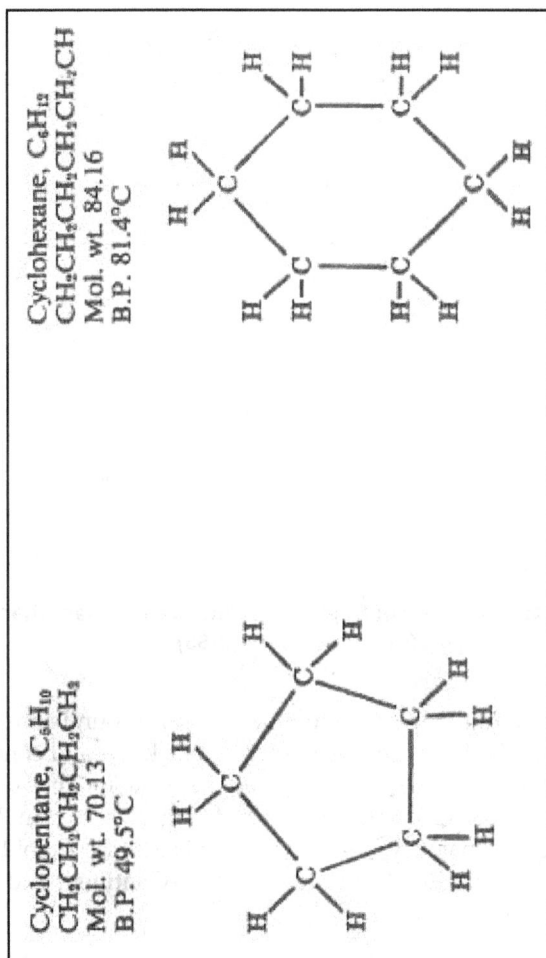

Figure 2.3: Structures of Cyclopentane and Cyclohexane
(*Source:* Levorsen, 1967)

Cyclopentane, C_5H_{10}
$CH_2CH_2CH_2CH_2CH_2$
Mol. wt. 70.13
B.P. 49.5°C

Cyclohexane, C_6H_{12}
$CH_2CH_2CH_2CH_2CH_2CH$
Mol. wt. 84.16
B.P. 81.4°C

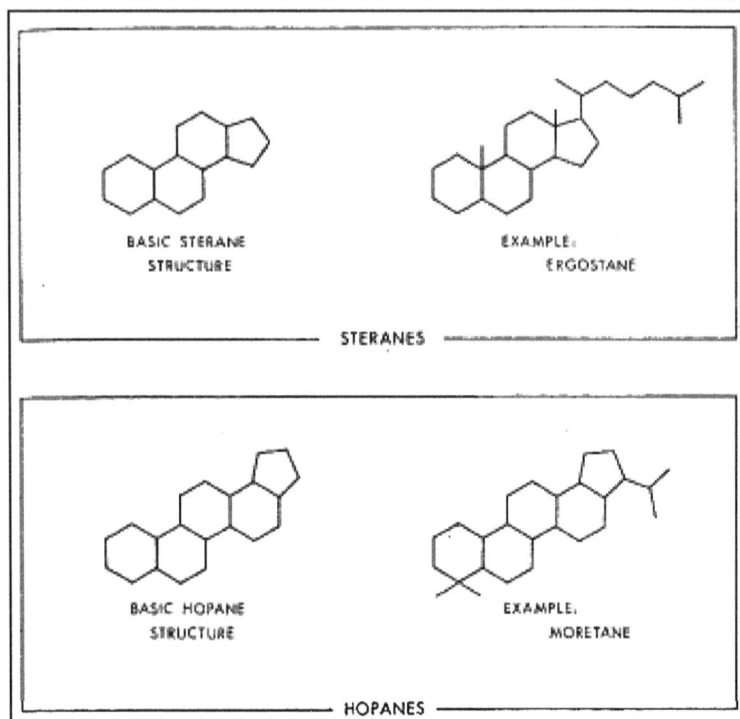

Figure 2.4: Structures of Selected Cycloalkanes (Naphthenes)
(*Source*: Barker, 1985)

Crudes dominated by the naphthenic compounds are called asphalt base oils. They are also known as black oils and constitute only a small fraction of the total oil (about 15 per cent). They occur in Venezula, Mexico, parts of California and Gulf coast, and in Russia. However, great majority of crudes are mixed base (naphthene–paraffin) and occur in all Middle East, Mid-Continent and in the North Sea (North, 1990).

2.2 Unsaturated Hydrocarbons

2.2.1 Alkenes (Olefins)

Alkenes commonly known as olefin compounds are very reactive compared to alkanes since they are unsaturated having double and treble bonds between carbon atoms. Olefins are uncommon in crude oils because they are readily converted into saturated compounds in the sediments. This is illustrated by the reaction of propylene with either hydrogen or H_2S to yield propane or 2- propyl thiol respectively (Figure 2.5). However olefins formed in refinery processes are largely used as basic materials in petrochemicals. Isoprene (C_5H_{10}) a diolefin is the basic building block for many hydrocarbon structures such as isoprenoids, terpenes, sterols, pigments and vegetable oils.

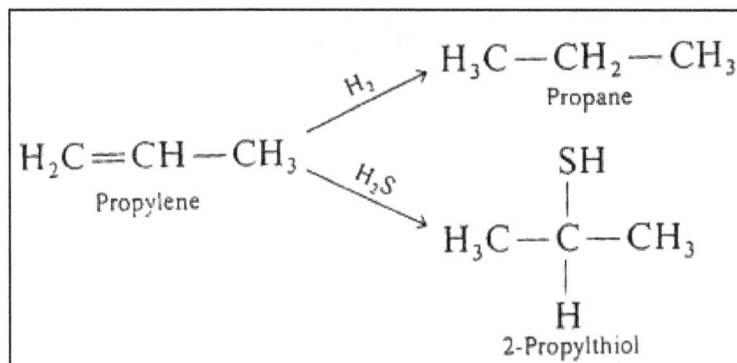

Figure 2.5: Reactions of Propylene with Hydrogen and Hydrogensulphide
(*Source*: Hunt, 1996)

2.2.2 Aromatics

They occur in many oils in small amounts (0.1 per cent to max.10 per cent). The molecular structure of aromatic hydrocarbons is based on a hexagonal ring of 6 carbon atoms as in benzene, the smallest member of the family. Toluene is the most common aromatic compound followed by xylene and benzene. They are in general, liquids at STP conditions and present in relatively minor fraction (about 10 per cent) in light oils. However, they increase in quantity to more than 30 per cent in heavy oils. Aromatic compounds with more than one ring system have been reported in many crude oils (Figure 2.6). They range from naphthalene and diphenyl upto compounds with five or six fused rings (perylene).

2.2.3 Naphtheno-aromatics

Compounds containing both aromatic and saturated ring systems are classified as naphtheno-aromatics. They constitute

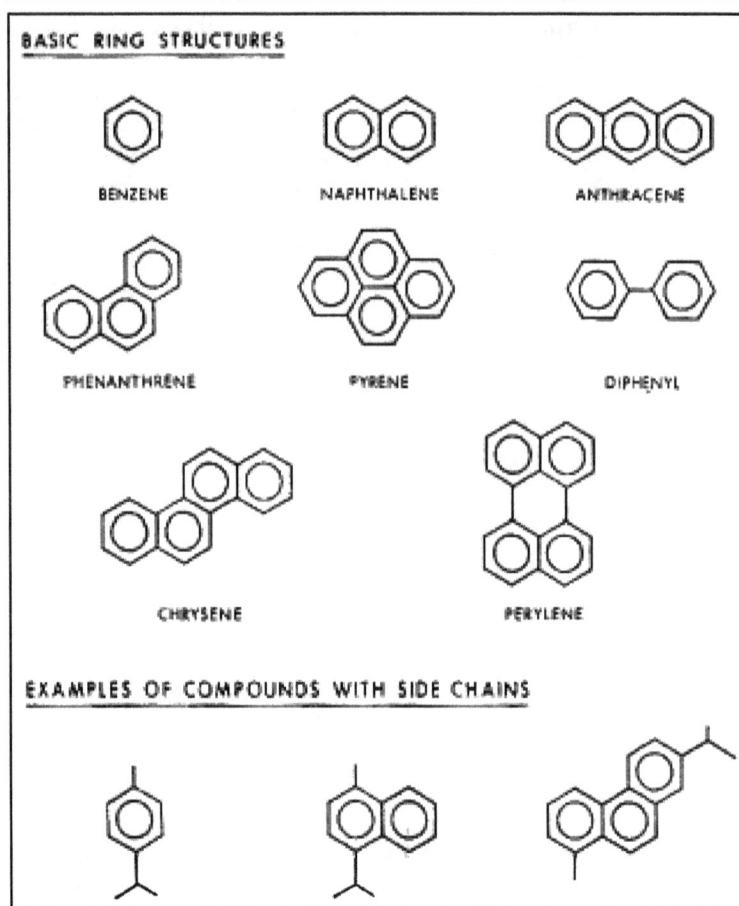

Figure 2.6: Structure of Selected Aromatic Hydrocarbons
(*Source*: Barker, 1985)

another group of compounds in crude oils particularly abundant in higher boiling point fractions (Figure 2.1). It is often possible to relate their particular structure to biological precursors which provide further impetus to their study. Some of the important compounds identified so far in oils include Sesequi- (C_{15}), Di-(C_{20}), and tri-(C_{30}) terpenes (Figure 2.7). These are typical biological markers and many of them with slight modification are recognized in shale extracts and crude oils. For example, the diterpenoid phytol, yields pristane

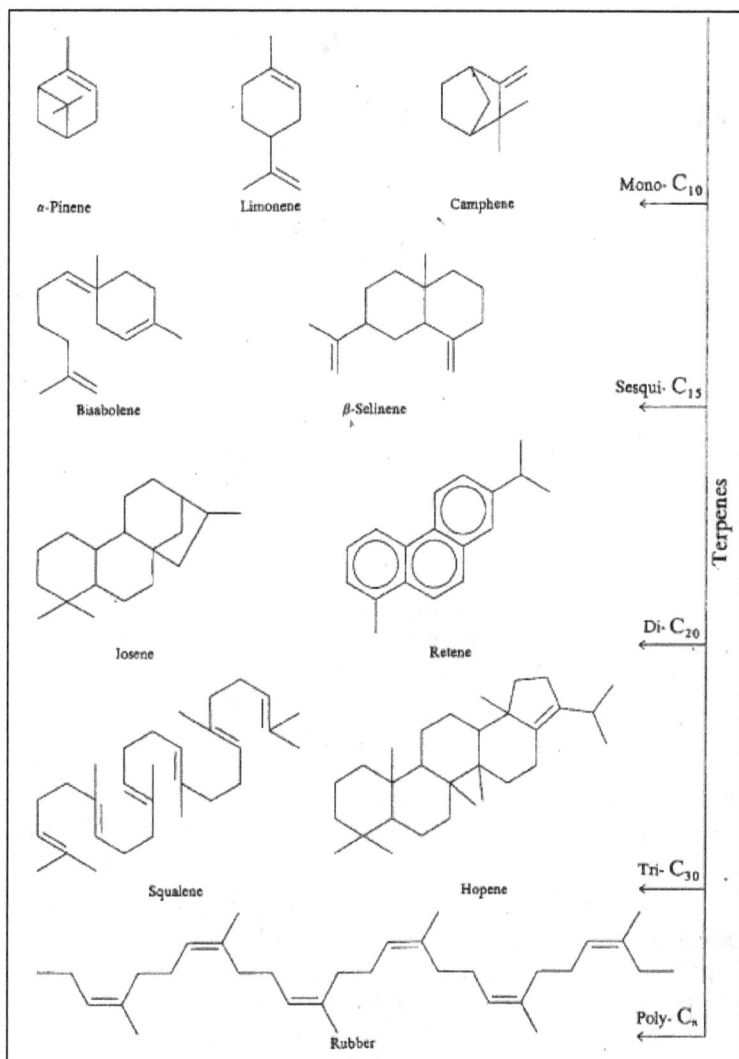

Figure 2.7: Typical Examples of Naphthene Aromatics, Sesequi and Triterpenes
(*Source*: Hunt, 1996)

and phytane (2.1.2). Squalene is a biological precursor of sterols (cholesterol, sitosterol, ergosterol etc.) and their counterparts such as tetracyclic triterpenes (steranes) and petacyclic triterpenes (hopanes etc).

2.3 Compounds with Hetero (NSO) Atoms

Crude oils comprise many compounds that contain other than carbon and hydrogen (hydrocarbons). They are in general, called as hetero or NSO compounds. The principal elements are nitrogen, sulphur and oxygen along with metals such as vanadium, nickel, iron, copper etc. They commonly account 10 per cent of the total crude oil, though in extreme cases it can be much higher. The percentage of hetero compounds is often higher in shallow degraded and immature crudes than in deep and more matured oils. They are distributed unevenly among the boiling point ranges being predominant in the residuum. Heavy distillate contains intermediate amounts and the lighter fraction are the poorest in hetero atoms (Figure 2.1).

2.3.1 Sulphur

Sulphur is by far the most important element of the hetero compounds. The sulphur containing compounds in crude oils vary on average from 0.1 to 7.0 per cent by weight excluding H_2S gas and native sulphur. A small portion of elemental sulphur occurs along with H_2S in young, shallow oils at temperatures greater than 100° C. Major part of sulphur is bonded with carbon mainly in poly-aromatic compounds such as (*i*) mercaptans (*ii*) disulphides and polysulphides of alkanes and (*iii*) cyclic aromatic sulphides or thiophenes.

Crude oils containing as low as 0.1-0.2 per cent sulphur, often termed as sweet crudes are concentrated in African basins (Algeria, Angola, Nigeria). Low, intermediate and high sulphur crudes contain <0.6 per cent, 0.6-1.7 per cent and >1.7 per cent sulphur by weight respectively. The highest percentage is found in reservoirs of dolomite-anhydrite facies, mainly in Middle East, Iran etc. In general, high sulphur crude oils have lower API (American Petroleum Institute) gravities (North, 1990).

2.3.2 Oxygen

The oxygen compounds range from 0.06-0.4 per cent by weight

in most crudes. They include acids, esters, ketones, phenols and alcohols. Acids are especially common in young immature oils and in fatty acids, isoprenoids, naphthenic and carboxylic acids. The presence of steranes in some crudes is an important indication of their biological origin.

2.3.3 Nitrogen

Nitrogen in crude oils is related primarily to the asphalt content. Nitrogen compounds range from 0.01-0.04 per cent by weight in most crudes. They include amides, pyridines, indoles and pyrroles. More than 0.2 per cent of nitrogen is considered as high. Such concentrations occur in some oils in the LosAngles, Maracaibo and Tampico basins. High nitrogen content is usually associated with high helium, carbon dioxide and/or sulphur content (North, 1990). Nitrogen containing compounds are classified as basic if they can be extracted with dilute mineral acids, or neutral if they can not be extracted. The ratio of basic to neutral compounds in any boiling point range is approximately constant even between different crude oils. This suggests that nitrogen compounds are not influenced by lateral alteration processes (7.5.0) unlike the sulphur compounds.

2.3.4 Trace Metals

Most crude oils contain 1 µg/g (ppm) to 1 ng/g (ppb) of metals, the notable among them are nickel and vanadium occurring as porphyrin complexes. Other commonly reported trace metals include Fe, Co, Cr, Hg, Cu and As. (Hobson and Tiratsoo, 1975). They in general, occur in shallow, young and degraded crudes while almost rare in old and deep marine oils. Metals tend to associate with resins, sulphur and asphaltene fractions of crude oils. Of particular interest is almost ubiquitous presence of nickel and vanadium that occur as organometallic compounds with porphyrins. Unlike other trace metals, the average concentration of vanadium and nickel in crude oils are 63 and 18 ppm respectively (Tissot and Welte, 1984). The known maximum values are 1200 ppm vanadium and 150 ppm nickel in the Boscan crude of Venezuela (Selley, 1998).

2.3.5 Resins and Asphaltenes

The content of resins (4-30 per cent) and asphaltenes (0.2–2.0 per cent) vary widely among different crude oils. The abundance of resins always exceeds asphaltenes, although their ratio may vary considerably in different crudes. They are separated from residuum obtained during distillation by first treating with liquid propane wherein oils dissolve, while resins and asphaltenes precipitate together. The latter fraction is then separated from each other by treating with n-pentane. The portion soluble in n-pentane is called as resins and the insoluble portion as asphaltenes. Resins are light to dark coloured, thick viscous liquids or black crystalline solids while the asphaltenes are solids or black powders. They commonly occur as colloidal particles in crude oils and their elemental composition show considerable amounts of sulphur, nitrogen and oxygen. The stability of asphaltic suspension depends on the ratio of resins to asphaltenes. When the stability of asphaltenes is disturbed by pressure, temperature or composition of crude oil, they precipitate and cause blockage problems in production and transport operations. The particle size and the molecular weight of asphaltenes are higher than resins in many crudes. For example, the former have molecular weight in the range 900–41,000 but appears to form aggregates with molecular weight over 1,00,000.

2.3.6 Diamondoids

Diamondoids are saturated, polycyclic organic compounds having a structure similar to a diamond lattice. Examples of diamondoids include adamantane, diamantane and tetramantane. Similar to formation of waxes and asphaltenes, the deposition of diamondoids in petroleum reservoirs is associated with changes in pressure, temperature and composition of reservoir fluids. However, their deposition is not as common as asphaltenes or waxes. Major difference between asphaltenes and diamondoids is that the former are chemically much more complex than the latter. Diamondoids form under high pressure and temperature conditions and provide novel method for measurement of thermal maturity (10.2.5) of oils and condensates (Dandekar, 2006).

2.4 Natural Gases

Natural gases present in crude oils can be broadly divided into two types: (i) hydrocarbon gases and (ii) non-hydrocarbon gases.

2.4.1 Hydrocarbon Gases

Methane is the most abundant gas (90-95 per cent) among the hydrocarbon gases. Ethane, propane and butane are present in small quantities (5-10 per cent). The gas composed almost entirely of methane is called 'dry gas'. If the proportion of other hydrocarbon gases exceeds 4-5 per cent, it is called as 'wet gas'. Methane is also known as marsh gas if it occurs in wetlands and in coal mines at shallow depths. Natural gas consisting largely or wholly methane has three origins: (*i*) extra terrestrial (mantle derived), (*ii*) bacterial degradation, and (*iii*) thermal maturation of organic matter in sediments or source rocks (Selley, 1998). Geochemical and stable isotope analysis (6.4.0) can differentiate the origin of methane from the above sources. It is sparingly soluble in water under STP conditions. However, the solubility increases with increase of pressure. Hence the gas accompanying oil in the reservoir will be in the form of solution under the pressure and temperature conditions of the reservoir. If the concentration of the gas exceeds its solubility in oil, it forms a gas cap over the oil.

Methane occurs as gas hydrate (12.7.0) in the form of a solid when frozen under favourable conditions of temperature and pressure. It also occurs in coal mines commonly known as coal bed methane (CBM, 12.6.0).

2.4.2 Non-hydrocarbon Gases

They mainly consist of nitrogen (6-14 per cent), carbondioxide (1-9 per cent) hydrogen sulphide (0.2-5 per cent), and trace amounts of hydrogen and inert gases (helium, argon and radon).

(*i*) Nitrogen

Nitrogen gas is commonly associated with both inert gases and hydrocarbon gases. Its origin is basement or more specifically igneous rocks. It is predominantly of inorganic origin, although organic processes such as bacterial degradation, atmosphere and coalification processes (CBM) also may account to a significant extent. Biogenic nitrogen is likely to be produced only in shallow conditions, whereas in nature it occurs in deep hydrocarbon reservoirs. Further, some atmospheric nitrogen gas might have been trapped in the sediments during deposition and occur in connate water. Thermal metamorphism of bituminous carbonates could generate both nitrogen and carbon dioxide (Hitchon, 1963). Nitrogen gas mixed with CO_2 and steam is used as an injectant in recovery of light oils.

(*ii*) Carbon Dioxide

Major subsurface CO_2 accumulations are common in areas of extensive volcanic activity and earthquakes. These include: Sicily, Japan, New Zealand, and the Cordilleran chain of North America from Alaska to Mexico (Selley, 1998). Both inorganic and organic processes generate significant amounts of CO_2 in the earth's crust. It may also be generated by leaching of carbonate rocks by meteoric waters or may be associated with connate water in the oil fields. Carbon dioxide is produced by the thermal maturation of kerogen and bacterial degradation of organic matter in sediments. Much of CO_2 generated in the subsurface remains in solution as carbonic acid (H_2CO_3) because of its greater solubility in water particularly at higher pressures. It is being increasingly used now a days to enhance oil recovery in the reservoirs (Taylor, 1983).

(*iii*) Hydrogen Sulphide

Hydrogen sulphide (H_2S) occurs both as free gas and in solution with oil and in oil field brines. It is commonly expelled together with SO_2 from volcanic eruptions. H_2S is highly corrosive and attacks production pipes, flow lines in the oil or gas fields. Even small amounts of H_2S in oils and natural gas create problems of its removal during refining. It is produced in subsurface by bacterial reduction of metallic sulphides, particularly of iron. It is associated with evaporite minerals such as anhydrite, calcite and lead-zinc sulphide ores (Davidson, 1965; Dunsmore, 1973).

(*iv*) Inert Gases

Among the inert gases, helium (He) is a common minor accessory in many natural gases found in the subsurface. It is associated with mines, hot springs and fumaroles. Helium rich natural gas (about 2 per cent) is found in oil fields in Texas (Panhandle), Peace River and Sweet water arches, foot hills of Rocky Mountains in Canada. Other regions containing helium-enriched natural gases include: Poland, Alsace and Queens Land and in Australia (Lee, 1963; Hitchon, 1963). Helium is known to be produced by the decay of radioactive elements such as uranium, thorium and radium. It is believed to have emanated in oil fields from the deep seated basement rocks especially granite or from petroleum source rocks derived from granites. Helium rich gases mostly derived from Pre-Tertiary reservoirs are usually rich in nitrogen gas. Helium is of considerable economic importance because it is lighter than air, inert and safer than hydrogen for use in dirigibles. Argon (Ar) and radon (Rn) are radioactive decay products of potassium and radium respectively. They are believed to have the same origin as that of helium. Radon is of considerable health hazard because its inhalation causes lung cancer. Its health hazards are highest in areas of basement in general, and granites in particular (Durrance, 1986).

(v) Hydrogen

Free hydrogen gas rarely occurs in the subsurface, partly because of its reactivity and partly because of its mobility. It is commonly dissolved in subsurface waters and in petroleum in trace quantities. It is probably produced in subsurface by the thermal maturation of organic matter.

2.5 Physical Properties of Crude Oils

Physical properties of crude oils reflect their chemical composition in a complex and largely unpredictable way. They are particularly important in handling, transporting and refining oils, and are important factors in establishing price. Some of the important physical properties of crude oils are: density (expressed as API gravity), colour, viscosity, fluorescence, pour point, cloud point and flash point.

2.5.1 Density

Crude oil density is commonly expressed in (degrees) API gravity, which is defined as

$$API\ gravity = \frac{141.5}{SG} - 131.50 \qquad (2.3)$$

where,

SG: Specific gravity or density at 60°F (15.6°C).

By general convention, oils with API gravities higher than 30° are considered light, those between 30° and 22° are medium, and those between 22° and 10° are heavy. API gravities less than 10° are characterized by extra heavy crudes (tars). This convention is however, not universal. It is difficult to distinguish extra heavy crudes by API gravity but viscosity provides a more useful discrimination. Oils greater than 50° API gravity are not really liquids but are rather condensates or distillates. As the API gravity increases, the density decreases. The API gravity of two oils may differ considerably even though they appear to be closely related but produced from different reservoirs within the same field.

Most of crude oils are less dense than water. However, when API gravity falls below 10°, the oil density exceeds that of water and they no longer rise by buoyancy in oil field waters. A typical histogram of API gravities of large number of oils (Figure 2.8) shows that they commonly have values 35-45°, but spread with decreasing frequency to lower and higher values. The world wide average value is 33.3°. Most favoured crudes (API gravity about 37°) are common in the Middle East, the Mid-Continent and the Appalachian province of USA, Alberta, Libya and the North Sea. Very light crudes (>40° API)

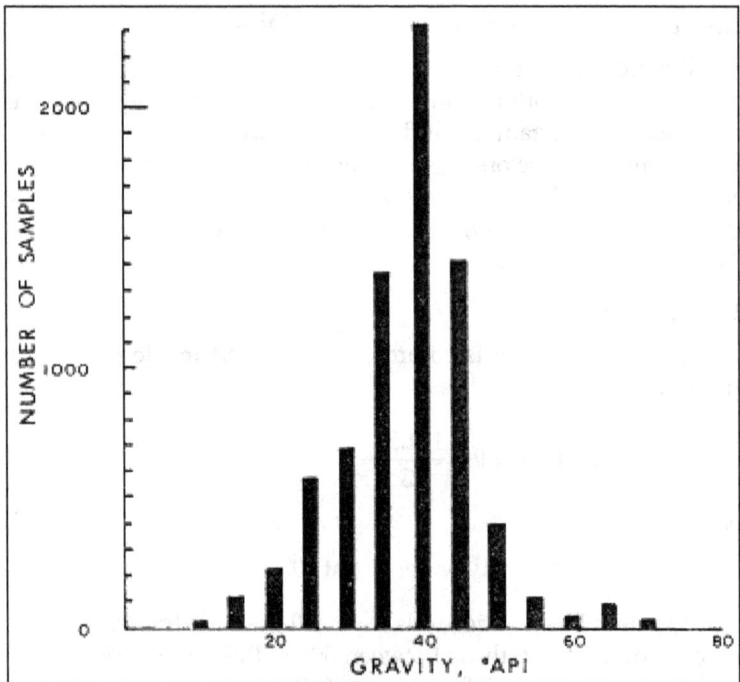

Figure 2.8: Histogram Showing Frequency of
Occurrence of Oils of Various API Gravities
(*Source*: Barker, 1985)

occur in large quantities in Algeria, south eastern Australia, in some Indonesian and Andean fields. Very heavy crudes occur in California, Mexico, Venezuela and Sicily oil fields (North, 1990).

2.5.2 Colour

The colour in crude oils is largely caused by the hetero (NSO) compounds. Most of the pure hydrocarbons in petroleum are colourless. The colours of paraffin oils in general, vary from yellow to brown. Asphalt base oils are commonly brown to black; many of them are known as black oils. However, dark brown and reddish tinged oils are more in common. The lower the API gravity of a crude oil, the darker is its colour. In general, there is no simple relationship between the chemical composition and the colour of crude oil.

2.5.3 Viscosity

Viscosity is an inverse measure of the ability of a liquid to flow, the greater the viscosity, the less readily it flows. Crude oils vary widely in viscosity; at one extreme they grade into highly viscous tars that are almost immobile, and at the other end into highly mobile condensates. Viscosity of a crude oil is a function of the amount of gas dissolved, its composition, density, temperature and pressure. The effect of dissolved gas on both viscosity and the API gravity (Figure 2.9) indicates that while the viscosity decreases with increase of dissolved gas (solubility), it increases with increase of API gravity. Further, viscosity decreases with increase of temperatures while it slightly varies with increase of pressure. In general, crude oils of lower API gravities (higher densities) have higher viscosities. Crude oils with high asphaltene content may be too viscous to be transported through a pipe line even in hot climate. Boscane crude from Western Venezuela and several Californian crudes are typical examples of this category (North, 1990).

2.5.4 Fluorescence

All oils exhibit more or less fluorescence, the aromatic oils being the most fluorescent. The fluorescent colours of crude oils in reflected light range from yellow through green to blue. Fluorescence can be observed under ultra violet (UV) light. Even a nacked eye allows the detection of 10 ppm of oil in UV light when dissolved in a solvent like carbon tetrachloride (CCl_4).Fluorescence is commonly used in the logging of wells to locate oil showing in the cores, cuttings and

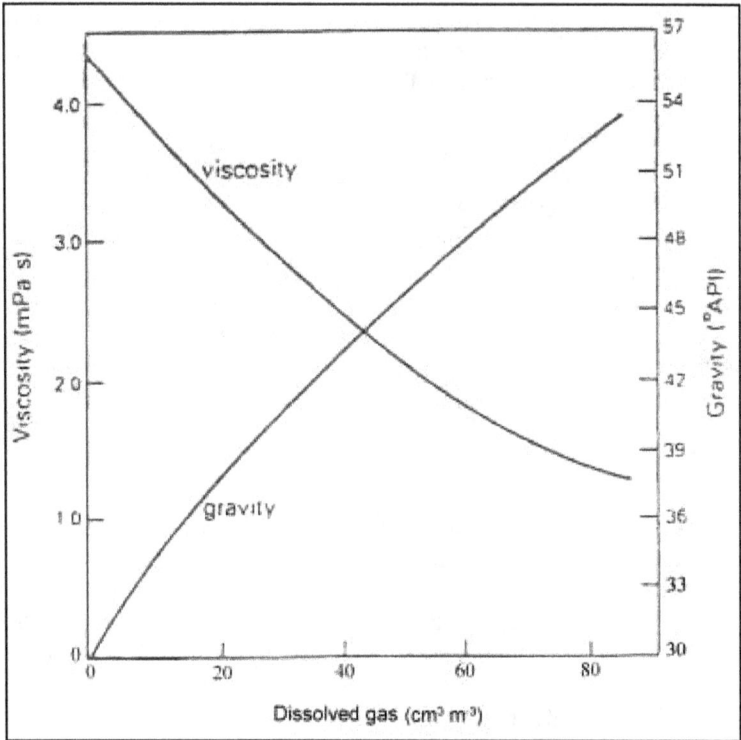

**Figure 2.9: Effect of Dissolved Gas on
Viscosity and Gravity of Crude Oil
(*Source*: North, 1990)**

drilling mud. It rapidly decreases its intensity with ageing. Thus fluorescence can be used to distinguish a fresh sample from the old containing traces of oil.

2.5.5 Pour Point

As the crude oil temperature is lowered, a point is reached at which an oil will no longer flow. This is called the pour point. By definition, the pour point is the lowest temperature at which the oil move, pour or flow when it is chilled without disturbance under definite conditions. Pour point values of crudes vary from below −50°F (−45.5°C) to over 100°F (37.8°C). In fact the pour point of an oil when used in conjunction with the temperature gives a better indication of oil in the reservoir than the viscosity. This combination gives an indication of the mobility and the ability of an oil to flow under reservoir conditions.

The pour point should be as low as possible so that the oil can be in a liquid state even at low reservoir conditions. Typical examples of this category include Middle Eastern and African crudes (–36° C). On the other hand, high pour point indicates that the crudes are immobile and can not move out under the reservoir temperature conditions. In such cases injection of steam is required to raise the temperature above the pour point of crude oil to facilitate its flow from the reservoir. High pour point oils are in general, rich in paraffin wax. Examples of this type include oils from Brazil, Gabon, Libya, Sumatra and China. Many deltas including the Niger, Mackenzie and Mahakam, Assam and Mumbai High have waxy, high-pour point crudes (Barker, 1985).

2.5.6 Cloud Point

The temperature at which a cloud or haze appears when the oil is cooled is called the cloud point. The cloud point is related to unsaturation of an oil. In general, the higher the unsaturation, the lower is its cloud point. The cloud or haze formed while cooling the oil is due to the settlingout of the solid paraffin wax crystals. Contrary to this, wax–free naphthenic oils show no cloud point. It is desirable that the cloud point of an oil should be as low as possible.

Cloud point is of particular importance in storage and handling of oil in outdoor bunkers and ship's tankers especially in cold weather. This is because the wax crystals formed can block filters and small–bore fuel lines.

2.5.7 Flash Point

Flash point is the minimum temperature at which the vapours of an oil when heated will ignite with a flash of a very short duration. When the oil is heated further to higher temperatures, it will ignite and burn with a steady flame at the surface.

Flash point is important primarily from a fuel handling point. Too low a flash point will cause fuel to be a fire hazard subject to flashing and possible ignition and explosion. A low flash point in general, may indicate contamination by more volatile and explosive fuels like gasoline. It is therefore important that the flash point of a crude oil should be as high as possible (35-66° C).

2.6 Thermal Alteration of Oil in the Reservoir

Composition of petroleum in the reservoir changes in response to changing conditions. Increase of temperature accompanied by increase of depth of burial leads to the alteration of crude oil composition. In this process large molecules are thermally cracked to smaller fragments with an increasing trend of formation of lighter fraction in the sequence from crude oil to lighter oil, to wet gas and finally to dry gas (Connan, *et al.*, 1975). This trend is maked by decrease in the concentrations of nitrogen, sulphur and metals (Ni and V). Further, physical properties of oils such as API gravities increase, and pour points and viscosities decrease. Asphaltenes which become less soluble in oil precipitate out as a residue during cracking in the reservoirs. This process is commonly known as "natural deasphaltening" (Milner *et al.*, 1977). The Composition of typical crude oils before and after deasphaltening (Table 2.4) shows that the deasphaltend oil is of higher API gravity. In nature, the quality of the produced oil improves while the asphaltic residue left in the reservoir adversely affect the porosity and create plugging problems. Further, the gases (CO_2, H_2S, etc.,) generated during thermal cracking of oil *in situ* may also assist precipitation of asphaltenes in the reservoir. Figure 2.10 shows the compositional changes and the trends produced by thermal maturation and biodegradation (9.9.3).

2.7 Classification of Crude Oils

Crude oils as is evident from the above discussion cover a wide range of chemical composition and physical properties. As such it is often important to be able to group them into broad categories of related oils. The classification scheme employed depends on the application-whether for establishing commercial value (price) or nature of the organic source material. Refiners (chemical engineers) are most concerned with the relative abundances of various distillation fractions (Figure 2.1) and their chemical characteristics. On the other hand, geologists and geochemists are usually more interested in relating oils to their source rocks or in establishing their degree of thermal maturity or degradation. In this context, the characteristic biomarkers that are only minor components, are often very useful (10.1.0).

Table 2.4: Changes in Chemical Composition which Accompany Deasphaltening of Selected Oils

	Murban	Arabian Light	Buzurgan	Boscan
Original oil				
Specific gravity	0.982	1.003	1.051	1.037
S (per cent)	3.03	4.05	6.02	5.90
N (ppm)	3180	2875	4500	7880
Ni (ppm)	17	19	76	133
V (ppm)	26	61	233	1264
Asph. (wt. per cent)	1.2	4.2	18.4	15.8
Oil deasphaltened with propane				
Specific Staiity	0.924	0.933	0.945	0.953
S (per cent)	1.80	2.55	3.60	4.70
N (ppm)	300	1200	930	1600
Ni (ppm)	1	1.0	2.0	6.0
V (ppm)	I	1.4	4.0	10.0
Asph. (wt. per cent)	0.05	0.05	0.05	0.05

Source: Barker, 1985.

2.7.1 Classification Based on Distillate Fractions

The earliest schemes developed by U.S. Bureau of Mines (Smith, 1927; Lane and Garton, 1935) classified crude oil into paraffinic, naphthenic and mixed types according to their distillate fractions at different temperatures and pressures. Subsequently asphaltic and aromatic types are also included in this classification (Sachenen, 1945). Two fractions namely the first one collected between 250-275° C at one atmosphere (760 mm Hg) pressure, and the second fraction collected between 275-300°C at a pressure of 40 mm of Hg, are used for this classification.

2.7.2 Classification Based on Geological Occurrence

An empirical classification taking into account geological occurrence was proposed by Biederman (1965). Based on arbitrarily defined age and depth, he classified oils into four types. These include:

Type 1: Mesozoic and Cenozoic oil at less than 600 m (young-shallow)

Type 2: Mesozoic and Cenozoic oil at over 3000 m (young–deep)

Type 3: Paleozoic oil at less than 600 m (old–shallow)

Type 4: Paleozoic oil at more than 3000 m (old–deep)

A detailed study of many oils showed a number of chemical and physical differences among the four groups as detailed below:

(*i*) Type 1 oils tend to be heavy and viscous containing relatively high sulphur, low paraffin content, and rich in aromatics.

(*ii*) Type 2 oils on the other hand, are less viscous, higher API gravity, more paraffinic and low in sulphur content.

(*iii*) Type 3 oils are broadly comparable to Type 2 in API gravity, viscosity and paraffinic content. However, they tend to be relatively high in sulphur content.

(*iv*) Type 4 oils tend to have the lowest viscosity, sulphur content and highest API gravity.

This classification has many limitations. For example, it is now well known that oils apart from age and depth of their occurrence, widely vary with the nature and composition of source rocks and the degree of their degradation during migration and in the reservoir.

2.7.3 Classification Based on Different Oil Components

This scheme proposed by Tissot and Welte (1984) is based on the relative amounts of various compound types in oil topped at 210° C and one atmosphere pressure (Figure 2.11). It relates to the ternary system with normal + isoalkanes (paraffins), naphthenes, and aromatics + NSO compounds as the end members (aromatics also include naphtheno–aromatics). It is evident from the figure that oils are unevenly distributed among the classes. Most abundant types are the paraffinic, paraffinic- naphthenic, and the aromatic-intermediates. The advantage of this classification is that it can be used to demonstrate the maturation and degradation path ways (Figure 2.10) of crude oils in the subsurface.

2.7.4 Classification Based on Geochemical Characteristics

Based on geochemical characteristics (Chilingar *et al.*, 2005),

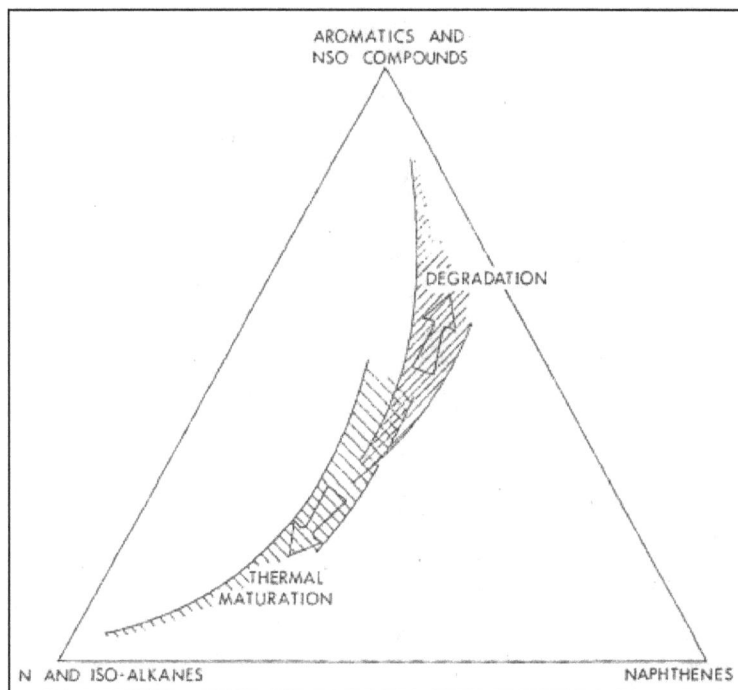

Figure 2.10: Effects of Thermal Maturation and Degradation on Crude Oil Composition and Hence Classification
(Source: Barker, 1985)

crude oils can be divided into 3 types–Type 1: catagenetically altered; Type 2: hypergenetically altered; and Type 3: migrationally altered.

Typical characteristics of Type –1 oils are (*a*) lowering of crude oil density, (*b*) lowering of resinous and asphaltene content, (*c*) simplification of complex hydrocarbon structure, (*d*) lowering of metal- porphyrine complexes and (*e*) increase in the aromatic content. Infra red (IR) studies indicate decrease in the long chain paraffin-naphthene fraction, and increase in short chain fraction. However, in some cases the number of long chains did not decrease. In general, catagenetically altered crude oils are rare.

Hypergenetic alteration (Type-2) affects the crude oil composition in the accumulation more profoundly. The geochemical parameters of this category include: (*a*) oxidation, (*b*) bacterial

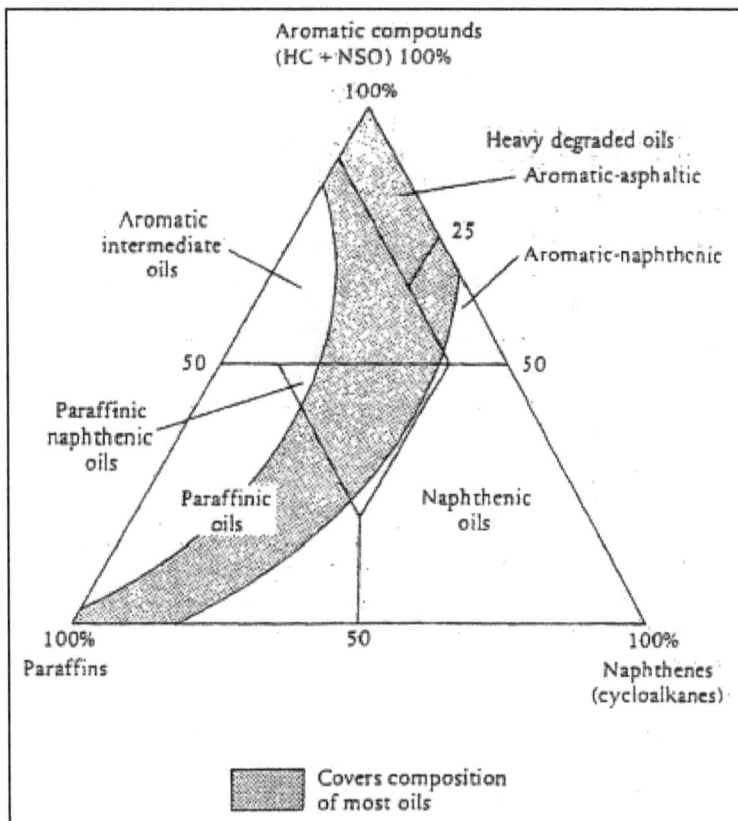

Figure 2.11: Ternary Diagram Showing the Classification of Oils
(*Source*: Selley, 1998)

reduction of sulphates, and (c) loss of light hydrocarbon fraction. Hypergenetically altered oils are usually heavier (lower API gravity), higher in resin and asphaltene content and lower in the gasoline fraction. Crude oils subjected to hypergenetic processes do not have IR absorption at 1710 cm^{-1} (carbonyl stretching frequency)

Alterations during migration (Type–3) result in changes of composition of crude oils depending on the type of migration. These alterations might be quite significant in secondary migration through sands and sandstones. They include increase in the number of paraffin structures and decrease in the naphthene cycles. The degree of cyclisation in naphthenes decreases and the aliphaticity (ratio of paraffins to naphthenes) drastically increases. Migration through carbonate rocks increases the density, the content of asphaltenes and resins, and naphtheno-aromatics.

Chapter 3
Occurrence of Petroleum in Sedimentary Basins

Sedimentary basins received much attention with the increased activities for petroleum exploration since 1950's (Huff, 1978). Sophisticated geological, geophysical and geochemical techniques made it possible to obtain vast amount of subsurface data which gave new insight to the mechanisms of various types of basin formations and their relationship with the different habitats of petroleum (Price, 1973; Watts and Rayon, 1976; Turcotte and Ahern, 1977). Before we go into the details of occurrence of petroleum in sedimentary basins, a brief account on their classification, basin types and hydrocarbon richness is presented in the following pages.

3.1 Definition of a Sedimentary Basin

Several definitions have been proposed to designate a sedimentary basin. It is defined as a low area in the earth's crust, of tectonic origin in which sediment has accumulated *e.g.*, a circular centrocline such as the Michigan Basin, a fault–bordered Intermontane feature such as the Bighorn Basin of Wyoming, or a linear crustal down warp such as the Appalachian Basin (Jackson, 1997). In petroleum geology sedimentary basins include parts of the earth's crust with significant thickness of sedimentary rocks deposited and preserved irrespective of their size, shape and structural style. A basin, therefore, represents a unit of geological structure which received sedimentary succession unique to it during a given span of geological time (North, 1979).

3.1.1 Classification of Sedimentary Basins

The basic concepts for classification of sedimentary basins have been refined, modified and characterized from time to time since 1950's. The most frequently quoted classification based on plate tectonic settings is that of Klemme (1980). According to this classification, sedimentary basins are divided into the following 8 types:

A. Intra Continental Basins on Cratonic Crust

1. Single-cyclic cratonic basins, mostly Paleozoic (*e.g.*, Willston).

2. Composite, multi cycle cratonic basins containing most Paleozoic oil (Ural–Volga, Western Canadian, Mid–Continent); some Mesozoic oil (West Siberian) and most gas reserves.

3. Graben or rift basins that may be Paleozoic (Dnieper–Donetz); Mesozoic (North Sea, Sirte, North East China); or Cenozoic (Suez).

B. Extra Continental Border Land Basins Formed on Intermediate Crust Associated with Plate Margins of Mesozoic and/or Cenozoic

4. Down warps into small ocean basins. They are of three sub types:

    4A. Basins closed by deformed belts (Persian Gulf, Orinoco, Caucasian).

    4B. Troughs between deformed belts (Upper Assam, Po Valley).

    4C. Open one sided basins (Gulf Coast, Alaska North Slope, North west Borneo)

5. Pull-apart (Stable Coastal) basins on trailing edges of continents, ultimate stage of type 3 basins (Gabon, Dampier-Rankin).

6. Subduction basins, second cycle, intermontane, small, mostly Tertiary; may be fore-arc (Talara); back arc (Sumatra); or non-arc, usually transform (California, Baku).

7. Median basins, also second cycle, intermontane, within compressed uplifted zones (Maracaibo,Gippsland).

8. Delta basins, especially Tertiary (Niger, Mississipi, Mahakan).

Each basin presents its unique characteristics and any classification is a guide to perceive the geological history and some regularities in its development to visualise zones for exploring petroleum traps.

3.1.2 Basin Styles and Sedimentary Fill

The basin types attractive for hydrocarbon exploration are closely related to the evolution of the earth's crust. The crustal genesis of a basin determines its subsidence and heat flow history, physical size and shape, structural style and the bottom configuration, and water depth at the time of deposition. Green (1985) presented an excellent review of basin settings and frame work in relation to filling of the basins. The genetic class of a basin and its tectonic history determines the provenance and the composition of the basin fill. The drift history determines the varying positions of the basin location with reference to surrounding drainage and provenance as well as the paleolatitudes. These physical factors set the stage for dynamic interactions with the atmosphere, the hydrosphere and the biosphere that create unique suite of depositional environments for each basin in space and time.

3.1.3 Environmental Controls

Profound temperature variations, ice cap melting, alteration of sea levels, rainfall volumes, ocean currents and winds–all influence the sediment fill. Major transgressions of sea upon the continents displacing the shore lines far from the interior contribute to the creation of restricted areas rich in benthic and phytoplankton productivity, and to enhance preservation of organic matter. Carbonate deposits emphasise the delicate inter-relationships of tectonics, eustatic sea level changes and climate, and evolution of reef-building life forms. Associated volcanic activity in some basin settings influence atmospheric carbon dioxide levels. Paleo-wind directions affect regional and local currents, areas of upwelling and high productivity as well as reservoir distribution along the deltaic systems. Water chemistry, another important aspect of paleoenvironment, influences the flocullation of fine clastics and interstitial fluids of sediments. The pore-water chemistry affects diagenetic activity, flushing and other factors which alter the reservoir properties.

In addition to providing an environment of high organic productivity, the basin setting must also provide an environment of its preservation. The sedimentation rates must be high enough to bury and preserve organic matter. The rate of subsidence has a pronounced influence on the reservoir characteristics. The greater the length of the time that a sedimentary basin subsides, the greater are the chances for favourable formation of rich source-rocks, reservoir rocks and effective seals.

Since each sedimentary basin presents its individual characteristics of evolution, sedimentary fill, structural style and habitat of hydrocarbons, the tectonic setting of a basin with reference to major crustal units provides indications in the initial stages the frame work for exploration. Regularities in basin development associated characteristics of hydrocarbon accumulations (Structural style of sediments, zones of preferred migration and entrapment) can be broadly understood by analysis of each basin (Kingston *et al.*, 1983 a, b). However, it should be noted that even within a single basin type several styles and modes of petroleum occurrences can coexist as in the case of polycyclic basins.

3.1.4 Relation Between Basin Type and Hydrocarbon Richness

Among the 700 sedimentary basins and sub-basins identified so far all over the world, about 400 are subjected to some exploration for oil and gas; about 150 of them are commercially productive. As much as 90 per cent of the world known oil and gas (excluding tar sands and oil shales) occur in only about 30 of them (Figure 3.1) indicating that hydrocarbon content among the basins is exceedingly variable. In fact they are pooled only in four regions: (*i*) along the Persian Gulf ; (*ii*) the Northern and Western sides of Gulf of Mexico and the southern side of the Caribbean; (*iii*) along the two flanks of the Ouachita-Marathan Mountains in the southern USA; and

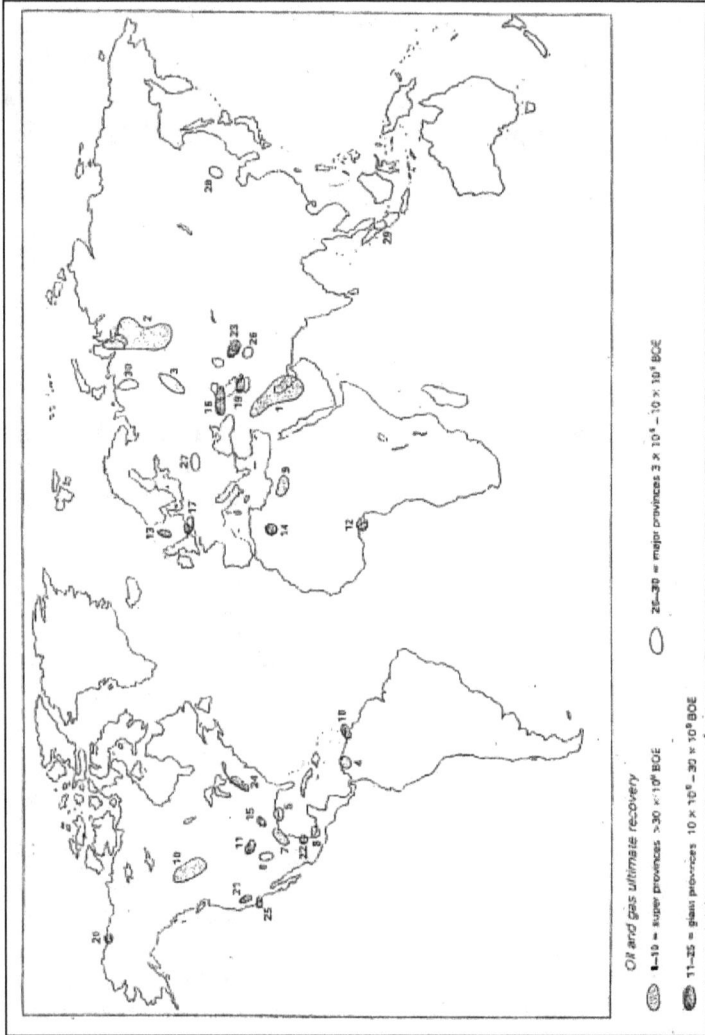

Figure 3.1: The World's Giant Petroleum Provinces Ranked in Order of Ultimate Estimated Recovery of Oil and Oil Equivalent Gas (*Source*: North, 1990)

(*iv*) along the Ural Mountains in the former Soviet Union. The recoverable hydrocarbon richness of a basin can be expressed in terms of its volume per unit area or per unit volume of sedimentary rock. In general, the average figures of oil richness have been in the range from 30,000 to 60,000 barrels per cubic mile of sedimentary rock (about 1,200 to 2,400 m^3 per cubic kilometer). Based on the production capacities in oil equivalent (m^3. km^{-3}), they can be broadly categorized into the following five types:

(*i*)	Ultra rich basins	> 10,000
(*ii*)	Rich basins	3,600-10,000
(*iii*)	Average basins	2,400-3,600
(*iv*)	Lean basins	1,000-2,000
(*v*)	Very lean basins (but productive)	< 1,000

Assignment of hydrocarbon richness to a particular situation within plates and their margins permits only two conclusions, namely:

(*i*) Nearly all Paleozoic oil reserves are in intra cratonic platform basins and nearly all Cenozoic oil reserves are close to plate boundaries while Mesozoic oil reserves are more equally divided between the platforms and their boundaries;

(*ii*) Nearly all gas reserves of all ages are in intra cratonic basin settings.

Though these conclusions confirm the tectonic control of hydrocarbon occurrence to the present basins, it is evident that most ultra rich basins are down warps at continental margins, either still open towards oceanic crust (like the Gulf Coast Basin) or closed by deformations and converted into foredeeps (the Persian Gulf Basin). These basins (Type 4 as mentioned in 3.1.1) contain about 50 per cent of the world's known oil reserves (Carmat and Bell, 1986). The most commonly productive basin type is the composite interior basin or cratonic foredeep (Type 2) on every continent except Antarctica. They contain about 25 per cent of the world's oil reserves and well over half of the gas reserves. Many young deltas (Type 8) are not oil productive, contributing only less than 5 per cent of the world's oil reserves. However, they tend to be gas prone, presumably because of their dominant terrestrial organic carbon input. The least productive basinal type is the simple cratonic basin within a Precambrian Shield. For further details on basin classification and hydrocarbon prospects, the reader may consult several articles (Dickinson, 1976; Klemme 1980; Kingston *et al.*, 1983 a,b) and books (Klemme, 1975; Bally and Snelson, 1980; North, 1990; Selley, 1998).

3.2 Classification of Indian Petroliferous Sedimentary Basins

Based on different criteria, the Indian sedimentary basins were classified by several researchers. Sastri *et al.*, (1974) classified basins according to their ages of the earlier sedimentary fill or the time of the basin formation in the Phanerozic basin map of India. Classification of Indian sedimentary basins based on tectonic criteria was first made by Raju (1979). Stratigraphy and occurrence patterns, of the petroliferous basins were described by Bhandari *et al.*, (1983). Based on the genetic relationships to the mechanism of formation and plate tectonic setups, Biswas *et al.*, (1993) classified Indian sedimentary basins into three main categories: (*i*) Interior Basins (*ii*) Rifted Settings and (*iii*) Orogenic Settings. Each category is further subdivided into different types (Figure 3.2). According to this scheme, Biswas *et al.*, (1993) classified them into 38 sedimentary basins (Table 3.1) and estimated their resource potential.

Based on geothermal criteria (Panda, 1985), the petroliferous sediments were classified into three types (Figure 3.3):

(*i*) *Low prospects*: Heat flow in the range of 70 to 100 m w m^{-2}

(*ii*) *Medium prospects*: Heat flow in the range of 100 to 130 m w m^{-2}

(*iii*) *High Prospects*: Heat flow range > 130 m w m^{-2}

About 29 prospective areas were identified of which 7 are of high prospects, 9 are of medium prospects and 13 are of low prospects. Most of the areas are associated with 6 major geothermal provinces namely, Himalayan tectonic belt, Cambay-Bombay graben, Krishna-Godavari graben, Son-Damoder graben, Andaman Volcanic zone, and Haridwar-Delhi ridge.

Kuldeep Chandra *et al.*, (1994) reported a genetic classification of basins based on geochemical perspective (hydrocarbon charge).

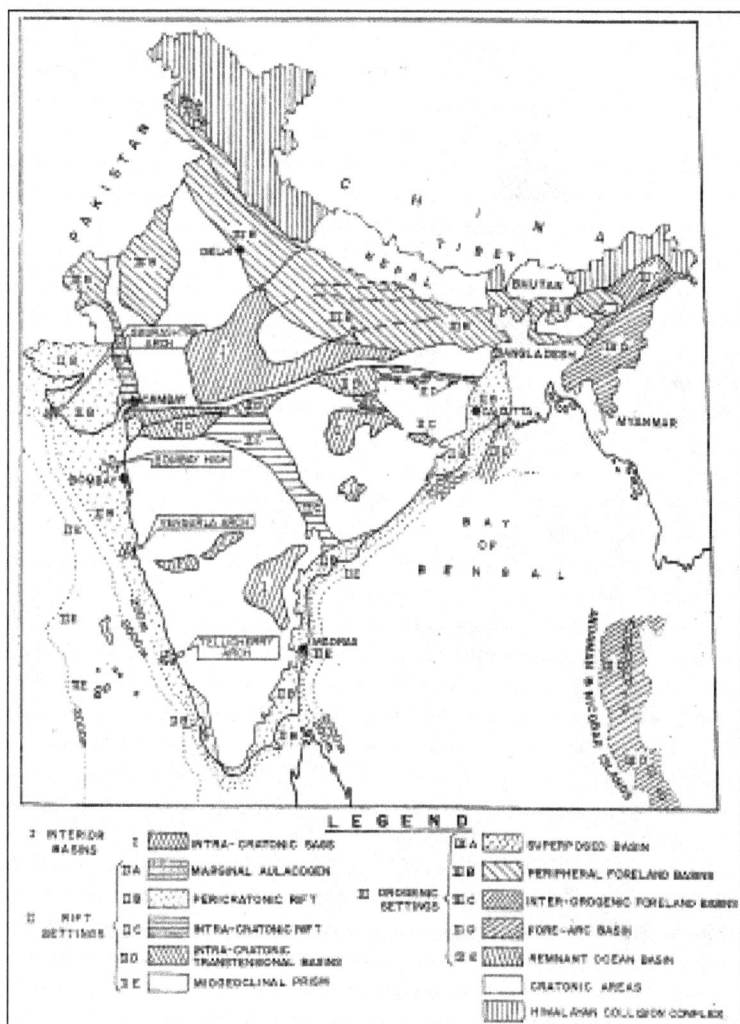

**Figure 3.2: Sedimentary Basins of India Based
on Tectonic Classification
(*Source*: Biswas *et al*., 1993)**

According to this scheme, the basins are categorized into 4 types
based on the magnitude of their charge (SRI)-A: Super (>1000); B:
Adequate (1000-500); C: Moderate (500-200); and D: Poor (<200).

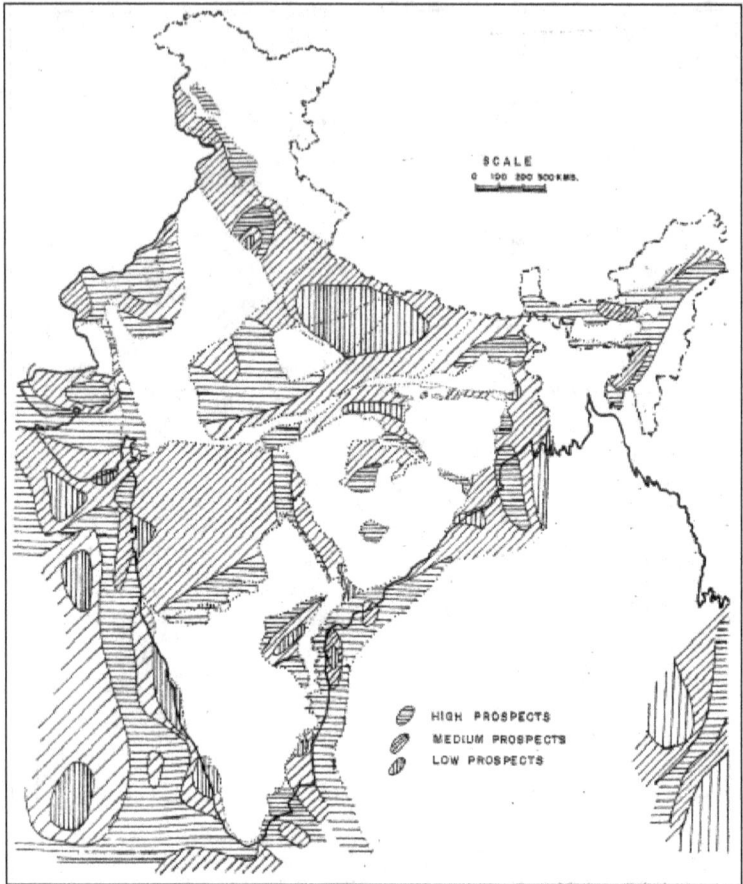

**Figure 3.3: Petroleum Prospects of Sedimentary Basins of India
Based on Geothermal Criteria
(*Source*: Panda, 1985)**

Petroliferous sediments are associated with a source-rock index (SRI) of 1000 or more.

Based on the generative concept of Demaison (1984), petroleum system approach developed by Demaison and Huizinga (1991) and reviewed by Magoon and Dow (1994), Kuldeep Chandra *et al.*, (2001) identified eleven petroleum systems within the five proven petroleum basins namely Cambay Basin, Bombay Offshore, Cauvery and

Krishna-Godavari Basins and Assam Shelf. Following are the eleven petroleum systems in the five basins:

(*i*) *Cambay Basin*: Cambay-Hazad and Cambay Kadi/Kalol systems

(*ii*) *Bombay Offshore Basin*: Panna-Bombay system.

(*iii*) *Cauvery Basin*: Andimadam-Bhuvanagiri, Andimadam-Nannilam and Andimadam-Neravi systems.

(*iv*) *Krishna-Godavari Basin*: Krishna/Gollapalli –Konakollu; Palakollu-Pasarlapudi; and Vadaparru-Ravva systems.

(*v*) *Assam Shelf*: Kopili and Barail-Tipam systems.

Table 3.1: Classification of Indian Sedimentary Basins

	Category	Type	Basins	
I	Interior Basins	Intra-Cratonic Sags	1. Vindhyan Basin	
			2. Chattisgarh-Bastar Basin	
			3. Cuddapah Basin	
			4. Kaladgi Basin	
			5. Bhima Basin	
II	Rifted Settings	II$_A$ Marginal Aulacogen	1. Cambay Basin	
		II$_B$ Pericartonic Rift	1. Kutch Basin	
			2. Saurastra Shelf	West
			3. Surat Basin	Coast
			4. Bombay Ratnagiri Basin	Basins
			5. Konkan Basin	
			6. Kerala Sasin	
			7. Western Bengal Basin	East Coast
			8. Mahanadi Basin	Basins
			9. Krishna-Godavari Basin	
			10. Palar Basin	
			11. Cauvery Basin	
			12. Mannar Basin	

Contd...

Table 3.1–Contd...

Category	Type	Basins
	II$_C$ Intra-Cratonic Rift	1. Pranhita Godavari Basin
		2. Mahanadi Graben
		3. Purnea Graben
	II$_D$ Intra-Cratonic Transtensional Basins	1. Satpura Basin
		2. Son Basin
		3. S. Rewa Basin
		4. Damodar Basin
		5. Narmada Basin
	II$_E$ Miogeoclinal Prism	1. Deep water Basins of west coast
		2. Deep water Basins of east coast
III Orogenic Settings	III$_A$ Superposed Basin	1. Karewa Basin
	III$_B$ Peripheral Foreland Basins	1. Rajasthan Basin
		2. Punjab Basin
		3. Ganga Basin
		4. Lower Assam Basin
	III$_C$ Inter-Orogenic Foreland Basin	1. Upper Assam Basin
	III$_D$ Fore-Arc Basins	1. Kohima Fore-Arc Basin
		2. Tripura-Mizoram Fore-Arc Basin
		3. Andaman Basin
	III$_E$ Remnant Ocean Basin	1. Eastern Bengal Basin

Source: Biswas *et al.*, 1993.

Based on typical rock types, structural fabric and depositional style, 26 sedimentary basins have been recognized and they have been divided into four categories (Figure 3.4) on the basis of their degree of prospectivity as presently known (DGH, 2006).

Category (I) Basins

Proven petroliferous basins with commercial production. They include Cambay, Assam Shelf, Mumbai Offshore, Krishna-Godavari, Cauvery, Assam- Arakan Fold Belt, and Rajasthan.

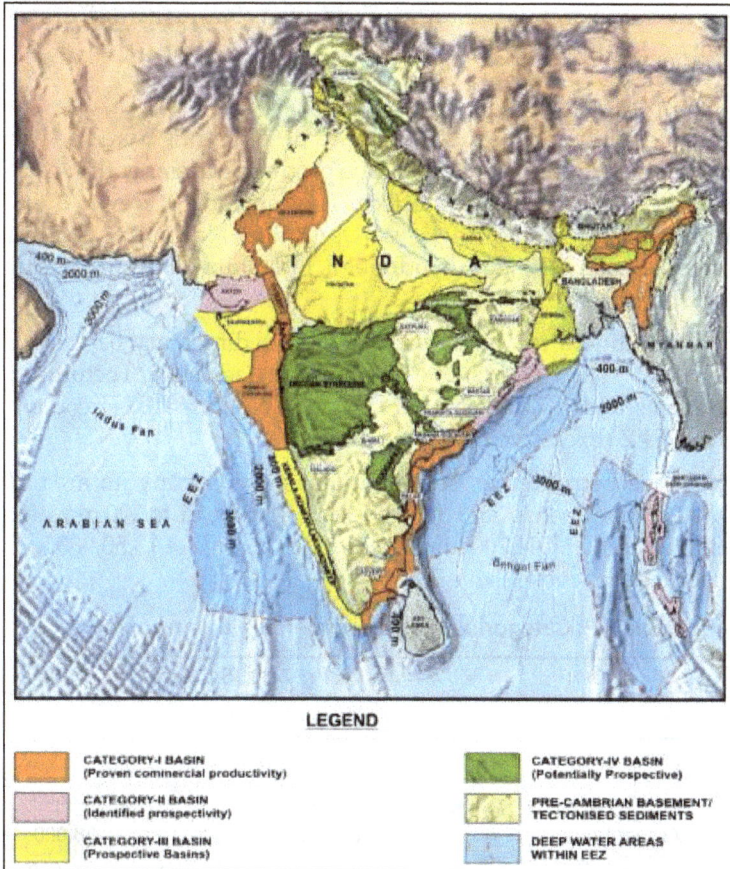

Figure 3.4: Sedimentary Basins of India
(*Source*: DGH, 2006)

Category (II) Basins

Identified prospectivity. They are basins of known hydrocarbon occurrence but no commercial production has yet been obtained. They include: Kutch, Mahanadi- NEC and Andaman –Nicobar

Category (III) Basins

Prospective. These are basins in which significant shows of hydrocarbons have not yet been found, but which on general geological grounds can be considered to be prospective. They include:

Himalayan Foreland, Ganga, Vindhyan, Saurashtra, Kerala-Konkan-Lakshdweep, and Bengal.

Category (IV) Basins

Potentially prospective. They are petroliferous basins which on analogy with similar hydrocarbon producing basins in the world can be considered as prospective. They include: Karewa, Spiti-Zanskar, Satpura-South Rewa-Damodar, Narmada, Deccan Syneclise, Bhima-Kaladgi, Cuddapah, Pranahita-Godavari, Bastar and Chhattisgarh.

In addition, two more categories of sedimentary basins are identified. They are: (*i*) Pre-Cambarian Basement/Tectonised sediments, and (*ii*) Deep water area (Kori –Comorin 85° E Norcodam) within EEZ (Exclusive Economic Zone).

These petroliferous sedimentary basins occupy an area of 13,90,200 km² on land and 3,94,500 km² Offshore upto 200 m water depth totalling 17,84,700 km². Besides this an area of 13,50,000 km² exists in deep water (Table 3.2)

Table 3.2: Categorisation of Sedimentary Basins of India

Category*	Basin	Basinal Area (Sq. Km.)		Total
		Onland	Offshore	
Up to 200M Isobath				
I	Cambay	51,000	2,500	53,500
	Assam Shelf	56,000	–	56.000
	Mumbai offshore	–	116,000	116,000
	Krishna Godavari	28,000	24,000	52,000
	Cauvery	25,000	30,000	55,000
	Assam-Arakan Fold Belt	60,000	–	60,000
	Rajasthan	126,000	–	126,000
	Sub Total	**346,000**	**172,500**	**518,500**
II	Kutch	35,000	13,000	45,000
	Mahanadi-NEC	55,000	14,000	69,000
	Andaman-Nicobar	6,000	41,000	47,000
	Sub Total	**96,000**	**68,000**	**164,000**

Contd....

Table 3.2–Contd...

Category*	Basin	Basinal Area (Sq. Km.)		Total
		Onland	Offshore	
III	Himalayan Foreland	30,000	–	30,000
	Ganga	186,000	–	186,000
	Vindhyan	162,000	–	162,000
	Saurashtra	52,000	28,000	80,000
	Kerala-Konkan-Lakshadweep	–	94,000	94,000
	Bengal	57,000	32,000	89,000
	Sub Total	**487,000**	**154,000**	**641,000**
IV	Karewa	3,700	–	3,700
	Splti-Zanskar	22,000	–	22,000
	Satpura-South Rewa-Damodar	46,000	–	46,000
	Narmada	17,000	–	17,000
	Deccan Syneclise	273,000	–	273,000
	Bhima-Kaladgi	8,500	–	8,500
	Cuddapah	39,000	–	39,000
	Pranhita-Godavari	15,000	–	15,000
	Bastar	5,000	–	5,000
	Chhattisgarh	32,000	–	32,000
	Sub Total	**461,200**	**–**	**461,200**
	Total	**1,390,200**	**394,500**	**1,784,700**
Deep Waters				
	Kori-Comorin			
	85°E	–	–	1,350,000
	Narcodam			
	Grand Total	**–**	**–**	**3,134,700**

* Categorization based on the prospectivity of the basin as presently known. The four recognized categories are basins which have:

I: Established commercial production.

II: Known accumulation of hydrocarbons but no commercial production as yet.

III: Indicated hydrocarbon shows that are considered geologically prospective.

IV: Uncertain potential which may be prospective by analogy with similar basins in the world.

This categorization will necessarily change with the results of further exploration.

Source: DGH, 2006.

3.2.1 Resources and Reserves of Petroleum

The hydrocarbon resource assessment of Indian petroliferous sedimentary basins has been carried out from time to time by ONGC. Rai *et al.*, (1994, 1998) estimated the total hydrocarbon potential (as on 1991) of 21,600 MMT of oil (O) and oil equivalent gas (O+OEG). As per the 2005–2006 estimates, the total hydrocarbon resources, inclusive of deep waters were around 28,085 MMT (million metric tones). Production of oil (O), gas (G) and (O+OEG) during the year 2005–2006 were 32.189 MMT, 32.20 BCM (billion cubic metres) and 64.389 MMT respectively (DGH, 2006).

Among the four categories of petroliferous sedimentary basins as mentained above (3.2.1), only Category I Basins where commercial qualities of oil and natural gas have been established are briefly described in this chapter.

3.3 Cambay Basin

Cambay Basin is an intra cratonic N-S trending rift graben which developed in the north-western part of Indian Peninsula and covers an area of 53,500 km^2 including the area within the Gulf of Cambay. From the Gulf area, the basin extends 450 km inland to the north of Sanchor where it is separated from the Rajasthan Basin by a basement swell. In the south, it merges with the Surat depression of the Mumbai Offshore Basin across the Gogha–Aliabet basement arc (Figure 3.5).

Stratigraphy, structural style and petroleum geology of the basin have been dealt extensively by several researchers (Mathur *et al.*, 1968; Raju 1968, 1979; Raju and Srinivasan, 1983,1993; Rao, 1969; Chandra and Chowdhury, 1969; Sudhakar and Basu, 1973; Bhandari and Chowdury, 1975; Kaila *et al.*, 1980; Biswas 1982,1987; Biswas *et al.*, 1993; Dhar and Bhattacharya, 1993).

The Cambay Basin has six large depressions separated by basement controlled uplifts and faults as seen in Landsat imageries (Lakshman Singh, 2000). This has led to the identification of five structural zones (Table 3.3) in the basin. They are Mehsana Block, Ahmedabad Block, Tarapur Block, Broach Block and Narmada Block (Raju and Srinivasan, 1993). In the Gulf of Cambay, the major structural elements comprises a NE-W trending arch (Gogha Aliabet arch), in the south a N-S trending uplift (Median ridge), and a depression (Gulf Low) in the east. The Gulf Low is the main kitchen

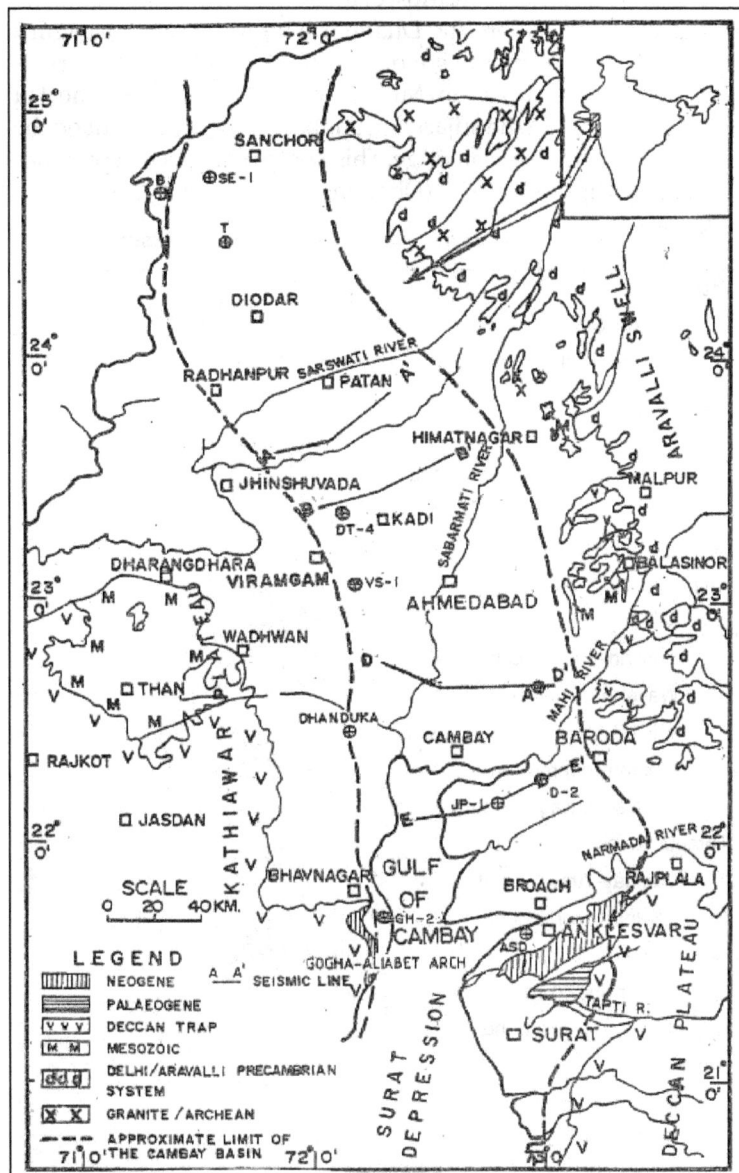

Figure 3.5: Location Map of Cambay Basin
(*Source*: Dhar and Bhattacharya, 1993)

for hydrocarbon generation. Most of the structural prospects lie along the Median ridge where the Dhadhar (Oligocene) and Babaguru (Early Miocene) formations would constitute the main reservoir facies. The Hazad Formation (Middle Eocene) which houses most of the hydrocarbons in the adjacent Gandhar Field pinches out on the eastern flank of the Median ridge. This updip pinchout belt provides excellent stratigraphic traps (Dhar and Bhattacharya, 1993).

Table 3.3: Structural Elements of Cambay Basin

Mehsana Block

    Unawa Cross Trend

    Western and Eastern basin rise zones

    Mehsana uplift

    Warosan depression

    Linen–Sobhasan cross trend

    Sobhasan uplift

    S. Kadi–Nandasan cross, trend

Ahmedabad Block

    Kalol uplift

    Sanand–Jhalora uplift

    Wavel depression

    Ahmedabad Eastern step fault zone

    Western depression

    Dholka–Nawagam cross trend

Tarapur Block

    Cambay–Aravali uplift

    Tarapur depression

    Western depression

    Eastern basin rise

    Kathana–cross trend

Broach Block

    Devla–Malpur uplift

    Tankari depression

    Jambusar rise

    Dadhar–Dabka cross trend

Contd...

Table 3.3–Contd...

Broach depression
Eastern basin rise
Aliabet uplift
Narmada cross trend
Gandhar Terrace
Dahej uplift
Narmada Block
Ankleshwar uplift
North Tapti–Kosamba uplift
Kim depression
Dumas depression
SE basin rise

Source: Raju and Srinivasan, 1993.

In the Cambay Basin, the major source rocks (Cambay Shale), reservoir rocks (Kadi, Kalol and Hazad formations) and the regional cap rocks (Tarapur Shale), successively lying one above the other and occurring in between the two basin wide unconformities (top of Olpad Formation and Tarapur Shale) constitute one mega petroleum system. This can be divided into two systems namely, Kalol Petroleum system to the north of the Jambusagar–Broach Block, and the Anklesvar petroleum system in the south. They are based principally on non-contiguity and difference in litho-association of the reservoir facies, and variation in qualifers (Table 3.4) such as major reservoir rocks charge factor, migration–drainage style, entrapment style and physical property of oil. (Demaison and Huizing, 1991; Dhar and Bhattacharya, 1993; Kuldeep Chandra *et al.*, 1994, 2001).

Several organic rich potential source rock sequences have been identified across the Cambay Basin. These include :

(i) Kalol Formation and its equivalent (Middle Eocene)

(*ii*) Cambay Shale Formation (Lower Eocene) and

(*iii*) Top and base of Olpad Formation (Paleocene)

Oligocene–Miocene sequence source rocks dominated by humic-sapropelic facies (Dadar Formation to Tarakeswar Formation) are

Table 3.4: Characteristics of Kalol and Anklesvar Petroleum Systems

	Kalol Petroleum System	*Anklesvar Petroleum System*
Major reservoir rocks	Deltaic sandstones intercalated with coals (Kadi and Kalol formations)	Deltaic sandstones nonassociated with coals (Hazad Formation).
Charge factor (Source Rock Index, SRI = thickness of source rocks × organic carbon contents per cent) for Cambay Shale	1500 (Ahmedabad–Mehsana block) to 1000 (Cambay–Tarapur block)	1875 to 2375 (Broach and Narmada blocks)
Migration drainage style	Vertical migration dominated over lateral in both, but lateral migration was comparatively more in the Anklesvar system than in the Kalol system.	
Entrapment style	Moderate impedance (resistance to dispersion)	High impedence.
Physical property of oil	Heavier oil (sp.gr.0.80–0.83)	Lighter oil (0.70–0.80)

Source: Dhar and Bhattacharya, 1993.

marginally mature; Paleocene–Upper Eocene sequence rocks dominated by sapropelic–humic facies (upper part of Olpad Formation to Anklesvar Formation) are sufficiently matured to produce hydrocarbons while Paleocene sequence source-rocks containing predominantly humic facies (lower part of Olpad Formation) are amenable for generation of gaseous hydrocarbons (Koshal, 1993).

Maturity modeling of potential source rocks in Kalol area indicates that they attained peak oil generation (Ro = 0.75 per cent) at top and base of Olpad Formation during 5 –45 MYBP (Millions of Years Before Present). Source rocks at top of Cambay Shale formation have entered the oil window (Ro = 0.5 per cent) in recent past whereas source rocks at the base of Kalol Formation have not entered in oil window. In Gandhar area source rocks at the top of Cambay Shale to base of Olpad have attained peak oil generation (Ro = 0.75 per cent) during 3-50 MYBP (Singh *et al.*, 1995; Banerjee *et al.*, 2000; 2002).

Cambay Basin is a rich hydrocarbon bearing province with estimated resource of 2050 MMT. Exploration in this basin was commenced in 1956 by ONGC and the first gas well was discovered in 1958. It was followed by the discovery of major oil in Ankalesvar Field in Narmada-Tapati Block during 1960. Subsequently several oil and gas fields have been discovered starting from Patan in north to Hazira in south (Figure 3.6). Maximum number of oil fields and large quantities of hydrocarbons occur in Kalol/Ankalesvar Formations of Middle to Upper Eocene age. The annual production (2005-2006) by ONGC of oil, gas and oil + oil equivalent of gas (O+OEG) in the Cambay Basin were 6.208 MMT, 2166.48 MMSCM (million metric standard cubic metres) and 8.375 MMT respectively. Production from Private (PVT) and Joint Venture (JV) companies during this period were 0.099 MMT oil, 2677.44 MMSCM gas and 6.14 MMT of O + OEG (DGH, 2006).

Detailed geochemical studies of oils in Cambay basin have shown dominant terrestrial organic matter input with Type III (vitrinite) kerogen and oxic environment of deposition. Majority of oils in the basin are paraffinic in nature. The oils of southern Cambay (Ankalesvar Formation) are thermally more evolved, updip migrated, better preserved and generated from more matured source rocks than northern Cambay basin (Kalol Formation) which are relatively heavy,

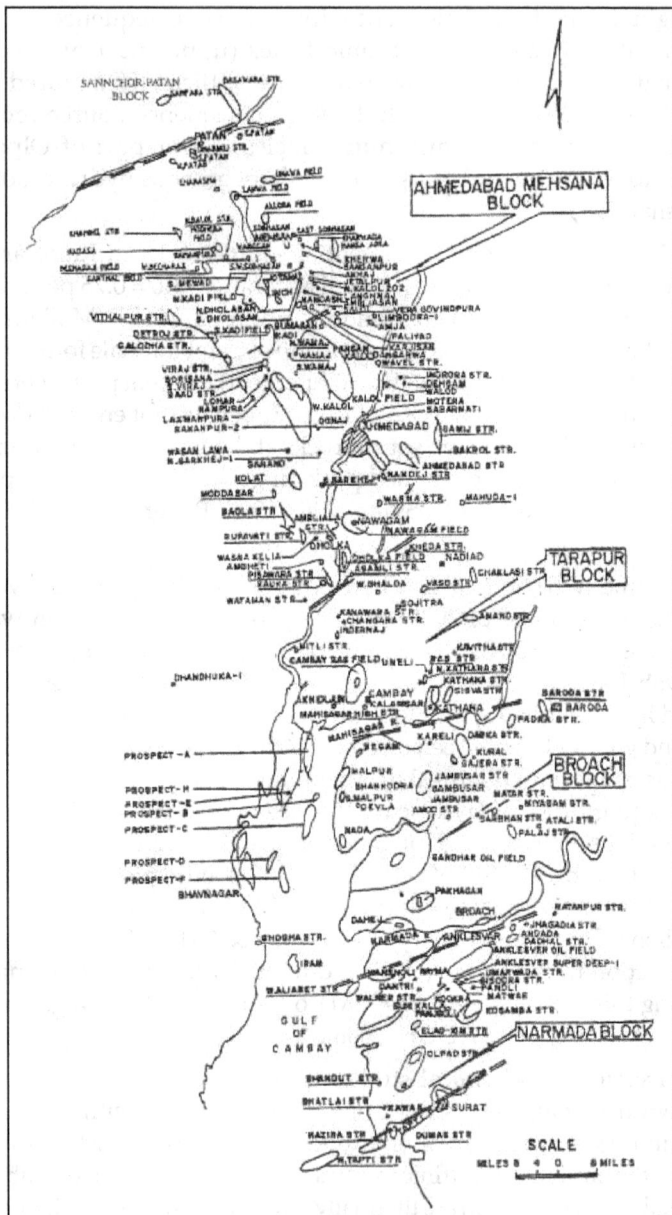

Figure 3.6: Oil and Gas Fields of Cambay Basin
(*Source*: Lakshman Singh, 2000)

less matured and altered (Kumar *et al.*, 1984; Gupta *et al.*, 1984; Singh *et al.*, 1987; Kuldeep Chandra *et al.*, 1994).

Chandra *et al.*, (1994) have established the oil–generation threshold *i.e.* the boundary between the diagenetic–catagenetic regimes at 1800 m in the Ahmedabad–Mehsana block of the Cambay basin based on several maturation parameters such as Rock–Eval Production idex, percent hydrocarbons and saturate/aromatic hydrocarbon ratio in EOM, Ts/Tm of 27 hopanes and methyl phenanthrene index (MPI-1). Natural gases from Mehsana and Mandhali pay sands of Sobhasan oil field of north Cambay (Mittal *et al.*, 2002) indicated thermogenic origin. The isotopic composition of ethane and propane gases of this oil field suggests their common origin and source which is in conformity with the biomarker study of oils in this area (Pande *et al.*, 1993).

3.4 Mumbai Offshore Basin

The Mumbai Offshore Basin is the central part of Western Continental margin of India. This margin, trending NW-SE and extending along the west coast upto Cape Comorin has the following major elements (Figure 3.7) from east to west.

(*i*) Shelf al Horst-Graben

(*ii*) Kori-Comorin Depression

(*iii*) Kori-Comarin Ridge

(*iv*) Arabian Abyssal Plain

The Mumbai Offshore Basin, mostly within Shelf al Horst-Graben Complex, is the most important petroliferous basin in the west coast sedimentary province of India. Virtually most of the commercially exploitable hydrocarbon fields discovered so far are located in this basin. It covers an area of 1,16,000 km² upto 200 m isobath. The basin is bounded by Saurashtra Arch in the north, Vengurla Arch in the south, Kori Arch in the west and the Western Ghats in the east. It is a divergent passive continental margin basin comprising three major structural units with carbonate dominant stratigraphy and three contiguous major depressions with clastic dominated fills. Tectonicallly, the basin is a horst-graben complex and divided into 12 structural units, each having several well marked sub-units (Figure 3.8). Table 3.5 summarises these units. (Mathur and Nair, 1993).

Table 3.5: Structural Units in Mumbai Offshore

Sl.No.	Structural Units	Subunits	Location	Max Sed. Fill (Km)	Classics/ Carbonates	Remarks
1.	Saurashtra Arch	–	NW of Saunisiltra Basin	0.8 2.5	Clastics and carbonates	Separates Kutch Basin from Saurashtra Basin
2.	Diu Arch	–	Between Surat Depression and Saurashtra Basin	4-5	Clastics	Possibly provided sealing to Surat Depression in Paleocene-Eocene
3.	Ratnagiri Arch	–	Between Murud depression and Rajapur Depression	3-4	Pelagic carbonates and mudstone	Extends to shelf and deep sea
4.	Vengurla Arch	–	Separates S. Ratnagiri block from Kerala–Konkan shelf	Thin	Carbonates	
5.	Kori Arch	–	Deep Sea Basin	3-4	Claystone with minor carbonates	
6.	Surat Depression	–	North and NE of Heera-Panna block	6-7	Clastics	Located on Narmada rift-Cambay Basin rift junction

Contd...

Table 3.5–Contd...

Sl.No.	Structural Units	Subunits	Location	Max Sed. Fill (Km)	Clastics/ Carbonates	Remarks
7.	Saurashtra Basin	–	North of Bombay platform	6–7	Clastics and carbonates	– Max. subsidence in post E. Oligocene – Merges westward with Shelf Margin Basin – Several carbonate banks present
8.	Shelf Margin Basin	– SM area west of Bombay Platform		6–7	Clastics	Max. subsidence in post E. Qliaocene
		– Murud Depression	South of Bombay Ratnagiri Arch	7	Mixed	–Muddy carbonates and claystone in Palaeocene-Eocene – Clayslone in Oligocene-Holocene
		– Rajapur Depression	South of Ratnagiri Arch	6	Pelagic carbonates and clayst, in Paleocene-Mid Miocene Clayst in post- Mid. Miocene	Starved during Paleoceae-Mid. Miocene

Contd...

Table 3.5–Contd...

Sl.No.	Structural Units	Subunits	Location	Max Sed. Fill (Km)	Classics/ Carbonates	Remarks
9.	Bombay Platform	– Deep Continental Shelf	Platform margin	4–5	Carbonates	Tons of Paleogene clay in south DCS.
		– Homoclinal Area	Between DCS and Bombay High	2.5–3.5	Carbonates	Eocene-E. Oligocene sequence wedgeout
		– Bombay High	Basement high between Homocline and Heera-Bassin block	2–2.5	Carbonates in L. Oligo-Mid. Miocene. Clastics in Mid. Miocene-Holocene	
10.	Heera-Bassein Platform	– Heera-Baasin Block	East and SE of Bombay High	2.5–3.5	Mixed	Contains several hydro carbon fields.
		– Central Graben	Between Carbonate platform and Eastern Homocline	3.5–4.5	Shale with minor carb.	Southward extension of Surat Depression
		– Eastern Homocline	Between coast line and Central Graben	0.5–3	Muddy Carb. and Clay stone	

Contd...

Table 3.5-Contd...

Sl.No.	Structural Units	Subunits	Location	Max Sed. Fill (Km)	Classics/ Carbonates	Remarks
11.	North Ratnagiri Platform	– Srivardhan Horst	Between SM and Vijaydurg Graben	2.5–3.5	Carbonates	Western flank dissected by numerous faults
		– Vijaydurg Graben	Between Srivardhan Horst and Jaygarh Homocline	3–4.5	Shales and Carbonates	
		– Jaygarh Homocline	Between coastline and Vijaydurg Graben	0.5–3	Muddy carb. and Claystone	
12.	South Ratnagiri Platform		Between Ratnagiri Arch and Vengurla Arch.	2–3.5	Eocene-E. Miocene carbonates	

Source: Mathur and Nair, 1993.

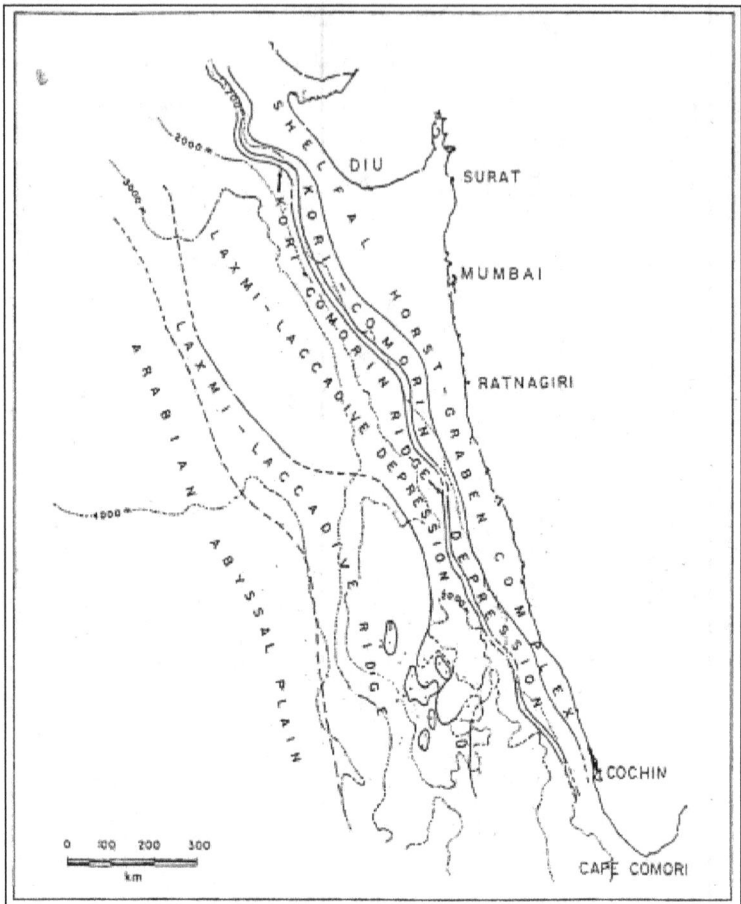

Figure 3.7: Tectonic Framework of Western Offshore Basin
(*Source*: Lakshman Singh, 2000)

Tectonic frame work, stratigraphy, structural style and depositional history of the Mumbai Offshore have been reported by several workers (Rao and Talukdar, 1980; Naini and Talwani, 1982; Basu *et al.*, 1980, 1982; Biswas, 1982, 1987, 1989, 1991; Biswas *et al.*, 1993, 1994; Mohan *et al.*, 1982, Mitra *et al.*, 1983; Sahi, 1986; Jokhan Ram *et al.*, 1998).

The Mumbai Offshore Basin has a tertiary sedimentary fill exceeding 5000 m directly overlying Deccan Trap floor. It varied

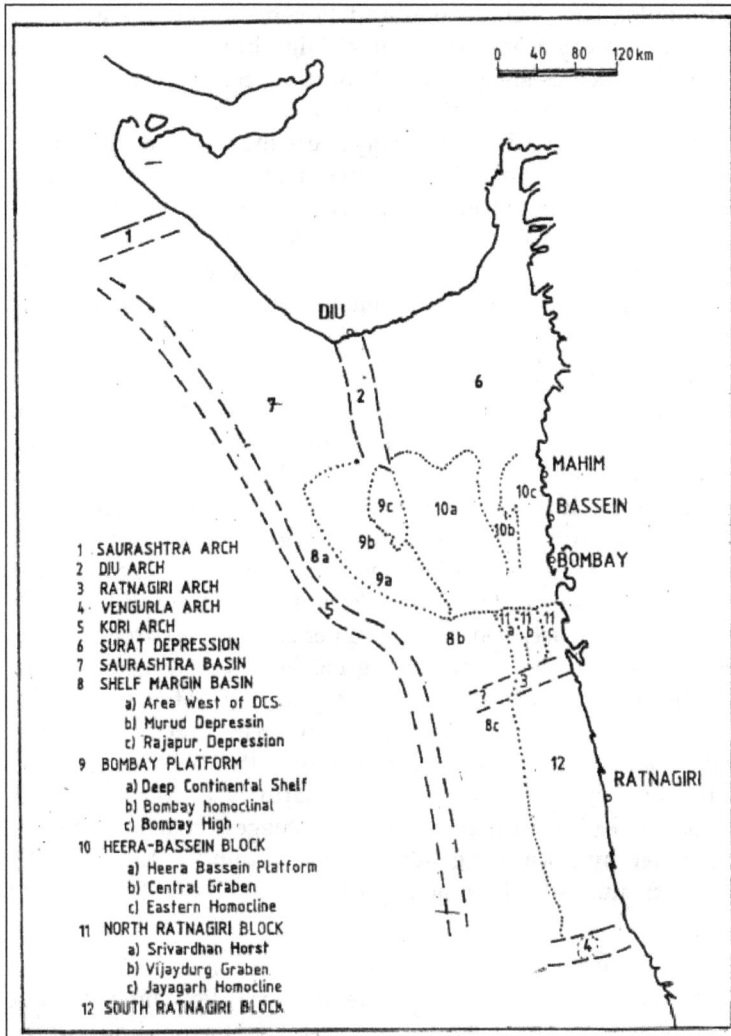

Figure 3.8: Structural Units of Mumbai Offshore Basin
(*Source*: Mathur and Nair, 1993)

widely as regards to its age, thickness and facies of its lithic fill. The thickness of Tertiary sediments over the Mumbai High structure is considerably less (1,800–2,000 m) as compared to its thickness in depocenters due to thinning of Oligocene and absence of Eocene sediments. These Oligocene and Eocene sediments are well developed away from the Mumbai High but occur in different lithofacies such as shales towards north in Surat depression and towards west of Mumbai platform, and as carbonates on Ratnagiri-Heera Ridge. The sedimentary sequence in the Mumbai High can be broadly classified into three lithological units–the lower unit composed of basal sand, lignite and clay of Paleocene age, the middle unit composed of thick limestone in the lower part and limestone–shale alterations in the upper part (Late Oligocene to Middle Miocene), and the upper unit composed of shale and claystone of post Middle Miocene age.

Synrift Paleocene-Early Eocene Panna Formation is the main source rock which has contributed to a large extent, for hydrocarbon generation and accumulation. This formation is spread over the entire Mumbai Offshore Basin except in Paleohighs. It is represented by sandstone at the bottom overlain by a section of coal shale alterations, suggesting coastal marginal marine (paralic) environment of deposition. However, the upper limit is represented by marine shales. In the Panna Formation, TOC (Total Organic Carbon) ranges between 0.5. to 20.4 per cent and Hydrogen Index (HI) ranges between 50 to 300 mg HC/g TOC. The sedimentary organic matter is in the catagenetic stage of maturation (T_{max} > 440° C). TOC and Rock-Eval pyrolysis (6.5.0) data indicates the presence of good and efficient source rocks in Upper Panna Formation. The burial history and progression of maturity at Middle Panna Formation (Kuldeep Chandra et al., 2001) suggests that this sequence has started hydrocarbon generation (GT) around 20 MY (Ro = 0.6 per cent), and critical moment (CM) around 10MY (Ro = 0.8 per cent).

Both limestones and sandstones are the main reservoir rocks in the Mumbai Offshore Basin. In the Mumbai High, limestones of Miocene sequence are the main reservoirs. The Bassein limestone of Middle Eocene to Oligocene age is an important reservoir in structures along Ratnagiri-Heera Ridge. Thin sandstone basal reservoirs of local significance occur in Mumbai High towards north. The sandstones of Oligocene-Lower Miocene age are important as gas reservoirs in Tapati Fields. The post Middle Miocene shales are the principal cap rocks throughout the Basin, even though Upper Oligocene shales provide cap rocks for all oil and gas fields in Bassein Formation. In addition, carbonates also act as seals in several cases.

The first Offshore well in Mumbai High the giant oil and gas field, was drilled in 1974. This was followed by several wells in Offshore Basin–Mukta, Panna, Bassein, Neelam, Heera, Tapati etc (Figure 3.9). Recent assessment (DGH, 2006) places the prognosticated hydrocarbon resource of Mumbai High at 9190 MMT. The annual production during 2005–2006 of oil, gas and (O+OEG) by ONGC were 16.31 MMT, 16, 943.1 MMSCM and 33.25 MMT respectively. Production by Joint Venture (JV) companies (BG-RIL-

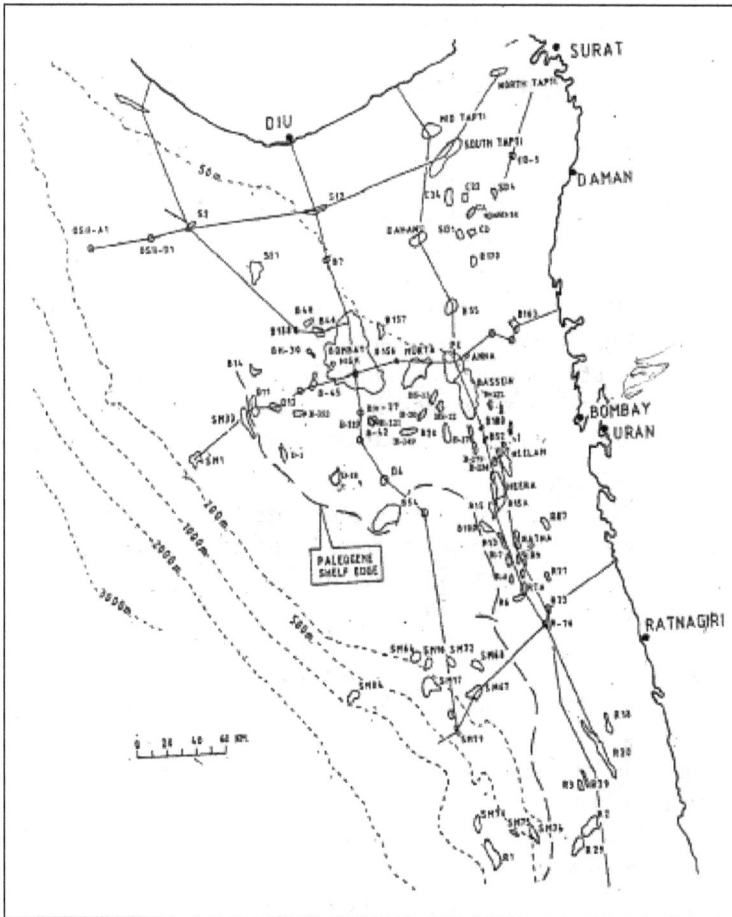

Figure 3.9: Map Showing Oil and Gas Fields in Mumbai Offshore
(*Source*: Lakshman Singh, 2000)

ONGC) during this period was 1.685 MMT, 3,688.31 MMSCM and 5.373 MMT respectively.

The Mumbai High oils are paraffinic and light (high API gravity) with relatively high wax content typically generated from organic matter of dominant terrestrial source. Their source may be either from single source rock or mixture of different source rocks. Biomarker study of Neelam Field oil wells and in the structures of deep continental shelf (DCS) revealed normal paraffin distribution and isoprenoid ratios typical of terrestrial organic source input deposited under oxic to peat swamp environment (Mathur *et al.*, 1993). Chemical composition of gases in general, indicate that they are typical associated gases generated from matured source rocks.

3.5 Krishna–Godavari (K–G) Basin

The Krishna-Godavari (K-G) Basin is located along the east coast of Indian Peninsula. It includes the deltaic plains of the Krishna and Godavari rivers and interdeltaic regions. Geographically the basin lies between Kakinada in the northeast and Ongole in the southwest. It extends southeast into deep waters of the Bay of Bengal (Figure 3.10). The basin covers an area of 28,000 km^2 on land and 24,000 km^2 in the Bay of Bengal upto 200 m isobath (DGH, 2006). Greater part of the onland basin is covered by alluvium of Krishna and Godavari rivers draining into the Bay of Bengal.

The basin was a major intra cratonic rift until the early Jurassic (Thomson, 1976). Since the Cretaceous, it has become a pericratonic basin (Biswas, 1992). During the Tertiary, the deltaic system generally prograded to the southeast, although some deltaic lobes have shifted in direction in response to changing rates of sediment flux and growth faulting (Ranga Raju, 1987). Murty and Ramakrishna (1980) have identified three sub-basins separated by two basement horsts. From the southwest, these are Krishna, West Godavari and East Godavari sub-basins separated by the Bapatla and Tanuku horsts respectively. The west Godavari sub-basin is further divided into the Gudivada and Bantumalli grabens, which are separated by the Kaza-Kaikalur horst (Kumar, 1983). Kommagudem and Mandapeta troughs are situated on either side of the Tanuku horst.

Exploration, hydrocarbon potential and petroleum systems of K-G Basin have appeared in several research publications (Rao, 1991, 1993, 1994, 2001; Rao and Mani, 1993; Venkatarengan and

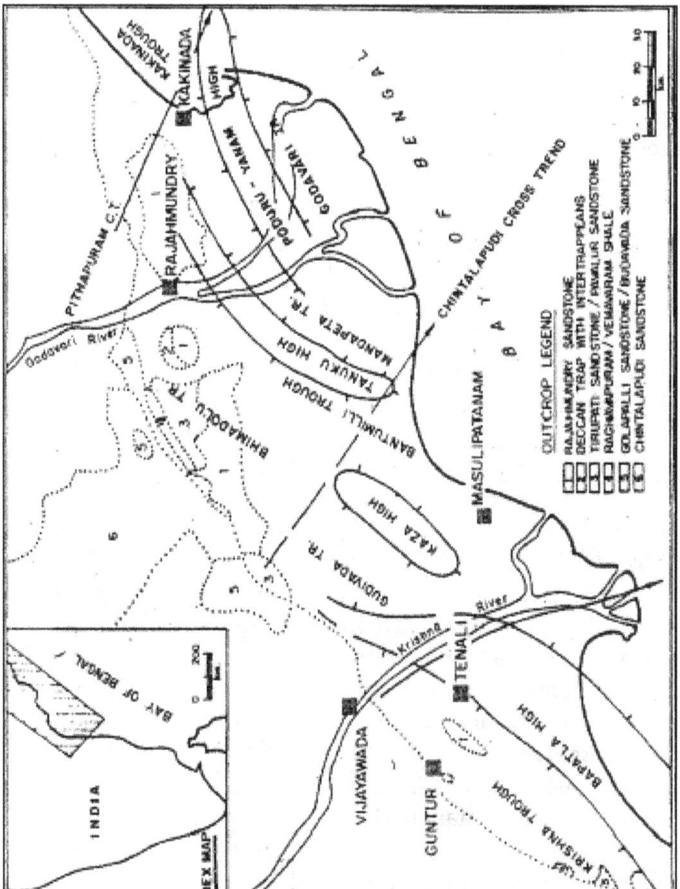

Figure 3.10: Geological and Tectonic Map of Krishna-Godavari Basin (*Source:* Lakshman Singh, 2000)

Ray, 1993; Gupta *et al.*, 1998; Kuldeep Chandra, 2001). While Kuldeep Chandra *et al.*, (2001) identified three major petroleum systems as mentioned earlier (3.2.1), Rao *et al.*, (2001) reported four systems in the K-G Basin. They are (*i*) Kommagudem–Mandapeta, (*ii*) Raghavapuram–Gollapalli, (*iii*) Palakollu–Pasarlapudi and (*iv*) Vadaparru (Ravva)–Godavari systems.

3.5.1 Kommagudem–Mandapeta Petroleum System

It is the oldest (Pre Cretaceous) petroleum system. Kommagudem Formation of Upper Early Permain is the principal source rock in the Mandapeta area. It has very good source potential and is in an advanced stage of maturation (T_{max} = 433–516° C, Ro = 1.0–1.3 per cent). However, the production index (PI; 6.5.0) is very low (0.05–0.15) as the Formation comprises source rocks of coal/sand shale alterations. Commercial quantities of gaseous hydrocarbons of thermogenic origin have been found in the Mandapeta Sandstone reservoir of Triassic age (Prasad *et al.*, 1995). Gas to gas and condensate to source extract correlations indicate that gas/condensates occurring in the Mandapeta Sandstones were generated from Kommagudem Formation and possibly migrated through the faults into the reservoir rock. The local seal of variegated clays (red beds) act as the cap rock in this petroleum system.

3.5.2 Raghavapuram–Gollapalli Petroleum System

In the West Godavari sub-basin, thin sands and limestones occurring within Konakollu Formation in Lingala-Kaikalur are the main reservoirs. In the East Godavari sub-basin, lenses of sands in the Chintapalli Claystone produce hydrocarbons (Rao, 2001). In the Gudivada and Bantumalli grabens of West Godavari sub-basin, potential source rocks are Krishna/Gollapalli Formations and Gajulapadu Shale. Krishna and Gollapalli Formations are inferred to have deposited in marginal marine environment whereas Gajulapadu (Raghavapuram) Shale was deposited in the middle shelf regime. Geochemical data indicates that Krishna/Gollapalli Formations are most likely the major source rock sequences. They are adequately rich in organic matter with TOC ranging 0.65–10.8 per cent. The Lower part of the Gajulapadu Shale has good potential (TOC>3 per cent). It is therefore possible that this sequence might have also generated hydrocarbons in the deeper part of the basin. These sequences are in catagenetic zone (T_{max} > 440°C and HI in the range 50-231 mg HC/g. TOC). Hydrocarbons in these reservoirs are probably generated from Krishna/Gollapalli Formations to a large extent, and migrated through the faults giving rise to this petroleum system (Kuldeep Chandra *et al.*, 2001). Raghavapuram Shale acts as cap rock for its sand as well as sands within Konakollu/Gollapalli Formations.

3.5.3 Palakollu–Pasarlapudi Petroleum System

This Paleogene system is the most prolific petroleum system in K-G Basin. Located south east of Matsyapuri–Palakollu Fault Zone (MPFZ) in the East Godavari sub-basin, the system contains abnormally pressured source rocks and normally pressured reservoir rocks (Rao and Mani, 1993). Anticlinal closures serve to entrap hydrocarbons. Palakollu Shale is the source for gas accumulations in thin Cretaceous reservoirs. Bantumalli Sandstone, Razole Valconics and Vadaparru Formation are the sources for the main Paleocene–Eocene hydrocarbon fields constituting this important petroleum system (Venkatarengan and Ray, 1993).

In the Pasarlapudi area, Palakollu Shale of Paleocene age shows fair to good source rock potential (Neerja *et al.*, 1997) with proclivity to generate dominantly gaseous hydrocarbons. It has TOC in the range 0.6 to 5.18 per cent and HI in the range 20-249 mg HC/g TOC. Good quality organic matter (OM) is concentrated in Mori-Razole-Chintalapalli-Tatipaka-Pasarlapudi belt. Maturity of sedimentary OM ranges from early phase to main phase of hydrocarbon generation. The OM is dominantly humic type rich in inertinite with about 10-20 per cent exinite macerals. Pasarlapudi Foramtion of Lower Eocene consisting of alterations of sandstone and shale with occasional limestone beds act as the main reservoir. It was deposited under cyclic conditions upto 30 m of paleobathymetry. Geochemical studies suggest that hydrocarbons have possibly migrated through vertical faults into the reservoir rocks. In Pallakollu-Pasarlapadi system, the intervening shales of Pasarlapudi Formation/Bhimannapalli Limestone are acting as cap rocks deposited under fluctuating inner shelf environment (Kuldeep Chandra, *et al.*, 2001). Subsidence history of Palakollu Shale suggests that the generation threshold occurred around 43 MY (Ro = 0.5 per cent) during Middle Eocene.

3.5.4 Vadaparru (Ravva)–Godavari Petroleum System

This Neogen Petroleum system is the most promising area for commercial hydrocarbon production in the Offshore. The abnormally high geothermal gradient has caused Lower Miocene sediments to mature and generate hydrocarbons (Rao, 2001). Ravva oil field is located in the Offshore shallow marine setup, south of Amalapuram. The potential source rock in this sub-basin is Vadaparru Shale of Eocene/Oligocene age. TOC of this sequence ranges between 1.1 to 6.7 per cent (ave: 4.2 per cent) and HI ranges between 50-250 mg HC/g TOC. The sedimentary OM is in the early stage of maturation (Neerja *et al.*, 1997) and it shows an increasing trend towards Offshore area. The OM in Vadaparru Formation is dominantly Type III (5.2.2) and it has potential to generate both oil and gas. Ravva reservoirs were deposited in geochemical events during sea level fluctuations confining the hydrocarbon accumulations in shelf/slope transition (Kuldeep Chandra *et al.*, 2001). Generation of threshold for Vadaparru Shale (Ro = 0.70 per cent) occurred during Lower Miocene (23 MY). Sand sequences of Matsyapuri/Ravva Formations in coastal and Offshore areas and sand bodies within the overlying Godavari Clay of Pliocene age are the potential reservoir rocks. Vadaparru Shale, intercalations within Matsyapuri and Ravva Formations, and Godavari Clay are the effective cap rocks.

Entrapment in the K-G Basin is mainly controlled by development of medium to high amplitude anticlines often dissected by a series of faults. The hydrocarbon accumulations both stratigraphic and structural are of limited vertical and lateral extent.

Extensive geochemical data have been generated to identify effective source rocks in the K-G Basin (Thomas *et al.*, 1991; Brahmaji Rao *et al.*, 1991; Philip *et al.*, 1991a, b; Benerjie *et al.*, 1994). Based on oil–oil, oil-source rock and gas-source rock correlations the drainage style was inferred as dominantly lateral (Chandra *et al.*, 1996).

Initial hydrocarbon exploration in K-G Basin was in thin Upper Cretaceous reservoirs in the Narsapur structure (Figure 3.11) of the East Godavari sub-basin. Subsequently several oil and gas wells were drilled in this basin. About 1130 MMT of hydrocarbon resource has been prognosticated (both onshore and offshore) of which over 0.216 MMT of oil, 1663.24 MMSCM of gas and over 1.879 MMT of O+OEG were produced by ONGC during the year 2005-2006. Cairn

KZ	- KAZA	TP	- TATIPAKA
NG	- MANDIGAMA	KP	- KESANAPALLI
KK	- KAIKALUR	MO	- MANDAPETA
BT	- BANTUMILLI	EM	- ENDAMURU
LG	- LINGALA	MM	- MUMMIDIVARAM
NP	- NARSAPUR	BP	- BHIMANAPALLI
RZ	- RAZOLE	BN	- BANDAMURULANKA
EL	- ELAMANCHILI	VA	- VADALI
MP	- MEDAPADU	MR-E	- MORI-EAST
PM	- PENUMADAM	KD	- KADALI
AC	- ACHANTA	MK	- MULLIKIPALLI
MN	- MANEPALLI	AD	- ADAVAPALEH
CP	- CHINTALAPALLI	LK	- LANKAPALEH
PO	- PONNAMANDA	PO	- PONNAMANDA
MR	- MORI	KW	- KESANAPALLI WEST
PS	- PASARLAPUDI		

Index

▦ Gas

▦ Oil

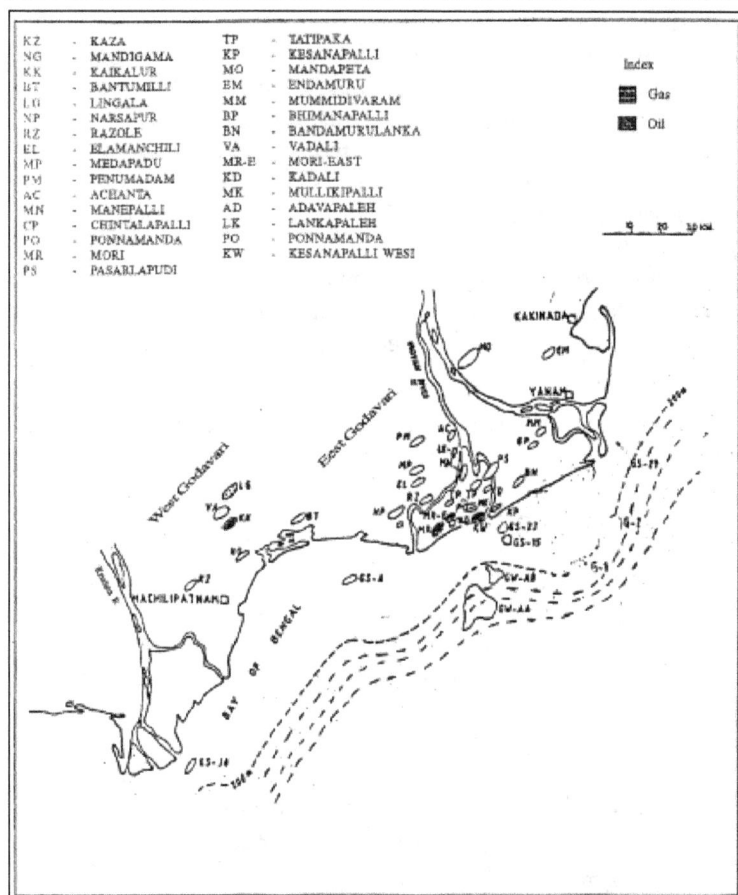

Figure 3.11: Map Showing the Location of Oil and
Gas Fields m K-G Basin
(*Source*: Lakshman Singh, 2000)

Energy, a PVT multinational company produced 2.452 MMT of oil and 911.625 MMSCM of gas in the K-G Basin Offshore during this period (DGH, 2006).

The Offshore oils of K-G Basin (Table 3.6) can be classified as high wax, low sulphur, low asphaltic and saturate rich crudes (Kumar *et al.*, 1985). The API gravities vary in the range 30-33° and C.P.I around 1.0 indicating normal crudes. Remarkable consistency in Pristane/Phytane ratio 5.8-6.4 shows a close genetic similarity among them. Further, the high Pristane/Phytane ratio (>3.0) suggests dominant terrestrial input of OM. Relatively high abundance of saturate hydrocarbons (67-79 per cent) shows that the oils are well preserved and have not been affected by biodegradation. Gases formed at Tatipaka, Pasarlapudi, Chintalapudi, Narsapur, Mandapeta and Kaza are in free state. They are in association with oils at Ravva in Offshore wells, and at Lingala, Kaikalur and Bantumalli in onland wells. Condensates are also produced with gas at Mandapeta, Pasarlapudi and Tatipaka wells. They have been derived from local sources in each sub-basin (Lakshman Singh, 2000).

Table 3.6: Physico-chemical Characteristics of Oils of K-G Basin

Characteristics	G–1–1 Oil (2160–65 m)	G–2–2 Oil (iii) (2026–31 m)	G–2–2 Oil (iv) (2010–14 m)
Sp. Gr (60/60°F)	0.8697	0.8743	0.8601
°API (Gravity)	31.20	30.34	33.01
Wax content (%))	20.1	9.2	11.0
Sulfur content (%))	0.15	0.46	0.19
Asphaltenes (%)	–	0.32	0.10
Group composition (%) wt			
Saturates	79.07	67.38	66.74
Aromatics	15.17	27.64	26.98
NSO's	5.76	4.98	5.20
n-Alkane range	C_{10}–C_{30}	C_{11}–C_{34}	C_{15}–C_{34}
C max.	C_{24}	C_{15}	C_{20}
CPI	1.00	1.14	1.18
Pristane/Phytane	6.20	5.76	6.36
Pristane/nC_{17}	2.20	3.33	4.47
Phytane/nC_{18}	0.32	0.59	0.66
Pr+nC_{17}/Phy+nC_{18}	2.20	2.78	3.10

Source: Kumar *et al.*, 1985

3.6.0 Cauvery Basin

The Cauvery Basin is a pericratonic basin located on the east coast of southern India, occupying most of the coastal plains of

Tamil Nadu and Pondicherry and extending Offshore into the Bay of Bengal (Figure 3.12). It occupies an area of 25,000 km² on the onshore and 30, 000 km² in the offshore. The southern part of the basin extends across Palk Bay and Gulf of Mannar into the western coastal area of Sri Lanka. Major tectonic units of the Cauvery Basin are parallel to the NE-SW Eastern Ghat trend and the basin consist of horst and graben morphology with the following tectonic elements from north.

1. Ariyalur–Pondicherry sub-basin
2. Kumbakonam–Madanam Ridge
3. Tanjavur sub-basin
4. Tranquebar sub-basin
5. Karaikal Ridge
6. Nagapattinam sub-basin
7. Vedaranyam Ridge
8. Palk Bay sub-basin
9. Ramnad-Palk Bay sub-basin
10. Mandapam Ridge
11. Mannargudi Ridge
12. Mannar sub-basin

The old sedimentary sequences encountered in the basin are of Late Jurassic age. At the beginning of the Cretaceous, sedimentation occurred in many sub-basinal depressions (Chandra and Venkataraman, 1988). Several transgressive phases are recognized during the Cretaceous age. By the end of Early Paleocene, sedimentation was wide spread on both the basement ridges and lows. Major regression in the Late Paleocene caused poor sedimentation during this period. Sedimentation was wide spread during Eocene and Oligocene times which witnessed progradation of deltas and development of turbidite fans. The basement highs also received considerable sediment cover during this period (Thomas and Sharma, 1993). Exploration history, geological setting, tectonic frame work and stratigraphy of this basin have been described by several researchers (Kumar, 1983; Sudhakar, 1991; Srivastava *et al.*, 1993; Prabhakar and Zutshi, 1993; Ranga Raju *et al.*, 1993).

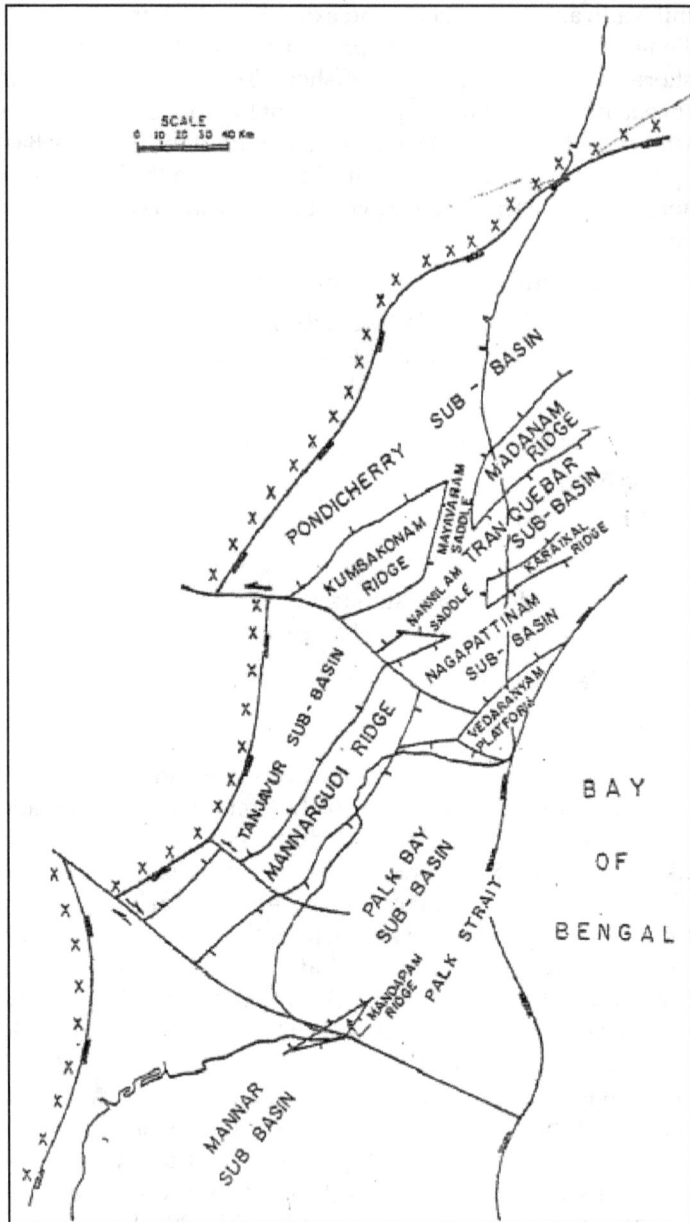

Figure 3.12: Tectonic Map of Cauvery Basin
(*Source*; Thomas and Sharma, 1993)

The postrift Cretaceous shales namely Porto-Novo, Kudavasal, Sattapadi and synrift Andimadam Formation are the potential source rocks in different sub-basins. Geochemical analysis indicates that Porto-Novo and Kudavasal Shales are lean in organic matter (OM) content and have very poor hydrocarbon generation potentials. Andimadam Formation has better source rock characteristics than Sattapadi Formation in wells nearer to depocentres. The OM in Andimadam Formation is mainly amorphous whereas Sattipadi Shale is rich in inertinite which indicates that the former is the main source rock near sub-basin depocentre (Kuldeep Chandra *et al.*, 2001). TOC in Andimadam Formation ranges from 0.35-10.7 per cent in different sub-basins namely, Tanjavur, Ariyalur-Pondicherry, Tranquebar and Nagapattinam. In Nagapattinam and Ramnad sub-basins, HI ranges between 176-416 mg HC/g TOC whereas in other sub-basins it ranges from 40-180 mg HC/g TOC. The OM in Andimadam Formation is in catagentic stage of maturation.

Based on source rock analysis and correlation studies, Khan *et al.*, (1995) envisaged the following petroleum systems (at an hypothetical level of certainty)

1. Andimadam–Bhuvanagiri system in Ariyalur–Pondicherry sub-basin.
2. Andimadam–Nannilam system in Tranquebar and Ramnad sub-basins.
3. Andimadam–Neravi system.

The main reservoir in the Ariyalur–Pondicherry sub-basin is the Bhuvanagiri Formation of Turonian age, deposited in an inner to middle shelf with drop in sea level near top. The deep marine Kudavasal Shale of Coniacian age is the seal rock.

Hydrocarbons bearing reservoirs in the Tranquebar and Ramnad sub-basins are mainly in the Nannilam Formation which is dated Santonian to Lower Campanian, and the cap rock is Komarakshi Shale of Upper Campanian age. Both reservoir and seal rocks were deposited in the middle shelf. No significant discoveries have been made in the Tanjavur sub-basin. In the Nagapattinam sub-basin, Neravi Formation of Upper Eocene to Lower Oligocene age forms the main reservoir deposited in a shallow inner shelf condition. The Shiyali Claystone of Upper Oligocene to Lower Miocene forms the cap rock. Besides the Neravi Formation, hydrocarbons have been discovered in the Bhuvanagiri, Nannilam and Kamalapuram Formations. Hydrocarbons also occur in Bhuvanagiri Formation in the Tranquebar sub-basin. Maximum hydrocarbon discoveries have been in the southern flank of Karaikal Ridge namely Narimannam, Thiruvarur, Kamalapuram, Adiakkamanglam and Nannilam areas.

Oil and gas pools in the Cauvery Basin (Figure 3.13) are dispersed in space and time. Accumulations are known in various geological formations from Pre-Cambrian fractured basement (Mattur and Pundi Fields) to Oligocene multistacked sandstone reservoirs (Narimanam Field). Most of the oil fields are in sandstone reservoirs in the depth range from 1080 to 3600 m in both structural and structural cum stratigraphic traps. The oils of Cauvery Basin are Paraffinic-Naphthenic in composition, mostly light to very light (37.5°-47.3° API gravity), and contain upto 5.8 per cent wax, upto 0.48 per cent asphaltenes and upto 1410 ppm sulphur (Table 3.7). These oils are derived from mixed organic source matter deposited in a reducing environment. The paleotemperature gradients observed are around 2.9° C/100 m in the depressions and 4.7° C/100 m on the highs (Kuldeep Chandra and Samanta, 1984). Based on biomarker and isotopic composition, the oils derived from various sub-basins reveal genetic differences except Oligocene oil of Narimanam and Eocene oil of Karaikal in Nagapattinam sub-basin (Pande, 1994). Gases accumulated in various structures are generated from localized source rocks at different maturities (Ro = 0.71-1.15 per cent) and they are genetically unrelated.

Exploratory efforts in the Cauvery Basin commenced from 1964. Since then several oil and gas wells were drilled both on land and offshore. The estimated hydrocarbon resource of this basin was 270 MMT offshore and 430 MMT on land totaling 700 MMT. The onland production figures in this basin by ONGC during the year 2005-2006 were 0.385 MMT oil, 906.33 MMSCM gas and 1.291 MMT of O + OEG. In addition, the private sector company (HARDY) produced 0.256 MMT oil, 68.246 MMSCM gas and 0.324 MMT of O+OEG from offshore during the same period (DGH, 2006).

3.7.0 Assam-Arakan Basin

Assam Basin, located in the alluvial foreland shelf zone known as Upper Assam Valley is the earliest known petroliferous basin of

Table 3.7: Compositional Parameters of Oils of Cauvery Basin

Sl.No.	Well Name	Sand	API Gravity	Pour Point °C	Wax Content % wt	Asphaltenes % wt	Sulphur ppm	Nickel ppm	H_2S % vol	Class*	Pr/Ph	Pr/nC_{17}	Ph/nC_{18}	$\frac{Pr+nC_{17}}{Ph+nC_{18}}$	Cmax
Group-1 Eocene Pasysands															
1.	NM-1	Eocene	44.7	<-12	3.6	0.01	330	1.24	Nil	PN	4.10	0.32	0.09	1.37	C_{17}, C_{19}
2.	NM-2	Eocene	41.5	<-9	3.4	0.23	600	1.46	Nil	P	3.20	0.48	0.17	1.44	C_{16}, C_{19}
3.	NM-3	Eocene	42.6	<-9	3.8	0.14	310	1.36	Nil	PN	2.61	0.41	0.18	1.35	C_{15}
4.	NM-4	Eocene	41.3	6	2.30	0.22	390	1.10	Nil	P	2.31	0.58	0.25	1.28	C_{19}
5.	NM-5	Eocene	43.3	6	5.8	0.26	500	ND	Nil	P	2.04	0.58	0.27	1.18	C_{19}
Group-II Oligocene Paysands															
6.	NM-6	Sand-1	47.3	<-9	2.8	0.01	1180	1.18	Nil	PN	2.46	0.57	0.24	1.31	C_{17}, C_{19}
7.	NM-7	Sand-1	45.7	<-12	2.3	0.15	700	1.08	0.02	PN	2.58	0.60	0.25	1.37	C_{16}, C_{19}
8.	NM-8	Smid-1	41.6	<-9	3.12	0.30	920	0.98	Nil	PN	2.51	0.60	0.27	1.42	C_{17}
9.	NM-9	Sand-1	44.1	6	2.32	0.05	610	1.1	0.01	PN	2.40	0.54	0.25	1-35	C_{15}
10.	NM-10	Sand-1	46.9	6	1.91	0.29	630	0.92	0.11	PN	2.56	0.58	0.25	1.40	C_{15}
11.	NM-11	Sand-1	44.1	9	3.64	Traces	520	0.36	Nil	PN	2.10	0.54	0.29	1.34	C_{17}
12.	NM-12	Sand-1	43.1	<-6	5.2	0.08	ND	ND	Nil	PN	1.79	0.51	0,31	1.26	C_{15}
13.	NM-13	Sand-1	45.9	<-6	ND	0.06	ND	ND	Nil	PN	2.29	0.66	0.32	1.38	C_{15}

Contd...

Table 3.7.—Contd...

Sl.No.	Well Name	Sand	API Gravity °	Pour Point °C	Wax Content % wt	Asphaltenes % wt	Sulphur ppm	Nickel ppm	H_2S % vol	Class*	Pr/Ph	Pr/ nC_{17}	Ph/ nC_{18}	$\dfrac{Pr+nC_{17}}{Ph+nC_{18}}$	Cmax
14.	NM-14	Sand-2	41.9	<-9	1.5	Traces	910	0.45	0.15	PN	2.40	0.58	0.26	1.37	C_{16}, C_{19}
15.	NM-15	Sand-2	42.3	<-9	3.9	0.17	800	0.91	0.09	PN	2.20	0.52	0.25	1.33	C_{16}, C_{19}
16,	NM-16	Sand-3	44.4	<-12	3.8	0.48	800	1.53	0.39	PN	3.00	0.45	0.17	1.45	C_{16}, C_{19}
17.	NM-17	Sand-3	39.5	<-12	3.01	0.10	1410	1.65	1.15	PN	2.14	0.96	0.43	1.17	C_{19}
18.	NM-18	Sand-3	47.0	<-9	2.53	0.15	1000	ND	0.30	PN	2.6	0.64	0.27	1.44	C_{14}
19.	NM-19	Sand-3	44.9	<-6	5.20	0.08	ND	ND	0.20	PN	1.93	0.54	0.31	1.31	C_{15}
20.	NM-20	Sand-4	43.8	-3	3.69	0.09	1140	ND	0.49	PN	2.40	0.60	0.28	1.40	C_{15}
Group-III Oligocene Paysands															
21.	NM-21	Sand-5	38.4	<-12	3.6	0.01	1180	1.54	0.01	PN	4.40	1.74	0.42	2.00	C_{19}
22.	NM-22	Sand-5	37.5	<-6	4.6	0.01	1165	1.71	0.04	N	2.14	0.96	0.43	1.17	C_{19}
23.	NM-23	Sand-5	38.1	6.2	0.03	0.03	1260	ND	0.09	N	2.14	1.16	0.67	1.60	C_{17}

*P: Paraffinic; PN: Paraffinic–Naphtenic; N: Naphtenic; ND: Not determined.

Source: Lakshman Singh, 2000.

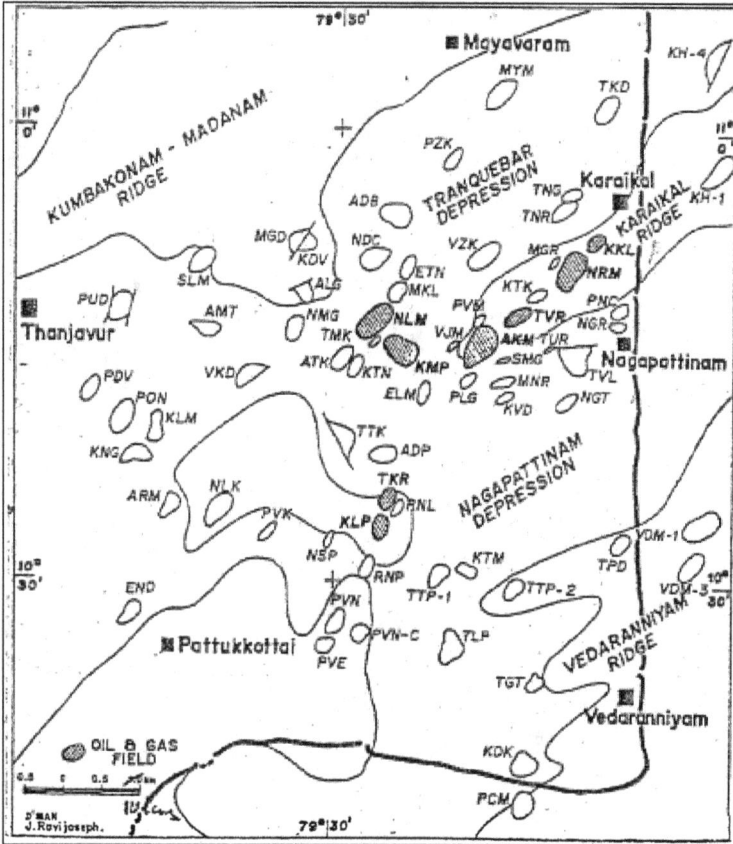

Figure 3.13: Map Showing Oil and Gas Fields of Cauvery Basin
(*Source*: Lakshman Singh, 2000)

India. Upper Assam Valley together with Assam-Arakan Fold Belt (AAFB) constitute the Assam-Arakan Basin. The basin covers the states of Manipur, Mizoram, Nagaland, Tripura, Assam and parts of Meghalaya. It occupies an area of 1,11,000 km² including 51,000 km² of Assam Shelf. The Upper Assam (Shelf) Basin extends from NE of Shillong to Mishmi Hills and includes the entire Brahmaputra Valley bound by Himalayan ranges in the north and Naga Hills in the south.

Following are the major structural elements (Figure 3.14) of Assam-Arakan province (Ahmed *et al.*, 1993).

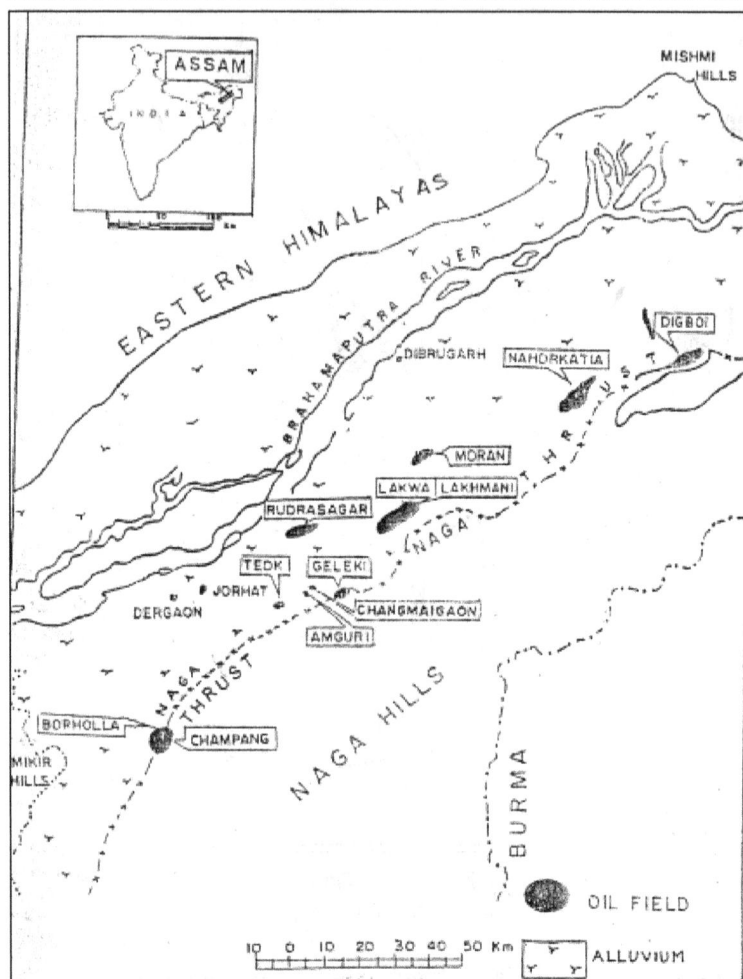

Figure 3.14: Index Map of Upper Assam Basin
(*Source*: Ahmed *et al.*, 1993)

1. Eastern Himalayas comprising south and south easterly moving thrust sheet of Tertiary rocks (including Mishmi Hills in the NE).

2. Naga Hills, consisting of over thrust masses (belt of Schuppen)

3. Assam Valley covered by alluvium.

4. Miker Hills and Shillong Plateau composed of Precambrian–metamorphics covered with thin patches of Tertiary rocks.

The basement in Assam Valley mainly comprises crystalline and metamorphic rocks. Shelf facies of Assam–Arakan basin are thin (upto 2000 m) in Shillong plateau and Mikir Hills while below the alluvium of Upper Assam they attain thickness upto 6000 m. The sediments ranging in age from Cretaceous to Recent are monotonous clastics except Sylhet Limestone of Eocene and some coal-shale of Barail group (BCS) of Oligocene age.

The geology, depositional environment, structural style and hydrocarbon prospects of Assam-Arakan Basin have been described by several researchers (Evans, 1932; Mathur and Evans, 1964; Raju, 1968; Bhandari *et al.*, 1973; Das Gupta, 1977; Murthy, 1983; Ranga Rao, 1983; Kumar, 1993; Samanta *et al.*, 1993; Ahmed *et al.*, 1993; Bastia *et al.*, 1993; Kuldeep Chandra *et al.*, 2001). Detailed geochemical analysis established the presence of two main petroleum systems namely (*i*) Kopili and (*ii*) Barail–Tipam system in Assam Shelf (Kuldeep Chandra *et al.*, 2001).

3.7.1 Kopili Petroleum System

In this system, Sylhet and Kopili Formations are potential source rocks. The Kopili of Middle to Upper Eocene has been recognized as one of the source rocks of Upper Assam Basin. Sediments have been deposited in fluctuating environments varying from marginal marine to shallow inner shelf and the organic matter is mainly terrestrial with some local marine input. TOC varied between 0.68-2.72 per cent (ave:1.5 per cent), 0.8-2.4 per cent (ave: 1.75 per cent) and 0.86-5.99 per cent (ave: 1.65 per cent) in Borholla–Changpang, Lakwa and Geleki areas respectively. The hydrogen Index (HI) varied between 30-200 mg HC/g TOC. Kopili Formation is in the catagenetic stage (Mangotra *et al.*, 1993) of maturity (Ro = 0.52 per cent). Maturity and richness of OM have been found to increase towards the Naga thrust (Srivastva *et al.*, 1993).

Main reservoirs within Kopili Formation are sandstones deposited in shallow marine marginal to marine–fluvial conditions.

However, Sylhet limestone was deposited in shallow inner shelf while Tura Formation was deposited in marginal marine and fractured basement. A reasonable correlation between oils and sediment extracts suggests that the former occurs in Eocene reservoirs probably generated *in situ* from Eocene source rocks. Basement oils from Borholla and Changpang Fields have close similarity with Eocene oil. This may be due to the fact that basement uplift of Archean granites and gneisses might have created prominent fractures along ancient lines of weakness which might have provided conduits for migration. The intervening shales within Kopili and Tura Formations act as cap rock.

3.7.2 Barail–Tipam Petroleum System

The Barail Formation consisting of sandstones and coal shale alterations deposited in deltaic system is an important source rock system in the area. Barail Coal Shale (BCS) sequence is rich in OM and has good hydrocarbon generation potential (Samanta *et al.*, 1996) due to the presence of Type II kerogen. TOC ranges between 0.59-7.18 per cent (ave: 2.5 per cent), 1.6-12.54 per cent (ave :4.5 per cent) and 2.54-9.86 per cent (ave : 4.5 per cent) in Borholla–Changpang, Lakwa and Geleki areas respectively. HI ranges between 50-300 mg HC/g TOC. Similarity, the BCS in Rudrasagar area is also found organically rich (TOC: 2.5 per cent). The BCS in drilled area (Figure 3.15) is in an early stage of maturation. Significant hydrocarbon generation and migration probably have taken place from the more deeper part of the basin.

Oil has been found in the fractured basement to Girujan Formation in Upper Assam Basin. However, the main reservoir rocks are Tipam (Miocene) deposited in a fluvial condition and Barail group (Oligocene) laid under marginal marine to fluvial environment. The Barail group contains rocks of excellent reservoir quality and are the producing formations in the fields of Nahorkatya, Kusijan, Moran, Raudrasagar and Geleki.

The Girujan Clay is the main cap rock for accumulations in Upper Assam Valley. However, local cap rocks are provided by shales associated with each of the sedimentary sequence. Anticlinal structures affected by faults mostly control the accumulations in the oil fields. Production from Tipam Formation is obtained in the Digboi, Duarmana, Bogapani, Hapjan and Lakwa fields (Figure 3.16). Various geochemical correlations indicate that the Tipam and Barail

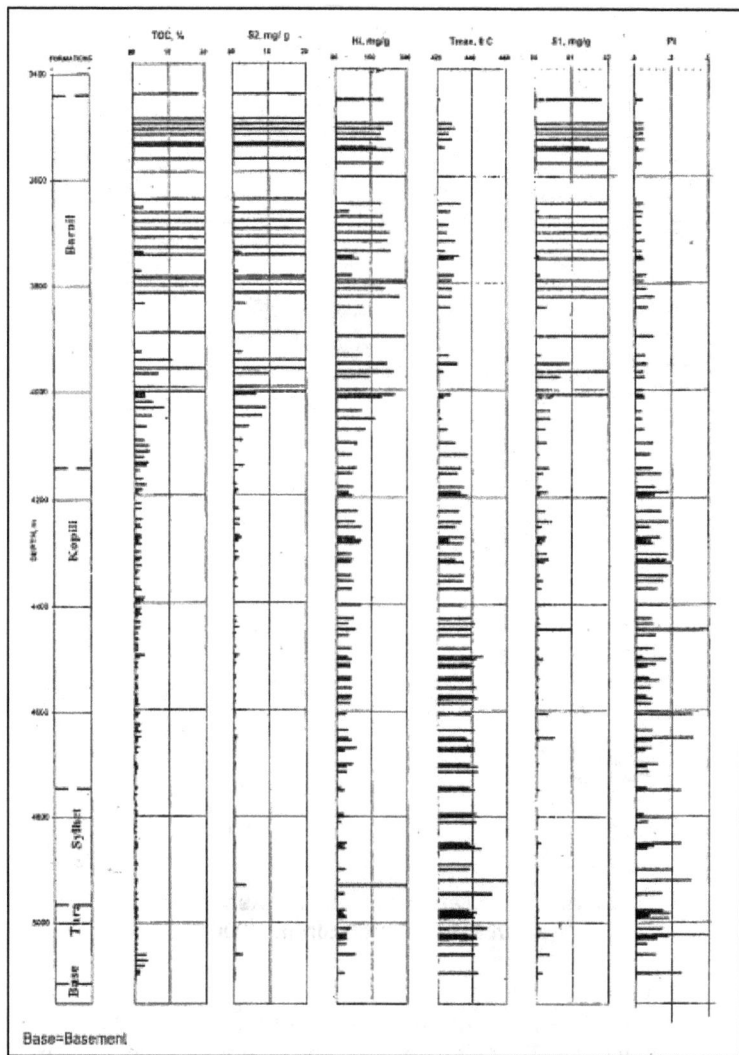

**Figure 3.15: Source Log of Deep Well of Upper Assam
(*Source*: Kuldeep Chandra *et al.*, 2001)**

group of oil were generated and migrated from Barail coal shale
(Dwivedi *et al.*, 1991; Singh *et al.*, 1993; Chandra *et al.*, 1995; Mallick
et al., 1997). Oils from Barail coal shale (BCS) and Barail main sand
(BMS) reservoirs from Demulgaon field of Upper Assam Shelf are

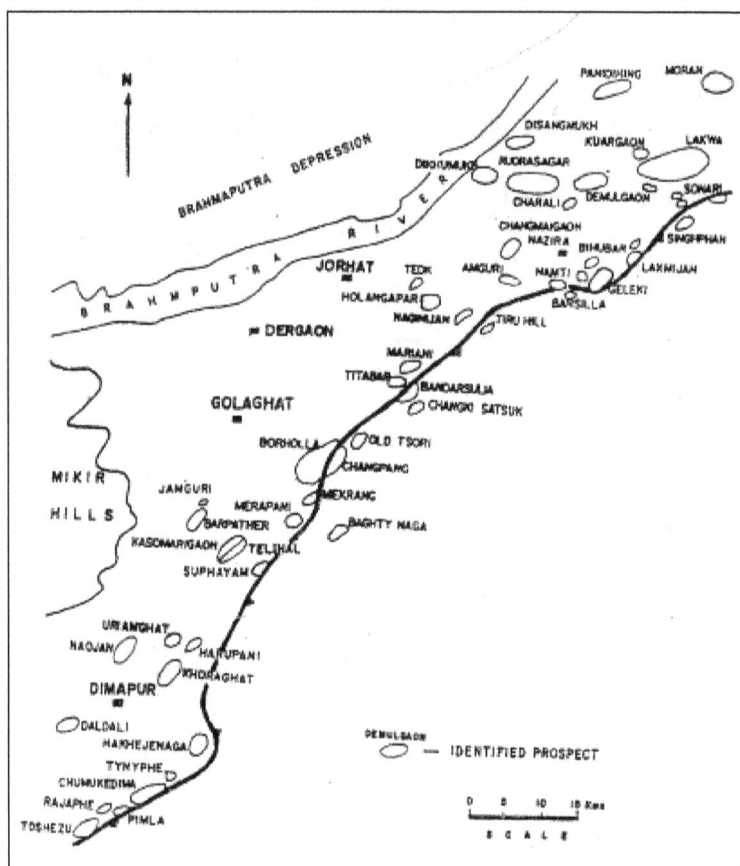

Figure 3.16: Oil and Gas Fields in Assam Shelf
(*Source*: Lakshman Singh, 2000)

paraffinic–naphthenic in nature (Table 3.8). They have high wax content (13.8–17.4 per cent), high pour point (27–30°C) and lower API gravity (30.9–37.1° C) exhibiting the characteristics of normal waxy crude oils. They are dominantly sourced from Type III kerogen (coaly) of terrestrial input (Singh *et al.*, 1993).

Major Oil Fields like Digboi, Changpang and Kharsang have been discovered in Schuppen Belt of Assam–Arakan Basin. The Schuppen Belt is bound by Naga thrust in the West, Disang thrust in the east, the Mishmi thrust in the north and Haflong Dauki compartmental fault in the south (Figure 3.17). Sediments ranging

Table 3.8: Gross Compositional Data of Oils from Demulgoan Field Upper Assam Basin

Sl.No.	Well No.	Formation	Depth (m)	API Gravity	Wax Content (wt%)	Pour Point (°C)	Asphaltene Content (wt%)	Resin Content (wt%)	Water Content (V/V %)	Specific Gravity 60/60°F
Block I										
1.	D#4	BMS	3736–44	30.7	13.8	30	0.4	5.8	0.2	0.8724
2.	D#5	BMS	3832–52	31.0	14.37	30	0.54	5.53	0.04	0.8708
3.	D#6	BMS	3826.5–49	30.86	14.4	30	0.45	5.5	10.0	0.8716
4.	D#9	BMS	3765–97	37.1	14.24	27	0.46	5.16	–	0.8390
5.	D#13	BMS	3820–27	33.23	17.4	30	0.82	9.36	1.0	0.8590
Block II										
1.	D#7	BCS*	3575–97	33.9	14.36	30	0.47	5.88		0.8525

* BMS: Barail Main Sand; BCS: Barail Coal Shale.

Source: Singh *et al.*, 1993.

Figure 3.17: Location Map of Schuppen Belt, Assam–Arakan Basin
(*Source*: Bastia *et al.*, 1993)

in age from Cretaceous to Recent have been encountered in this basin. About 24 prospective structures (notable among them are Singphan, Bhagtinga, Sukhovi, Bandersulia, Chanki-Satsuk) have been identified for drilling. Besides, numerous indications of oil and gas have been reported in outcrops as well as in drilled wells through out in this belt.

Oil exploration in Assam Basin has a long history which goes back to the discovery of Digboi oil field in 1889. Subsequently a large oil field was discovered near Nahorkatiya in 1952 by the Assam oil company, the precursor of the Oil India Ltd. (OIL). ONGC stepped up oil exploration and discovered a deep well in Rudrasagar structure in 1960. A number of oil fields like Lakwa–Lakshmani;

Gelki, Charli; Changmaigaon etc. were discovered in quick succession (Kumar, 1993). The area was established as a major petroliferous region of India with an identified hydrocarbon resource of Upper Assam (UA) 3180 MMT and Assam–Arakan Fold Belt (AAFB) 1860 MMT. Production of oil, gas and O+OEG by ONGC, OIL and other private and Joint venture companies in the two regions (UA and AAFB) during the year 2005-2006 were 3.575 MMT, 3152.33 MMSCM and 7.74 MMT respectively (DGH, 2006).

3.8.0 Rajasthan Basin

The Rajasthan Basin comprises an area of about 1, 26, 000 km² and geologically it forms the western part of the shelf zone of the Indus basin (Pakistan). This area is divided into four sub-basins (Figure 3.18) from north to south. These include:

1. Bikaner–Nagaur Sub-basin
2. Jaisalmer Sub-basin
3. Barmer Sub-basin, and
4. Sanchor Sub-basin.

The Jaisalmer Sub-basin is a pericratonic basin, whereas the other three sub-basins are predominantly intra cratonic type (Datta, 1983).

The tectonic setting of Indus Shelf including west Rajasthan (Figure 3.19) indicates that Bikaner-Nagaur Sub-basin is a part of westerly dipping platform flank of Indus Basin. It is mainly a Paleozoic basin with a relatively thin (1.5 to 2.5 km) cover of Mesozoic and Cenozoic sediments. Structurally the basin is bound by NNW–SSE trending–Pokhran Devikot–Nachna High in the south–west, Indus shelf in the south, Aravalli mountains in the east, and NW–SE trending Delhi-Sargodha ridge in the north (Das Gupta, 1974; Mitra *et al.*, 1993).

Jaisalmer Sub-basin having large gas fields constitutes the eastern contiguous shelf of Indus Basin and stretches to the west of the Aravalli ranges. Tectonically the shelf has been divided into a passive homocline represented by Kishangarh Sub-basin in the northeast and stepfaulted Shahgarh Sub-basin in the south-west. NNW-SSE trending Jaisalmer-Mari arch separates these two sub-basins. The basin hosts 7-9 km thick sediments ranging in age from Permian to Quaternary (Dhannawat and Mukherjee, 1997).

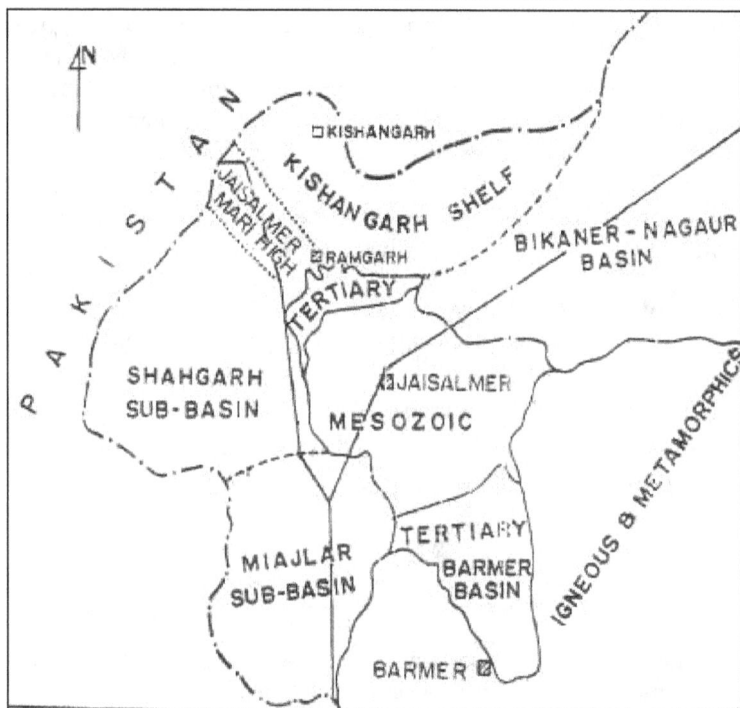

Figure 3.18: Generalized Geological Map of Western Rajasthan
(*Source*: Mitra *et al.*, 1993)

The Barmer Sub-basin is a narrow NW–SE trending graben separated from Jaisalmer Basin by an arcaute Fatehgarh Fault towards NW. It has a sediment thickness of 1.5 to 2.0 km and sedimentary sequence of Lower Jurassic to Lower Eocene age deposited under shallow marine to fluvial environment (Das Gupta, 1974). The Sanchor Sub-basin which geologically can be related to the northern continuation of Cambay Basin has a sedimentary sequence ranging in age from Jurassic–Lower Cretaceous to Quaternary with intervening Deccan traps. The sediments mostly represent fluvial deposits (Roy Chowdhury *et al.*, 1972).

Baisakhi and Pariwar Shales of Lower-Cretaceous are the regional source rocks containing Type III and Type II kerogen. They are genetically gas prone and attained sufficient maturity in the

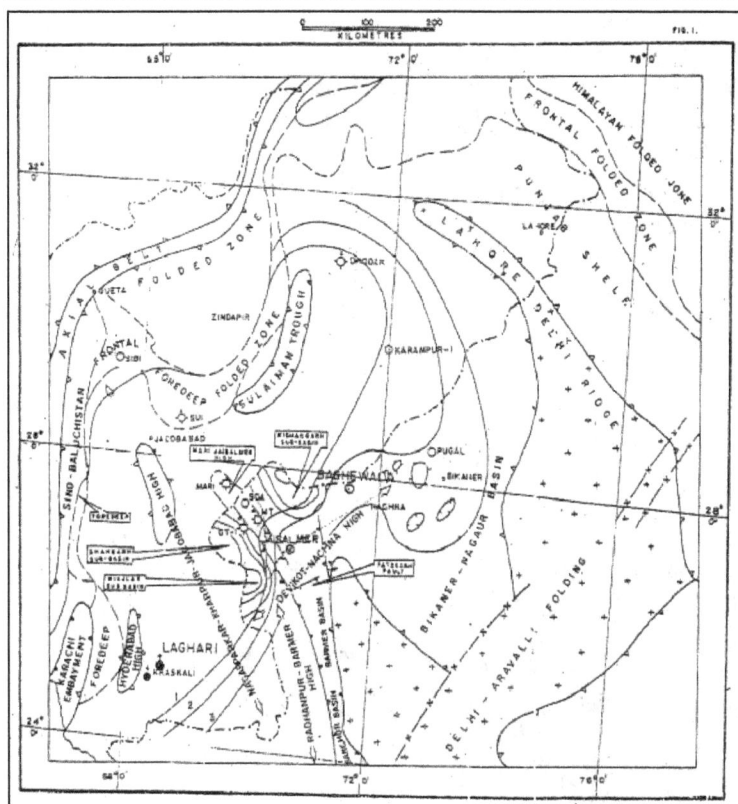

**Figure 3.19: Tectonic Setting of Indus Shelf
including West Rajasthan Basin
(*Source*: Mitra *et al.*, 1993)**

north-western parts of Jaisalmer Sub-basin. The main reservoir rocks
are sandstones in Goru, Sanu and Pariwar Formations. Limestone
reservoirs of Mid-Eocene age prevail in the deeper parts of Shahgarh
area (Banda Formation).

In the Jaisalmer Sub-basin argillaceous sequence in the upper
part of cretaceous and thick limestone sequence in the middle Jurassic
serve as good principal cap rock. The structures are controlled by
long through going master faults. In Bikaner-Nagaur Sub-basin good
reservoirs, both sandstone and carbonates are present ranging from
Early Cambrian to Jurassic age with a good indication of gas/oil

shows. Evaporite sequences of Lower Paleozoic, thick shales and limestones of Mesozoic–Cenozoic age provide good cap rocks. Deposition of thick carbonate–evaporite sequence of Cambrian age contributed in the formation of several large anticlinal structures in the Bikaner–Nagaur Sub-basin.

Geochemical data from Jaisalmer Sub-basin indicates that the OM is ubiquitously of terrestrial origin containing mostly Type III

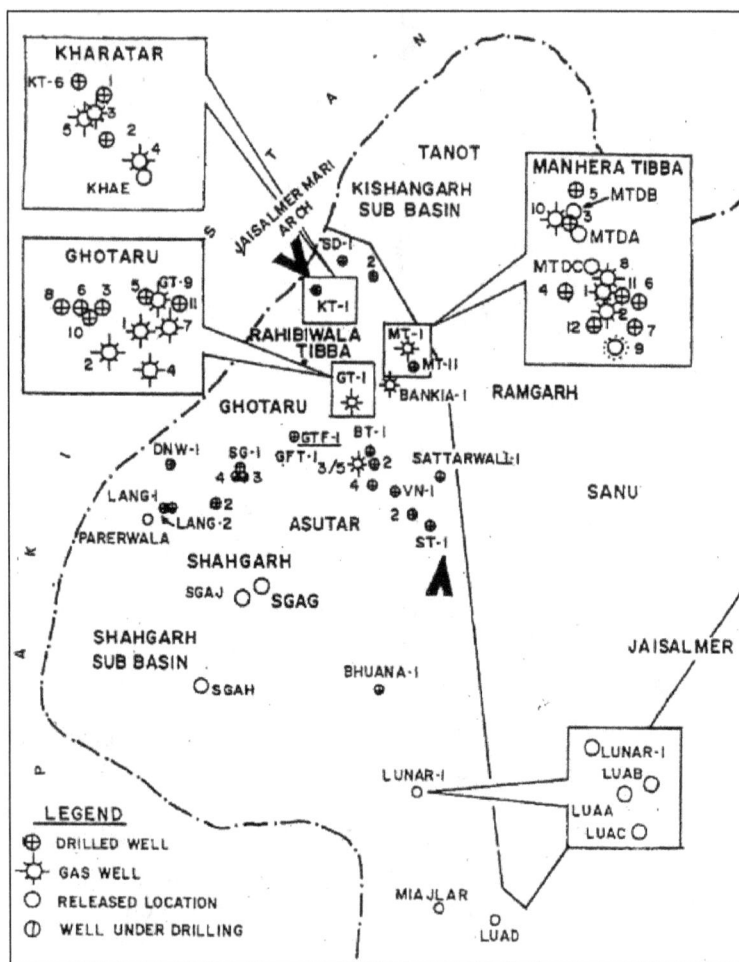

Figure 3.20: Gas Fields in Jaisalmer Basin
(*Source*: Mohan and Sangai, 1995)

but occasionally Type II kerogen. It shows fair to good maturity and generative potential of predominantly gas with occasionally oil showing (Dhannawat and Mukherjee, 1997). Several gas fields namely Bakhri-Tibba, Ghotaru, Kharatar, Manhera Tibba, Tanot, Dendewala have been discovered in Pariwar and Baisakhi-Bedesir Formations (Figure 3.20). Recently a large oil well (Mangala) was commissioned for production in the Barmer Basin. Non-degraded heavy oils with high wax content (33 per cent), high pour point (39°C) and almost devoid of distillate fraction (only 4 per cent of the total yield of liquid) were found in Ghotaru well of Goru reservoir. Heavy oil shows were also observed in Bilara and Jodhpur Formations. The gases of Ghotaru B well drilled to a depth of 1600 m in the Goru reservoir of Jaisalmer Sub-basin indicate high nitrogen (67.2 per cent), carbon dioxide (10 per cent) but relatively low methane (22.1 per cent) content as its major constituents along with minor proportions of other hydrocarbon gases (ethane, propane etc.) and He, H_2S. Such compositional characteristics of gases lend support to their deep seated origin and long distance migration (Kumar, *et al.*, 1998).

Initial exploration in the Rajasthan Basin can be traced back to 1956 leading to drilling of six wells discovering gas in Manhera–Tibba in Dungham Limestone. Sub-sequently several wells have been drilled in Jaisalmer, Bikaner-Nagaur and Barmer Sub-basin. Out of the total hydrocarbon resource (380 MMT) estimated in this basin, 15.39 MMSCM of gas and 0.015 MMT of O+OEG were produced in the year 2005-2006 (DGH, 2006).

Chapter 4
Origin of Petroleum

The origin of petroleum is a matter of debate since its discovery (Levorsen, 1967). Theories on its origin are broadly divided into two, based on the source material "whether it is organic or inorganic". Though early ideas lent support to inorganic (abiogenic), the present belief is mostly centred around the organic (biogenic) theory. Each theory has some evidence and reasoning in its support as well as against. Hence it is worthwhile to deliberate some of the salient features of these two theories.

4.1 Inorganic (abiogenic) Theory

This theory postulates that most of the hydrocarbons on the earth are in fact, primordial. Carbonaceous chondrites appears to have been the most abundant source rocks during the formation of earth. This type of meteorite contains a significant amount of hydrocarbons. As the earth formed, it would have acquired these hydrocarbons via accretions (bodies of roughly equal size clumping together through collisions) and later through meteorite impacts (formed by the reaction of meteorite carbon with hydrogen at high pressure and temperatures). As the earth gradually cooled, a solid crust was developed while the interior remained liquid or semisolid. Volatile substances expelled from the interior, after biological modification, would have yielded the present atmosphere. On the way up, it is presumed that hydrocarbon gases would have been trapped in suitable formations resulting in the formation of oil and gas fields, tar sands, oil shales, bitumens etc. all over the world (Gold and Soter, 1982). The historical development of this theory led to several variants, the most important of them are outlined below.

4.1.1 Early Proposals

When the incidence of fluid petroleum hydrocarbons was first recognized as being wide spread, two Russian scientists, Mendeleev and Sokoloff put forward their inorganic origin, in the late 19th century. Mendeleev (1877, 1902) proposed that metallic carbides deep within the earth reacted with water in hydrothermal fluids to produce acetylene, a familiar reaction that can be reproduced in the laboratory:

$$CaC_2 + 2H_2O \rightarrow C_2H_2 + Ca\,(OH)_2 \tag{4.1}$$

Sokoloff in 1889 proposed a cosmic origin for petroleum. According to him, the bitumina, meaning the whole range of hydrocarbons from petroleum to tar was precipitated as rain from the original nebular matter from which the solar system was formed, and subsequently ejected from the earth's interior into the surface rocks. Both these hypotheses have survived in a variety of forms (North, 1990).

4.1.2 Later Variants

Solid petroleum bitumens occur commonly in various igneous rocks in small amounts. Micro inclusions of petroleum hydrocarbons have been identified in alkaline carbonatite rocks. This is believed to be due to the reaction of carbon dioxide derived from the (rock) carbonate with hydrogen gas in the presence of rare earth minerals acting as catalysts. This hypothesis has subsequently proved to be valid by the Fischer-Tropsch reaction, a well known industrial process developed for production of synthetic hydrocarbons (Storch *et al.*, 1951). It involves the reaction of carbon dioxide and hydrogen in presence of a catalyst (haematite and magnetite) at a temperature of about 250° C.

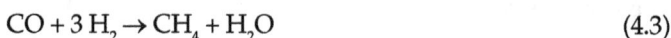

$$CO_2 + H_2 \rightarrow CO + H_2O \tag{4.2}$$

$$CO + 3\,H_2 \rightarrow CH_4 + H_2O \tag{4.3}$$

Robinson (1966) suggested that this process could produce methane and liquid petroleum in nature. Porfirev (1974) reviewed the occurrence of oil in the igneous rocks.

Studies carried out by several Russian Scientists revealed the association of petroleum hydrocarbons with hydrothermal systems. Gaseous homologs of methane, and aromatic and naphthenic oils occur in the fumaroles and injecta of many volcanic regions, particularly near mud pots or moderately hot hydrothermal fields. Hot springs of H_2S-CO_2 type commonly deposit some bituminous materials. For example, ozourite and asphalt occur in hydrothermally altered traps. A remarkable coincidence was pointed out between the major mud volcano regions of the world and the major oil producing areas such as Persian Gulf, Caspian Sea, Indonesia and Venezuela. Craters on the ocean floor have been reported from several areas such as Adriatic, North Sea, Gulf of Mexico, South China Sea and Baltic Sea (Samar Abbas, 1996). Sonar experiments in the North Sea revealed shallow circular ridges ranging from few metres to 200 m in diameter over an area of 20,000 km^2, roughly coinciding with the oil and gas producing fields. This is cited as an evidence for the sudden release of hydrocarbon gases through chimneys (gas vents) in oceans by sub- marine mud volcanoes.

Gold (1979), the principal proponent of volcanic origin of petroleum hydrocarbons invoked outgassing from the upper mantle via earthquake activity. According to him, large volumes of methane along with CO_2, H_2S and traces of noble gases are being degassed from the mantle. They subsequently migrate up through major faults to be trapped in sedimentary basins or to be dissipated at the earth's surface. Examples of such petroleum feeders include flinking faults of the Suez, Rhine, Baikal and Burguzin grabens. Based on the association of gas–oil deposits in the fractured crust/tectonically active regions in the Bay of Bengal, particularly in the Andaman Sea, Paropkari (2008) affirmed their origin to abiogenic at least to a partial extent. The earthquake outgassing of methane along faults gives rise to deep gas reserves which on subsequent polymerization produce oil at shallow horizons. This theory implies limitless quantities of oil and gas universally at great depths along the faults in areas of tectonic activity. However, this does not appears to be true in all cases.

Mass balance calculations indicate that methane is being released to the lithosphere, hydrosphere and atmosphere at a rate of 10^6–10^7 tonnes annually. The triggering mechanism is the earth's movement as testified by the sharp increase of radon gas content in the ground water and in the atmosphere following many earthquakes. According to Gold (1999), the existence of thermophile or extremophile bacteria in the earth's crust is responsible for the presence of biomarkers in petroleum or in source rock extracts. They have been found to survive at temperatures as high as $169°$ C in marine sediment cores at depths upto around 10 km, where the pressure is sufficient to permit water to remain in liquid state. This is verified by the discovery of microbial life in deep drilling wells at Alaska (4.2 km) and Sweden (6 km).

The extra terrestrial (cosmic) origin of Sokoloff has been resuscitated by many scientists including astronomers Hoyle (1955), chemists Robinson (1966) as well as geologists because of two critical discoveries. The first one is the presence of methane in the atmosphere of some celestial bodies such as Uranus, Neptune and Comets. Uranus atmosphere consist of hydrogen, helium and methane gas while the inner liquid shell is thought to consists of water, methane and ammonia (Lang and Whitney, 1991). Halley 's Comet was found to emit hydrocarbon gases, the core being black, presumably due to admixture of minerals, organic compounds and metals.

It is widely argued that the original atmosphere of the earth was reducing containing methane, other hydrocarbons, ammonia, hydrogen and water vapour. Various photochemical reactions mediated by ultraviolet (UV) light would have modified its atmosphere to oxidizing. Heavier hydrocarbon compounds would have builtup and accumulated in low temperature atmosphere as oily and waxy material of several metres deep. This layer, virtually a primordial oil slick, could have acted as the host for a variety of prebiotic compounds, possibly including the precursors of life.

The second critical discovery was that of carbonaceous chondrites (meteorites). They consist mainly of hydrous silicate minerals such as chlorite, carbonates and sulphates of calcium and magnesium; iron minerals; elemental sulphur and carbon compounds with chondruls of mafic minerals. They are distinguished within the chondrite class wherein all most all iron is present in the oxidized state (Fe^{3+}). This and the presence of hydrous minerals indicate an aqueous-oxidizing environment of the parent meteorite at temperatures far below than those involved in the formation of stony meteorite. The best studied carbonaceous meteorite, orgueil contains more than 6 per cent organic matter (not graphite) with $\delta^{13}C$ values substantially different from those of terrestrial carbon. Hydrocarbons identified include saturated paraffins, aromatics and hetero (NSO) compounds. A large polymeric material insoluble in ordinary solvents is also found. The solvent extractable (lipid) fraction includes optically active components with $\delta^{13}C$ values similar to those of crude oils and marine plants.

Kudryavtsev (1959) one of the strong proponents of abiogenic origin of petroleum hydrocarbons proposed a rule later named after him. According to this rule, hydrocarbons in at least small quantities are often present in horizons below any accumulation largely independent of the composition and mode of formation of horizons. This rule has been verified and confirmed in many locations including Oklahoma, Wyoming, Canada, Iran, Java, Sumatra and Russia (Earlich, 2001). The statistical mechanical model based on thermodynamic calculations predicts that methane and some other petroleum hydrocarbons are stable up to 30-40 k bar pressures encountered at depths around 150-200 km, and temperatures around 1000°C, mimicking the conditions of the mantle (Kenney et al., 2002). It is presumed that these hydrocarbons of primordial origin at high pressures are transported via cold eruptive processes into the earth's crust until they escape into the surface or trapped by impermeable strata forming hydrocarbon reservoirs.

In summary, it may be stated that a small proportion of hydrocarbons discovered earlier could be of abiogenic (inorganic) origin. Most oil and almost all gas generated before 400 MY (million years) having this origin entered into early oceans and incorporated in early sediments. Combination of methane, CO_2 ammonia and water vapour under ultra-violet radiation resulted in the formation of amino acids and proteins. Presence of isoprenoid and sterane hydrocarbons in Proterozoic sedimentary rocks long before the existence of plant and animal life may therefore represent extraterrestrial evolution and capture by the early earth. Hence a duplex theory combining features of both inorganic and organic theories may be the final choice (Robinson, 1963). On this hypothesis, the present stock of petroleum hydrocarbons would represent biogenic additions to a fundamentally primordial endowment. This would perhaps involve the enrichment of existing petroleum hydrocarbon deposits of biogenic source through a fundamentally abiogenic origin (Samar Abbas, 1996).

4.2 Organic (biogenic) Theory

The organic theory holds that the first stage in the genesis of petroleum involves marine plankton (phyto and zoo). After their death and decay in the sea water column, the unoxidised organic matter gradually sinks to the ocean floor and accumulates in the sediments. Over a geological (millions of years) time, it is transformed into a disseminated mixture of soluble petroleum like hydrocarbons (bitumen) and insoluble matter (kerogen) under increased over burden and consolidation of sediment leading to the formation of shale or source rock (Levorsen, 1967). Under favourable conditions of time and temperature, the kerogen trapped in the shale is transformed via cracking into petroleum and natural gas (Samar Abbas, 1996).

4.2.1 Analogies

Before world war II, the evidence favouring the organic theory largely rests on analogies. The chemical composition of petroleum closely allies with organic materials fundamental to life-proteins, fats, fatty acids, lipids etc. The carbon cycle intrinsically involve living and dead plants and animals. The fact that oil and gas were invariably associated with unaltered sedimentary rocks which commonly contain marine fossils lead to the analogy with aquatic life, typical examples being whale oil and fish oil. A further analogy was with coal known to originate in terrestrial plant material after burial in sediment analogous to petroleum in marine plant and animal source. This generalized (biogenic) concept was further strengthened by biochemical evidence-for example, the presence of bio-markers (porphyrins, terpanes steranes etc;), optically active compounds (cholesterol) in oils, and depletion of $\delta^{13}C$ values in oils, bitumen, coal and terrestrial plants when compared with limestone, magnetic rocks and meteorites.

There is a close association of coal and oil in their natural cycle of sedimentation either aqueous or marine. This is reflected by the fact that many rich oil bearing sandstones are members of coal measure successions. Typical examples are Carboniferous Donetz Basin, Cretaceous Western North America and the Early Tertiary of the Gippsland, Cambay, Upper Assam and Orinoco Basins (North, 1990). Coal and oil might have formed under the same conditions (of time, climate, tectonics etc.) but materials of different origin. The association of coal and gas is much more common than between coal and oil. The outstanding examples of the former are large gas deposits in the Permanian of Western Europe and of the Ukraine, and those in the Cretaceous of northwestern Siberia. Hence we can conclude that coal and natural gas are derivatives of terrestrial vegetation whereas oil and wet/dry gas are derivatives of aquatic vegetation akin to the sapropelic (decomposed) mud.

Waxes and hydrocarbons similar to those in petroleum are formed in kelp and other marine algae from the organic compounds produced by photosynthesis. Non-marine algae are the principal contributors to the hydrocarbon content of oil shales. In a number of rich oil provinces of Tertiary age (Baku, California, Sakhalin), the oil is closely associated with diatomaceous sediments. Marine organism comprising phyto (diatoms) and zoo (foraminifera, radiolaria, copepods) plankton contain minute droplets of oil in their cells and they are widely represented in fossils in young oil bearing strata. Hydrocarbons have been isolated from bacterial populations. Thus marine life is the most important source of oil and gas in sediments (North, 1990).

Antia (2007, 2008, 2009) proposed a polymerisation theory for formation of hydrocarbons in sedimentary strata from methane and carbon dioxide. He developed a new model for stoichiometric thermogenic degradation of organic matter (OM). A mechanism has been outlined for hydrocarbon expulsion from low temperature (T = <20–150°C), low maturity (Ro = <0.6 per cent) and over pressured sediments (clays, shales and enclosed sands). The principal products from low temperature degradation of OM are a mixture of kerogen (<10 to >30 wt per cent), gases such as CO_2 CO, CH_4 H_2O, H_2 (<20 to >80 wt per cent), carbonyls and minor quantities of paraffinic oil. Methane migrates from deeper gas accumulations into shallow reservoirs containing saline porewaters and polymerises over a time to oil. The polymerisation could be a result of biogenic activity or abiogenic catalytic activity using Fischer-Tropsch synthesis, and its rate can be enhanced by clay minerals or zeolites. Thus the presence of oil in association with biogenic methane in some cases is explained in terms of low temperature catalytic polymerisation rather than the traditional explanation of thermogenic formation of oil from kerogen of matured source rocks (Waples, 1984).

Early objection to an organic origin for oils is centered upon the absence of adequate amount of life in the constitution of most petroliferous formations. But it should be recognized that oil being a fluid, is potentially migratory. Its source may be elsewhere than in its present host rock even if that is a fossiliferous marine limestone. Further, the original fauna and flora may not have been suitable for the formation of visible fossil remains.

4.3 Conclusion

Majority of petroleum geologists and geochemists today believe that petroleum is formed by thermal transformation (maturity) of organic matter (kerogen) preserved in fine-grained sedimentary rocks (shales). Gas chromatographic finger printing of organic matter in shale perfectly match with petroleum in adjacent reservoirs. Laboratory studies on thermal transformations of organic matter to petroleum-like materials; empirical studies on the depth and temperature at which the composition of sedimentary organic matter changes and at which petroleum and gas accumulations occur; the ability of time–temperature models of thermal generation of oil; the observed facts such as the occurrence of highly specific organic compounds of unquestionable biological affinity have all lend support to the theory of organic (biogenic) origin of petroleum (Waples, 1984).

Commercial and economic petroleum accumulations have originated from organic matter in sedimentary rocks but not in igneous and metamorphic rocks. If at all any petroleum occurs, it is only where the igneous rocks intrude or unconformity overlain by sediments. Further, it is apparent that unquestionable instances of indigenous oil in basement (impermeable hard rock) are rare and not commercially significant. Not only the volumes of the hydrocarbons trapped this way are insignificant, the reservoirs are impermeable unless fractured. Thus it is to be concluded that organic (biogenic) theory is currently favoured over inorganic (abigenic) theory on the origin of petroleum.

Chapter 5

Hydrocarbon Generation, Migration and Accumulation in Sedimentary Basins

---

Almost all life on the earth depends on the process of photosynthesis wherein the atmospheric carbon dioxide is taken up by plants (terrestrial and marine) to be converted to organic carbon so as to enter into their cells (tissues).

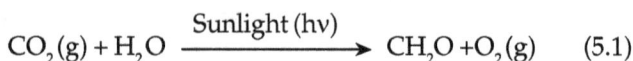

$$CO_2(g) + H_2O \xrightarrow{\text{Sunlight (hv)}} CH_2O + O_2(g) \qquad (5.1)$$

Animals (herbivores, carnivores etc.) eat the plants and give part of CO_2 during respiration in their life cycle. After their death, the organic matter undergoes microbial and chemical degradation to release most of CO_2 into the atmosphere.

5.1 The Carbon Cycle

The uptake and release of CO_2 in general, is represented as carbon cycle (Figure 5.1). All quantities in the cycle are expressed in billion metric tonnes (BMT). The numbers in parentheses represent stored quantities, while the numbers without parentheses are yearly fluxes. It is evident from the figure that small amounts of organic material produced escapes recycling, and is stored in sediments.

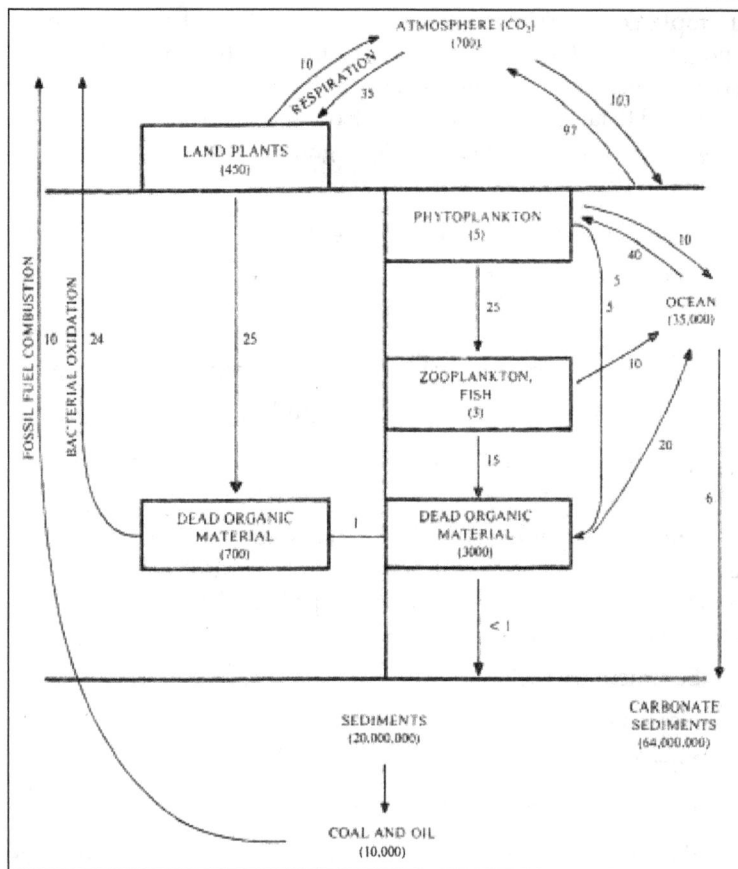

Figure 5.1: The Carbon Cycle
(*Source*: Bolin, 1970)

Some of the fossil organic material stored in sediment is converted into coal, oil and natural gas. It is apparent from the cycle that less than one BMT of organic carbon (OC) drops out of the carbon cycle each year. Though most of this is disseminated in fine-grained sedimentary rocks, only less than 0.01 per cent will be concentrated and appears as economically recoverable petroleum. The total amount of carbon (Figure 5.1) present at any one time in land plants (450 BMT) is much grater than the amount stored in marine phytoplankton (5BMT), but the amount of CO_2 taken up annually by phytoplankton (40 BMT) is actually greater than that taken up by land plants (35BMT). This is due to the fact that phytoplankton have extremely short life cycle and have high metabolic rates compared to land plants. The data also suggests that phytoplankton has a major contribution to the total organic carbon preserved in the sediments (Bolin, 1970).

5.1.1 Organic Matter in Sediments

The organic matter (OM) in sediments is often represented as organic carbon (OC). The interrelation between the two terms is:

$$OM \ (\% \ by \ weight) = OC \ (\% \ by \ weight) \times 1.22 \qquad (5.2)$$

$$OM \ (\% \ by \ volume) = OC \ (\% \ by \ weight) \times 2.50 \qquad (5.3)$$

Organic carbon (OC) some times is also called as total organic carbon (TOC)

The total organic carbon (TOC) content of all sediments and sedimentary rocks amounts to about $12,000 \times 10^6$ MMT (million metric tonnes). This is distributed among the principal natural sources as detailed below in units of 10^6 MMT

Sedimentary rocks	11,000
Coal and peat	15
Petroleum in non-reservoir rocks	265
Petroleum in reservoirs	1

It is evident that 1 in 11,000 parts of total organic carbon (about 0.01 per cent) has become petroleum accumulation since the beginning of life. In other words the generation and accumulation of petroleum are very inefficient processes (Hunt, 1996).

Organic matter (OM) content increases with decrease of grain size of the sediment. Thus fine-grained sediments like shales are enriched with OM compared to coarse-grained sediments like terrestrial sands. Organic rich sediments occur in anoxic, environments (where oxygen is almost absent, and virtually all aerobic biological activity ceases). Biochemical evidence (Brooks, 1981) suggests the deposition of organic rich source rocks in four main environments: (*a*) large anoxic lakes (Lake Tanganyika, Eocene Green river formation); (*b*) anoxic silled basins (Black Sea, Baltic Sea Lower Jurassic, Toarcin of the Paris Basin); (*c*) anoxic layers caused by upwelling (coastlines of California, Peru, Chile, western Australia, south west Austria); and (*d*) open ocean anoxic layers (Indian Ocean, Gulf of California, Atlantic Ocean of Late Jurassic, and Middle Cretaceous blackshales).

A sediment to be oil prone, it needs at least 0.5 per cent of OC by weight for a clastic while 0.3 per cent for a carbonate source rock. In general, though carbonate source rocks contain less OC than shales, the former are enriched with hydrocarbons than the latter. However, the actual percentage of OC required will depend on other conditions (6.8.0). Petroleum geochemists categorise in general, the sediment source quality based on its OC content (Table 5.1).

Table 5.1: Sediment Source Quality in Terms of Organic Carbon

OC (Per cent of Weight)	Quality
0.0–0.5	Poor
0.5–1.0	Fair
1.0–2.0	Good
2.0–4.0	Very Good
Above 4.0	Excellent

OC: Organic Carbon.

Source: North, 1990

5.1.2 Classification of Sedimentary Organic Matter

All organic matter in petroliferous sedimentary rocks may be classified into two types–sapropelic and humic (North, 1990). The term sapropelic refers to the decomposition and polymerization products of fatty acids, lipid organic materials such as spores and planktonic algae deposited in sub-aquatic muds (marine or lacustrine) usually under anoxic (oxygen deficient) conditions. Sapropelic organic deposits undergo maturation to form rich oil shales. The term humic refers to products of peat formations derived mainly from land plant material and deposited in the swamps in presence of oxygen (oxic environment). Humic matter derived from plant cellwall material composed mainly of lignin, cellulose and aromatic tannins. This type also includes charcoal and other oxidised plant matter. On maturation humic organic matter forms a good source of coal and gas (methane).

Another broad classification of sedimentary organic matter in petroliferous (source) rocks is based on its solubility and molecular size. The small size fraction is usually soluble and can be extracted into common organic solvents like chloroform, carbon disulphide etc. This extractable organic matter (EOM) fraction is called bitumen. The insoluble organic material comprising large sized polymeric fraction is called kerogen.

5.2 Genesis of Petroleum

Petroleum generally originates from source rocks containing organic matter through a series of successive inter-related processes (Brooks, 1981).

(*i*) Accumulation and preservation of organic rich fine grained sediments (source rocks) under anoxic conditions,

(*ii*) Maturation (thermal alteration) of organic matter during burial with transformation into petroleum like products,

(*iii*) Expulsion of oil (and/or gas) from the fine grained source rocks and their primary movement (migration),

(*iv*) Movement of petroleum after expulsion from the source rocks through the wider pores and more permeable, reservoir rocks (called secondary migration), and

(*v*) Accumulation via secondary migration, in permeable porous reservoir rocks in a trap.

The whole process of petroleum formation may be broadly summarised as follows: some hydrocarbons are formed from the living organisms during their life processes and undergo little change before being part of petroleum. Other hydrocarbons form from bacterial residues, and from early diagenetic reactions of lipids, proteins and carbohydrates. Most petroleum hydrocarbons, probably 80-95 per cent, form from the thermal transformation of kerogen with temperature and time. This process referred to as maturation or thermal alteration occurs during catagenetic stage (5.2.3). After petroleum formation, it undergoes continuing thermal alteration (maturation) in source and reservoir rock, eventually forming gas (methane) together with graphite during metagenesis and/or metamorphism (5.2.5). We now consider in more detail, the conversion of sedimentary organic matter by these processes.

5.2.1 Diagenesis

Tissot (1977) defined three major phases, namely diagenesis, catagenesis and metagenesis in the evolution of organic matter in response to burial. Diagenesis is a process of biological, physical and chemical alteration of organic matter (OM) in sediments prior to a pronounced effect of temperature. It occurs with burial over a depth range (below 1000 m) where the temperatures are too low (50-60° C) for significant cracking of large molecules to occur. The OM deposited in sediments consists primarily the biopolymers of living things: carbohydrates, proteins, lipids, lignins and subgroups such as chitin, waxes, resins, glucosides, pigments, fats and essential oils. Some of this material is consumed by burrowing organisms, some is attacked by microbes that use enzymes to degrade the biopolymers into simple monomers such as sugars, amino acids, fatty acids and phenols from which they were originally formed. Some degraded monomers condense to form complex high molecular-weight geopolymers. The random structure of geopolymers which are relatively stable and resistant to anaerobic microbial degradation, serve to preserve OM even in the presence of bacteria. Among several kinds of geopolymers (fulvic, humic acids etc.) formed, kerogen is most important for petroleum. In terms of petroleum exploration, diagenesis is an immature stage where mostly CO_2, water, some methane (marsh gas) and hetero compounds are generated.

5.2.2 Chemical Composition of Kerogen

Formation of kerogen in sedimentary rocks occurs in two successive stages–polymerization and re-arrangement. Polymerization involves the formation of geopolymer in a geologically short time, probably in few hundred or thousand years. However, the re-arrangement stage continues as long as the kerogen exists. Three types of kerogens are generally recognized based on differences in their chemical composition, and nature of the original organic matter (Tissot, 1977; Dow, 1977). However. a fourth type of kerogen is also known which is not important from the point of view of petroleum generation. Table 5.2 shows the chemical composition of the three basic types.

Type I kerogen is essentially algal in nature. It has higher proportion of hydrogen and lower proportion of oxygen than the other two types (II and III). This is reflected in the higher H: C (1.40) and lower O: C (0.08) atomic ratios respectively than other two types. Lipids are the dominant compounds in this kerogen, with derivatives of oil, fats and waxes. Type I kerogen is particularly abundant in algae such as *Bottryococcus* which occur in modern Coorongite and ancient oil shales. The well-known Green River Shales of Colorado and Utah belong to this type (Tissot and Welte, 1984; Hunt, 1996). Type I kerogen has a high genetic potential of petroleum.

Table 5.2: Chemical Composition of Kerogens

	Weight Percent					*Ratios*		*Petroleum Type*
	C	*H*	*O*	*N*	*S*	*H-C*	*O-C*	
Type I Algal (7 Samples)	75.9	8.84	7.66	1.97	2.7	1.40	0.08	Oil
Type II Liptinic (6 Samples)	77.8	6.8	10.5	2.15	2.7	1.05	0.10	Oil and Gas
Type III Humic (3 Samples)	82.6	4.6	14.3	2.1	0.1	0.67	0.13	Gas

Source: Selly, 1998.

Type II called liptonitic kerogen is of intermediate composition. Like algal kerogen, it is rich in aliphatic compounds with relatively low H: C atomic ratio (1.05) than Type I. Its organic matter consists of algal detritus and material derived from phyto and zooplankton, laid down in a confined environment. Many classical oil source-rocks belong to this type: Silurian of North America, Cretaceous of Middle East, Jurassic of Western Europe etc. Type II kerogen (like Type I) has a rich genetic potential of petroleum (Figure 5.2a)

Type III refers to humic kerogen having relatively low hydrogen and high oxygen content. Consequently it has relatively lower H: C (0.67) and higher O: C (0.13) atomic ratio than the other two types (I and II). The organic matter is derived from higher land plants, transported by rivers, and laid down either in non-marine or in deltaic or in continental margin environments. The woody plants contain lot of lignin and cellulose which decomposes to yield phenols. Type III kerogen has comparatively low potential for oil, but it can provide a good gas source at higher depths and temperatures (Figure 5.2b). Most coals can be classified into this category. The Lower Mannville Shales of western Canada, the Upper

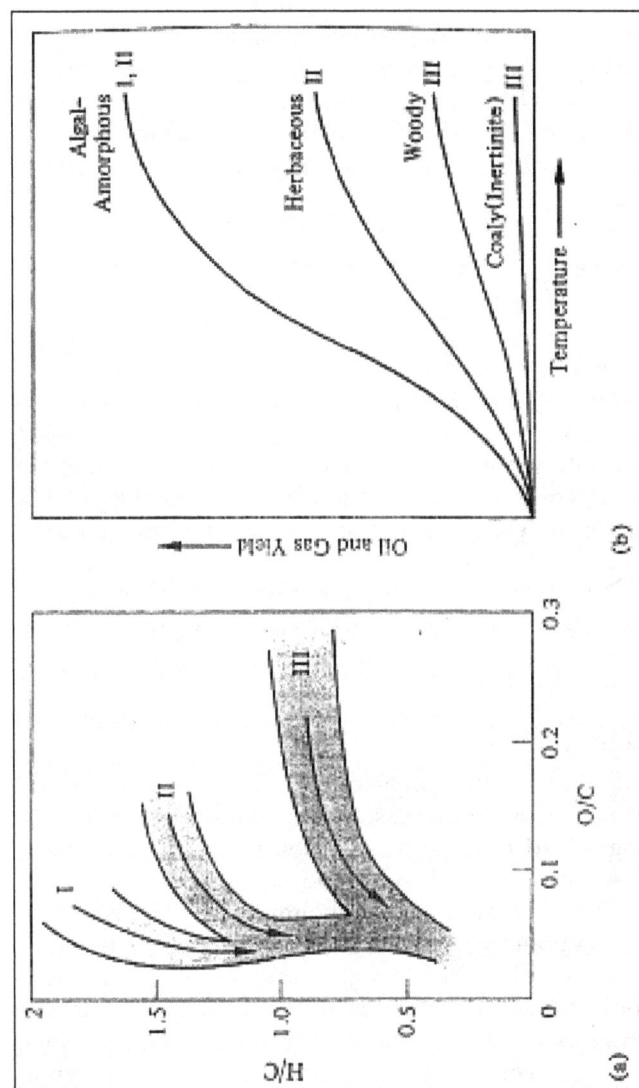

Figure 5.2: Effect of Increasing Temperature on Kerogen Type
(*a*) Change in hydrogen/carbon (H/C) and oxygen/carbon (O/C) ratios, (*b*) Relative yields of oil and gas
(*Source:* Hunt, 1996)

Cretaceous of the Doula Basin, Cameroom, are examples of sediments containing this kind of kerogen (Selley, 1998).

The main types of kerogens are conveniently distinguished by using Tissot or van Krevelen–type diagram where the atomic H/C and O/C ratios (resulting from the elemental analysis) of kerogens are plotted (Figure 5.2a). Each type of kerogen is located on the diagram along a specific line (indicated by solid arrows) called an evolution path. The increasing evolution of organic matter, with increasing depth of burial and rising temperature appears on the diagram as a change of composition along the same evolution path- the shallow immature samples have originally high H/C and O/C atomic ratios while the deeper mature samples have low ratios. Both of them approach origin in the diagram.

Chemical composition of kerogen is also reflected through microscopic examination. Based on the visual examination in a microscope under transmitted light, kerogen particles or macerals are classified into four types- alginite, exinite (sometimes also called liptinite), vitrinite and inertinite. The first three correspond to the three basic types of kerogen namely I, II and III respectively (Table 5.2). Alginite include materials of algal origin. Liptinite (exinite) include lipid rich materials (plankton). Humic (vitrinite) consists largely of woody and cellulose debris. Inertinite (not shown in Table 5.2 or Figure 5.2a) is thought to be the material of various origins which has undergone extensive oxidation or reworking prior to deposition (Waples, 1981).

5.2.3 Catagenesis

As the depth of burial of organic matter increases, porosity and permeability of sediment decreases while temperature increases. These changes lead to gradual cessation of microbial activity, bringing diagenesis to a halt. As the temperature rises however, thermal reactions become increasingly important where kerogen begins to decompose into smaller, more mobile molecules called bitumen. The process by which kerogen undergoes alteration due to the effect of increasing temperature is called catagenesis. In this process petroleum occurs in source rocks in the temperature range 60-120° C. Major geochemical reactions that take place during catagenesis include: dispropotionation and redistribution of hydrogen atoms and breaking of some carbon–carbon bonds giving rise to lower molecular-weight compounds of increasing volatility and hydrogen content (bitumen).

Catagenesis, similar to any cracking reaction requires energy called activation energy which is provided by geothermal heat. This is the reason why deep burial of sediment is vital to catagenesis. The average temperature in the earth's crust increases by about 2-4° C per 100 m depth (geothermal gradiant). A linear increase in temperature causes an exponential increase in reaction rate for most reactions involved in petroleum formation. It has been demonstrated that the rate of catagenesis approximately doubles with the increase of temperature by 10°C (Lopatin, 1971; Waples, 1980). For example, a sediment at surface at a temperature of 20° C when buried to a depth corresponding to a temperature of 120°C, the reaction rate is increased by 2^{10} (approximately, 1000) times. These observations are consistent with the predictions of chemical kinetics–specifically that the decomposition of kerogen is a first order process. Not only temperature, but also the time (age of the sediment) influence the rate of catagenesis. For example, if one million years was required to accomplish a certain transformation at 100° C, the same could be achieved at 90° C in approximately 2 million years or at 80°C in 4 million years. In other words, the rate of reaction varies linearly with time (6.6.2). Catalysts either increase or decrease the rate of a chemical reaction by providing alternate pathways that have lower or higher activation energies respectively. It seems that the role of catalysts are not uniform in all sediments. For examples, clays probably are more effective catalysts, and that catagenesis proceeds more rapidly in shales than in carbonate rocks.

During early stages of catagenesis, the rate of bitumen formation from kerogen dominates over its destruction and this results in the net increase of total bitumen formed over that present inherently in the source rock. However, during later stages when the rate of bitumen formation decreases to considerable extent, migration and destruction will dominate over its formation, resulting in the net decrease in total bitumen content of the source rock (Figure 5.3). The section of the curve in the figure from surface to point A is the zone of diagenesis where the rate of thermal reactions is too slow to be of any significance. The small amount of bitumen present in the source rock is inherited directly from the original source material, and is called diagenetic or inherited bitumen. The curve AB represents

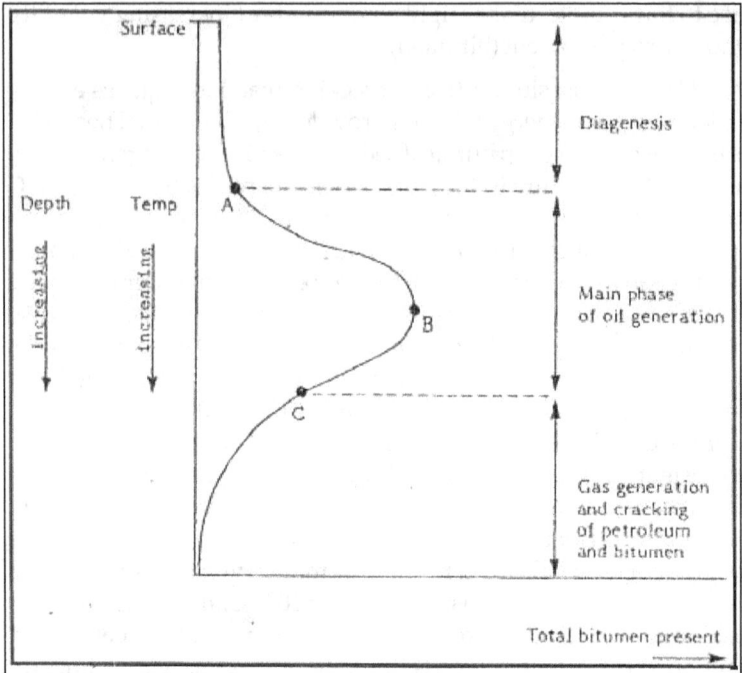

Figure 5.3: Generalized Plot of the Bitumen Concentration in a Source Rock as a Function of its Thermal Maturity
(*Source*: Waples, 1981)

(temperature range) the increase of bitumen production with little destruction or migration. On the other hand, curve BC represent net rate of decrease of bitumen production with simultaneous increase in its destruction and migration. Beyond point C, bitumen formation virtually ceases, leading to formation of gas and its migration from the source/reservoir rock. The depth at which catagenesis begins depends upon the age and geothermal gradient of the sediment (Waples, 1981).

5.2.4 Distinction Between Coal/Anthracite, and Bitumen/Petroleum

Both coal and bitumen are formed by maturation (thermal alteration) of kerogen under different conditions. While coal is formed by losing hydrogen under low energy conditions, bitumen is formed by gaining hydrogen under moderately-high energy (catagenetic) conditions (Figure 5.4). Bitumen contains the same group of compounds as petroleum but in different proportions (Figure 5.5). Crude oils are mostly enriched in saturated hydrocarbons (S), moderately enriched in aromatics (A) and depleted in resins and asphaltens (RA) compared to bitumen in source rock extract. A similar trend is evident either with shale sand sequence or with carbonate sequence.

5.2.5 Metagenesis

Metagenesis is a process that occurs at great depths and high temperatures (above 150°C). It is the late stage of catagenesis when the bitumen forming possibilities of kerogen have been exhausted. At this stage only dry gas (methane) is generated due to cracking of alkyl chains from the kerogen matrix. This process ultimately leads to the formation of large sheets of fused aromatic rings called graphitization, the end product being graphite. The late stage of metagenesis is also referred to as metamorphism.

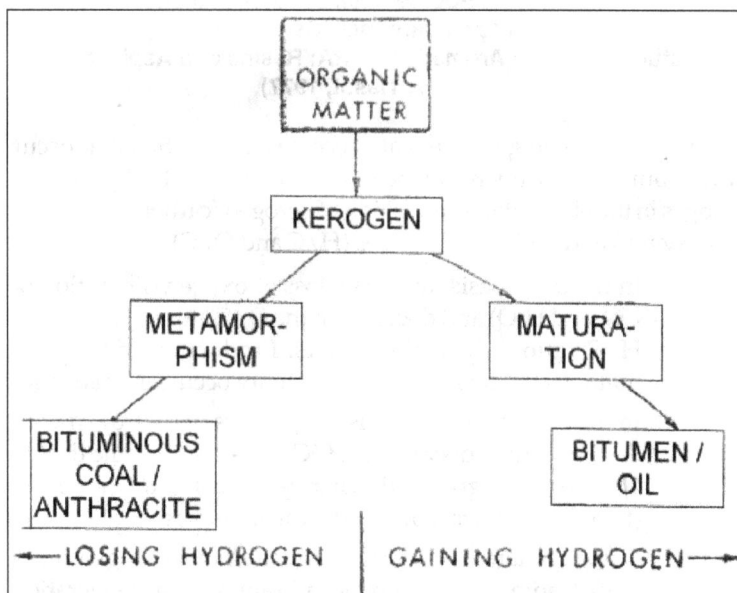

Figure 5.4: Schematic Representation of the Redistribution of Hydrogen from Bituminous Coal and Bitumen
(*Source*: Barker, 1985)

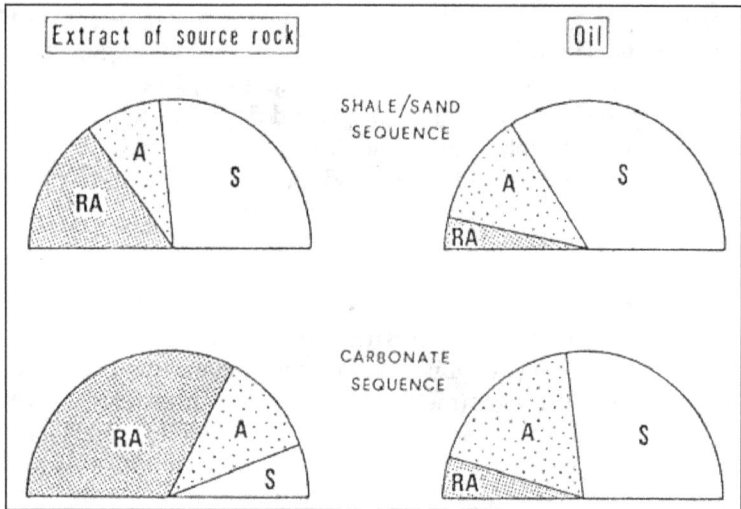

**Figure 5.5: Comparison of Gross Chemical Composition of
Crude Oils and Source Rock Bitumen in a Shale/
Sand and Carbonate Sequence
S: Saturated HC; A: Aromatic HC; RA: Resins and Asphaltenes
(*Source*: Tissot, 1977)**

The three main geochemical processes (Figure 5.6) that occur
during burial of sediment can be followed (Brooks, 1981) from the
changes in the elemental composition of kerogen (carbon, hydrogen,
and oxygen) and their atomic ratios (H/C and O/C).

(*i*) In the diagenesis, there is a loss of oxygen (liberation of
CO_2 and H_2O) and decrease in the O/C atomic ratio while
H/C ratio only slightly alters. Production of biogenic
methane by micro biological activity occurs in this stage.

(*ii*) In the catagenesis, there is a rapid loss of hydrogen with
corresponding decrease in H/C atomic ratio while the O/
C ratio undergoes little change. Petroleum generation
(together with gas) occurs during this stage.

(*iii*) In the metagenesis, total elimination of hydrogen takes
place leading to H/C atomic ratio almost zero. Generation
of dry methane gas occurs during this stage.

The general scheme of hydrocarbon formation as a function of
burial depth of source rocks (Figure 5.7) depicts generation of three

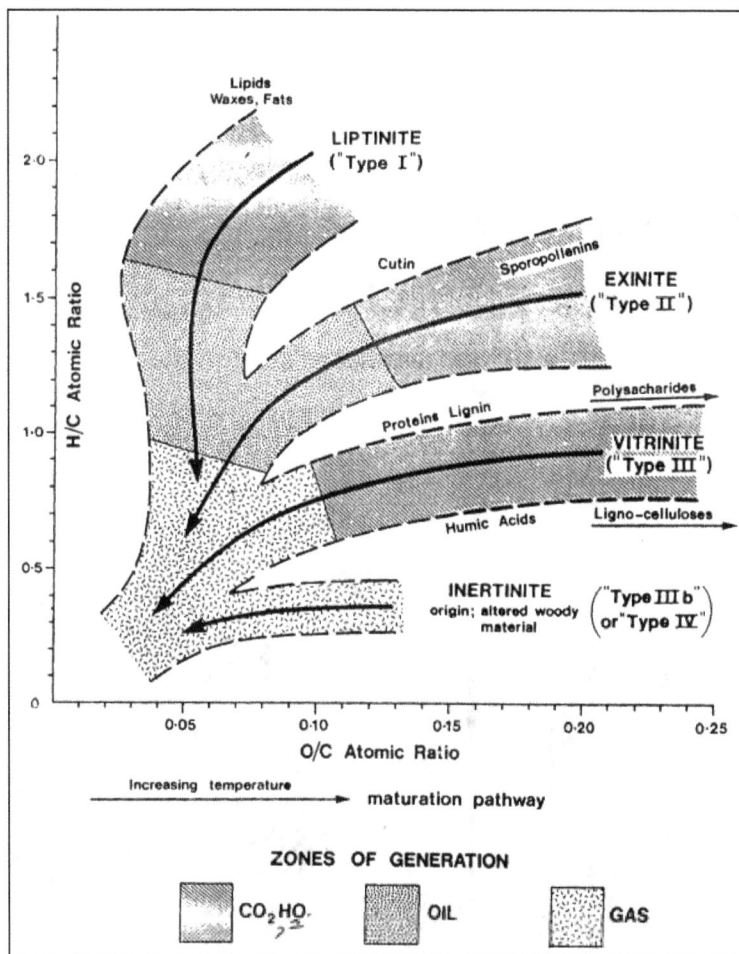

**Figure 5.6: Relationships Between Kerogen-Type, Elemental
Composition and Organic Maturation Pathways
(*Source*: Brooks, 1981)**

types of hydrocarbons in the three distinct zones namely biochemical
methane (marsh gas) in diagenetic (immature) zone, oil and wet gas
in catagenetic zone and dry gas (methane) in metagenetic zone. Thus,
the oil window is defined, overlain by the immature zone and
underlain by the gas zone. The shaded area in the figure representing
biologically derived bitumens steadily decreases with depth relative

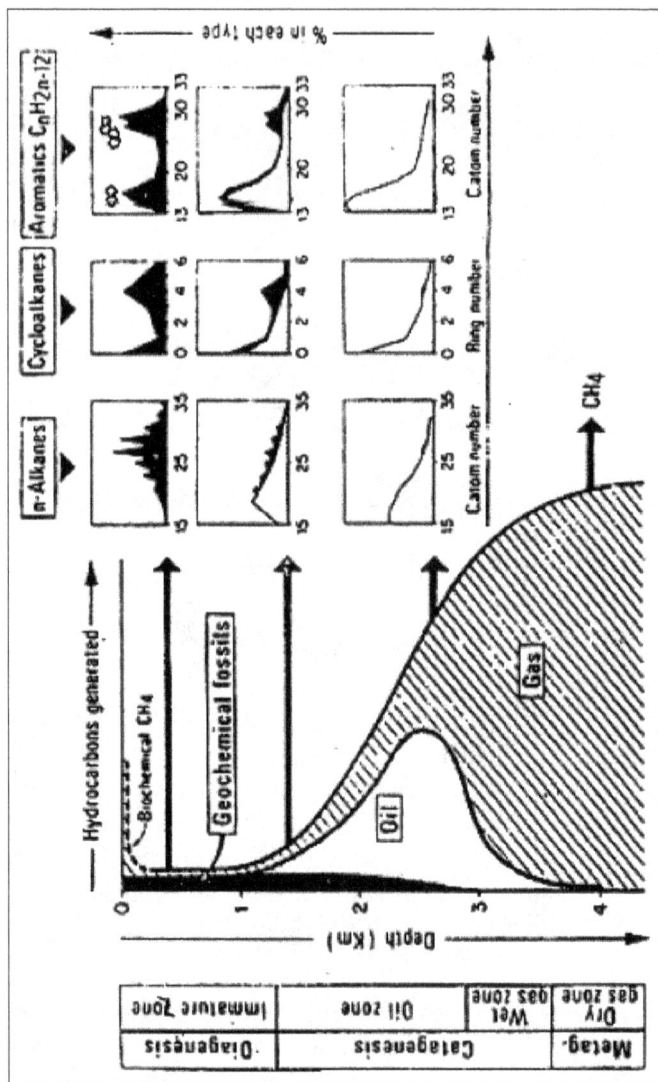

Figure 5.7: General Scheme of Hydrocarbon Formation; Depth Scale is Based on Examples of Mesozoic Source Rocks; Inset Shows Evolution of the Composition of the Three Structural Types (*Source:* Barker, 1985)

to thermally generated bitumens (oil and gas). The lower molecular weight compounds in the bitumen fraction that are generated from kerogen show none of the biological characteristics typical of compounds from recent sediments. This is because the large amounts of thermally generated bitumens at higher depths (increasing temperatures) overshadow the small amounts of bitumens preserved from recent sediments. Thus the bitumen fraction loses features such as odd–even predominance (CPI) in the long chain alkanes, optical activity and the predominance of four-and five-ringed compounds in the naphthene (cyclohexanes) fraction. Further, the normal alkane distribution in the shallow depths with marked odd-even predominance (especially in the longer carbon atom chains) moves progressively to smoother distribution of compounds with shorter chain lengths at higher depths. The multi ringed naphthenes initially showing high abundance for compounds with four and five rings steadily show a diminishing trend, until one and two ring compounds are most abundant in the deeper samples (Barker, 1985).

5.3 Migration and Accumulation of Hydrocarbons

It is evident (5.2.0) that hydrocarbon compounds are generated in appreciable quantities through geochemical alteration (maturation) of high molecular weight organic substances (kerogens) usually found in abundance in fine- grained sedimentary rocks. Further, some insoluble organic residue remains in the rock at least through the oil generating stage. However, petroleum accumulations generally found in relatively coars-grained porous and permeable rocks contain little or no insoluble organic matter. It is highly improbable that huge quantities of petroleum found in these rocks could have originated in them from solid organic matter of which no trace remains. Therefore, it can be concluded that the place of origin of petroleum is normally not identical with the location where it is found in commercial quantities. Evidently it has to migrate from place of its origin (source) to its present reservoir.

5.3.1 Concept of Migration

The concept that oil originates in fine-grained sediments (source rocks) and moves out into coarse-grained sediments (reservoir rocks) through pore fluids during compaction was putforth by Munn as early as in 1909. Since then extensive information was gathered on the fluid movement in the subsurface. Numerous hypotheses have been proposed from time to time as to how exactly petroleum is concentrated in reservoir sands. Nevertheless, the least clearly understood phenomenon among the overall process of origin, migration and accumulation is the mechanism of migration. It is a complex problem involving several simultaneous processes operating at different types of sediments. While recognising the importance of understanding the migration process, Illing (1933) posed the following pertinent questions. (*i*) How does the oil move from the source rock to the reservoir rock?; (*ii*) How is it trapped in the reservoir rock?; (*iii*) How does the water-oil mixture of the source rock in which water predominates become oil rich?; and (*iv*) How does the oil separate from water in the process of migration?

Before we proceed further, it is necessary to acquaint with definitions of some of the terms often we come across with migration.

Primary Migration

It is the movement of oil and gas out of the source rocks into permeable reservoir rocks.

Secondary Migration

It is the movement of fluids within the permeable reservoir rocks that eventually leads to the segregation of oil and gas in certain parts of the reservoir rocks.

Accumulation

It is the termination of secondary migration (arrested migration) in the reservoir. During this process, the migrated hydrocarbons are concentrated in a relatively immobile configuration where they can be preserved over longer periods of time.

Abnormal Pressure

Any departure from normal hydrostatic pressure is called abnormal pressure. Abnormal pressures (>12 k Pa) per metre are created and maintained by the inability of pore fluids to migrate within a reasonable geological time period when subjected to stress such as (*i*) rapid loading, (*ii*) thermal expansion of fluids, (*iii*) compression by tectonic forces, (*iv*) generation of petroleum hydrocarbons from organic matter in the rock matrix.

Capillary Pressure

It is the difference between the ambient pressure and the

pressure exerted by the column of liquid in rock pores. Capillary pressure increases with decreasing pore size or more specially pore throat diameter.

5.4 Mechanism of Primary Migration

Many theories such as migration of petroleum hydrocarbons in true solution as colloids (micelles), free oil as discrete droplets or globules, have been putforward from time to time. All of them are subsequently discarded because of several reasons against them. Only three transport mechanisms namely (*i*) diffusion, (*ii*) molecular solution, and (*iii*) oil/gas phase migration are presently believed to be the possible mechanisms to carry most of the hydrocarbons from source to reservoir rocks. Let us briefly examine these three mechanisms.

5.4.1 Diffusion Controlled Migration

In special geological settings, the diffusion of dissolved hydrocarbon molecules represents an effective process for primary migration of gas but not of oil (Leythaeuser *et al.*, 1982). In diffusion process, the aqueous phase would be stationary, *i.e.*, there is no need of flow of water. Only dissolved hydrocarbon molecules would move independently from places of higher concentration (higher chemical potential) to lower concentration (lower chemical potential). Transportation distances are generally short and always follow local gradients. Such gradients exists within the source rock near the contact to over or underlying porous carrier rocks, around fractures, or faults or in proximity to interbeded siltstones. Apart from them, the molecular diffusion by itself can account for transport of huge quantities of gas over geological times. For example, it was estimated that a 200 m³ volume of high mature gas-prone Jurassic Shale source unit in Western Canada yielded 1.5×10^9 m³ of methane (at STP) in 5,40,000 years by diffusive transport. Due to an exponential decrease of diffusion coefficient values with increasing carbon numbers of gaseous molecules, compositional fractionation effects occur during initial periods of diffusion of hydrocarbons through source rocks. In particular, primary migration by diffusion is controlled by three factors namely (*i*) concentration of initially generated hydrocarbons in the source rocks, (*ii*) relative differences in their diffusion rates, and (*iii*) the source rock thickness.

Diffusion is probably most effective in immature rocks where preexisting light hydrocarbons expelled out of the rocks prior to the onset of significant generation. The main problem with diffusion as an important mechanism of migration is that by definition, it is a dispersive force whereas accumulation of hydrocarbons requires concentration. It would therefore be imperative that diffusion has to be coupled to a powerful concentrating force to yield accumulations of appreciable size. During intense hydrocarbon generation, any contribution by diffusion will be over-whelmed by that from other expulsion mechanisms.

5.4.2 Migration in Solution

Molecular solution of hydrocarbons in pore waters has a serious drawback for primary migration because of limited solubility of most of them even at $100°C$ which appear to be required in water migration in peak oil generation. However, this mechanism seems to function best for the lighter hydrocarbon fraction (*e.g.*, methane, ethane and aromatics) because of their higher solubility in water. Based on the comprehensive treatment of molecular solution of hydrocarbons in water during primary migration, Price (1976) concluded that only a small percentage of the world's petroleum have undergone migration by this mechanism.

Hydrocarbon gases in deep wells are well known to carry oil in gaseous solution. The Russian scientists, Sokolov and coworkers (1963) were the most notable early advocates of primary migration of oils in gaseous solution. Gases such as methane and CO_2 formed during kerogen maturation enhance the solubilities of crude oils to a considerable extent under high pressures. For example, solubility of crude oils of the order of 1000 g m^{-3} are attained when temperature and pressure values reach within drilling depths (Figure 5.8). Paraffinic oils are more soluble in methane than naphthenic oils contrary to their normal trend of solubilities in aqueous solutions. The presence of water is critical to the solubility of crude oil in methane. Water also lowers the temperature and pressure necessary to achieve co-solubility in which the methane and the crude oil behave as a single phase mixture. Co-solubilities as high as 4000 g m^3 are then attained at about $200°C$ (a representative temperature) in deep wells (Price, 1981). However, methane is unlikely to be the most important solute gas in the transfer of oils out of source rocks because of its relatively low concentration compared to liquid

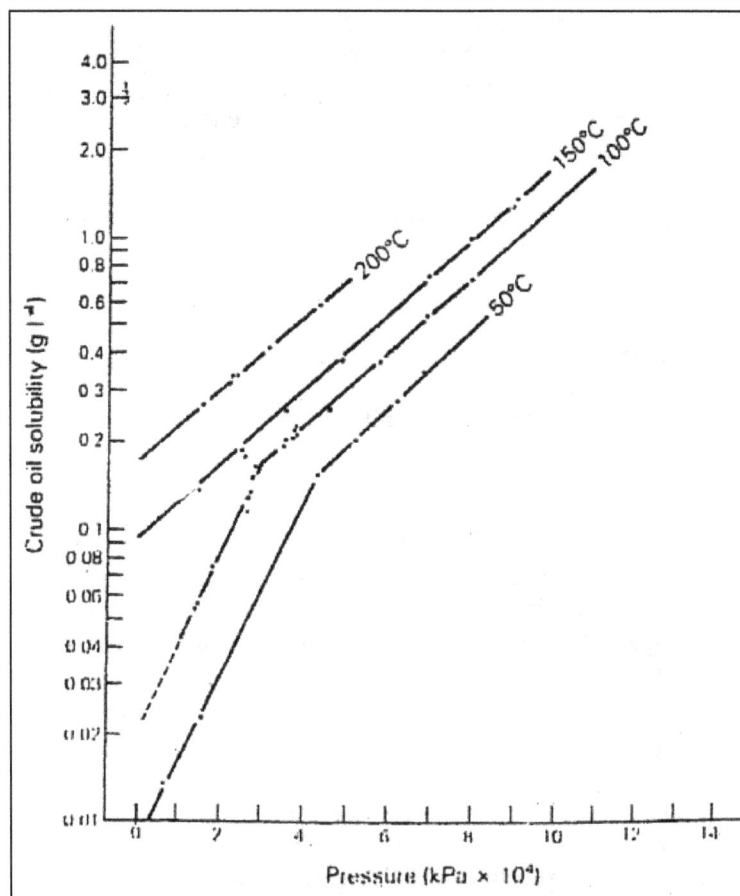

Figure 5.8: Solubility of 44° Gravity Crude Oil in Methane Over a Range of Temperatures Occurring within Drilling Depths (*Source*: Price, 1981)

hydrocarbons during the peak generation of oil. Bray and Foster (1980) observed that hydrocarbon molecules as heavy as C_{28} can be transported if the dominant solute gas is CO_2. At about 4×10^4 kP_a pressure, the quantity of crude oil carried in water saturated both with CO_2 and hydrocarbon gases (mainly methane) rises to 3 per cent by volume which is 100 times more than the quantity taken up into solution under mere rise of temperature (200°C). Nearly all the oil generated in the source rock-even tars and asphaltenes can be accommodated in CO_2–saturated water. Without CO_2 in the solute, however, the only hydrocarbons carried in the water are in the gasoline range, lighter than C_{10}. But the obvious question is a source of sufficient CO_2 in or near the source-rock. Generation of enough CO_2 to be an effective migration agent requires a burial depth in excess of 1000 m and minimum weight percent of kerogen (or TOC). However, some of the favourable properties of CO_2 are greater solubility in oil than in water, miscibility with both oil and methane and enhancing the oil mobility by lowering its viscosity.

Expulsion of oil dissolved in gas requires far greater amount of gas than that of liquid. It would be expected only in the late stages of catagenesis or in source rocks capable of generating mainly gas. Because neither case is of any significance for petroleum formation, we conclude that solution in gas is a minor mechanism for oil expulsion.

5.4.3 Hydrocarbon Phase Migration

By far the most popular mechanism invoked to explain primary migration is expulsion of hydrocarbons in a hydrophobic (oil) phase. There appears to be three distinct ways in which oil phase expulsion can occur.

One of the ways is when bitumen forms a continuous three dimensional interconnected net work that replaces water as the wetting agent in the source rock. Expulsion of hydrocarbons is facilitated because water-mineral and water –water interactions no longer need be overcome. These net works were observed in shales containing 1 to 6 per cent organic matter in the source rocks. The level of saturation required in organic matter in order for oil flow to occur would be from 2.5 to 10 per cent. This mechanism is applicable to very rich organic source rocks found in the Bakken Formation of the Williston Basin (Meissner, 1978).

The second way in which oil phase expulsion can occur is from very rich organic source rocks prior to the onset of strong hydrocarbon generation. As compaction reduces pore volume, the organic material begins to occupy substantial fraction of the total pore space with a reasonably rich shale (OC>5 per cent) and with a substantial compaction, the organic molecules would be free to move under the influence of physical stress. The organic matter expelled consists mainly of lipids that were present in the source rocks during sediment deposition and diagenesis. Therefore this early expulsion mechanism seems to be limited to rocks having very high content of lipids (Type II, kerogen). Khaveri (1984) has proposed this mechanism to account for some of the solidified bitumens in the Unita Basin of Utah. The source rocks for these bitumens probably lie in the extremely organic-rich Unita Green River and Wasath Formations.

The third and the most important way by which oil phase expulsion occurs is through microfracturing induced by over pressuring during hydrocarbon generation. When the internal pressure exceeds the shear-strength of the rock, microfracturing occurs, particularly along lines of weakness such as bending planes. Laminated source rocks may therefore expel hydrocarbons with greater efficiency than massive rocks. Once the internal pressure has returned to normal, the microfractures heal. The hydrocarbons within the pores then become isolated again because of the waterwet source rocks and over- pressuring commences anew. Many cycles of pressure built up, microfracturing expulsion of hydrocarbons and pressure release can be repeated over geological time in order to accomplish meaningful migration of hydrocarbons.

An important implication of microfracturing model is that expulsion cannot takes place until the shear strength of the source rock has been exceeded. Based on empirical evidence, Momper (1978) proposed a generation of 850 ppm of extractable bitumen (EOM) equivalent to 50 million barrels of oil per cubic mile as the threshold value. Although the exact threshold value varies considerably as a function of rock lithology and other factors, Momper's value has been widely accepted as a reasonable average. Once the threshold has been exceeded, most of the hydrocarbons are expelled, but a large proportion of NSO compounds and heavier hydrocarbons are left behind. This is responsible for the differences in composition of bitumen and petroleum (5.2.4, Figure 5.5).The expulsion efficiency estimated from the fraction of bitumen left in the source rock during microfracture induced expulsion is about 50 per cent.

5.4.4 Distance and Direction of Primary Migration

Distance travelled by hydrocarbons during primary migration are of the order of few centimetres to 100 metres. This is because petroleum is being forced through rocks having low matrix permeabilities. As soon as easier paths become available, they will be taken up by the migrating fluids. Such pathways can be continuous sands, unconformities, fracture-fault systems or any beds that are highly permeable compared to shales. In shales with normal compaction, the usual direction of fluid flow is upward because of the decrease in pressure. In compaction disequilibrium however pressure barriers may be created within the shale that prevent the vertical migration. Since the driving force for microfracture-induced primary migration is pressure release, hydrocarbons can be expelled in any direction that offer a lower pressure than the source rock. Because the source rock is over-pressured, expulsion can be lateral, upward or downward depending upon the carrier bed characteristics of the surrounding rocks. Thus the source rock lying between two sand beds on either side will expel hydrocarbons into both of them (Waples, 1985).

5.4.5 Timing of Primary Migration

Petroleum hydrocarbons mostly undergo primary migration in the burial depths of 1,500-3,500 m. This is also approximately the depth range for bitumen generation from kerogen. However, the time lag between generation and migration of hydrocarbons depends on the migrational mechanism. Philippi (1965) has suggested that the first generated bitumen is adsorbed on the kerogen sites and that migration can occur only when all adsorption sites have been filled. This view obviously suggests some time delay between generation and migration of bitumen. This constraint would also apply equally well to solution or continuous phase migration as explained earlier. Migration to microfractures would also require a delay because they will not occur until internal pressures (caused at least partly by catagenesis) have built up sufficiently. A comparison of time of oil generation and migration (Figure 5.9) indicates that the migration window (*e.g.*, the depth, temperature and time limits for migration) overlaps with that of generation. In fact the former is slightly narrower than the bitumen generation window. In most cases they do not differ by more than few million years (Waples, 1981).

Biogenic methane begins migration even in the early stages of compaction. Evidence of its migration in Pliocene sandstones at a depth of 400 m has been observed by Whelan (1979). Gaseous hydrocarbons will migrate from the fine-grained source rocks as they are generated at all depths. Migration, both in solution, and as

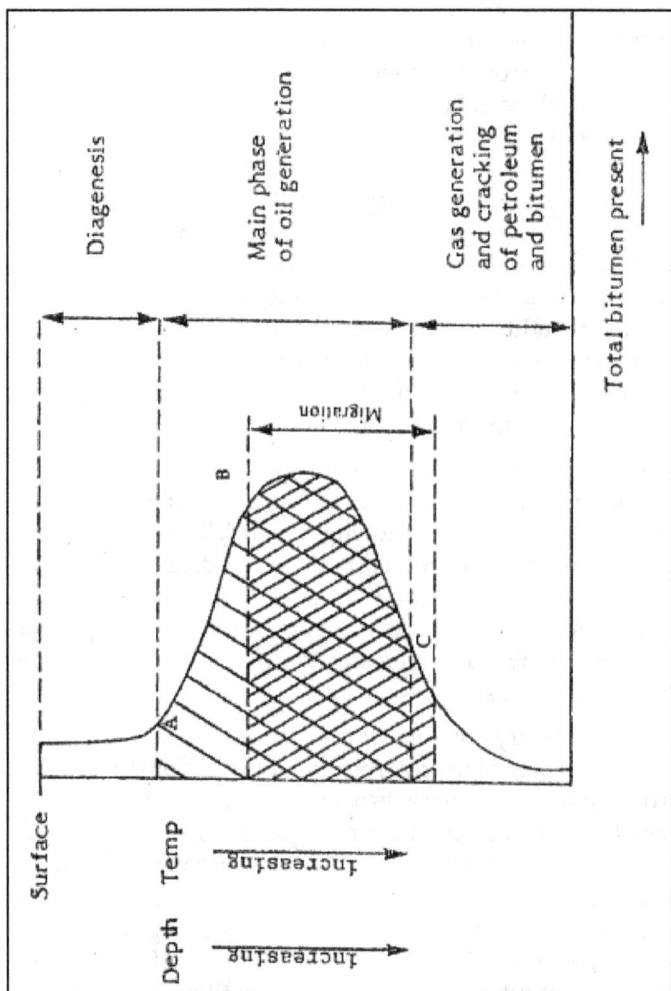

Figure 5.9: Comparison of Timing of Oil Generation and Migration
(*Source:* Waples, 1981)

an oil phase, would occur mainly during periods of compaction and fluid movement.

5.5 Secondary Migration

Once hydrocarbons are expelled by primary migration from source rocks in a separate phase into reservoir rocks, their subsequent movement occurs through secondary migration. Three important parameters control secondary migration and subsequent formation of oil and gas pools. They are: (*i*) buoyant rise of oil and gas in water saturated porous rocks, (*ii*) capillary pressure that determine the multiphase flow, and (*iii*) hydrodynamic fluid flow. As long as the aqueous pore fluids in the subsurface are stationary, *i.e.*, under hydrostatic conditions, the only driving force for secondary migration is buoyancy. If there is water flow in the subsurface *i.e.*, under hydrodynamic conditions, the buoyancy rise of oil and gas may be modified by this water flow. Capillary pressure in narrow pores, always oppose the buoyant pressure, thus stopping migration leading to hydrocarbon entrapment.

Hydrocarbons are less dense than formation waters and are more buoyant. They are thus capable of displacing water downward with the consequence of their moving upward. The magnitude of the buoyant force is proportional both to the density difference between water and hydrocarbon phase, and the height of the oil stringer. Coalescence of globules of hydrocarbons after expulsion from the source rock therefore increases their ability to move upwards through waterwet rocks.

The capillary entry pressure which opposes the buoyancy offers resistance to the entry of the hydrocarbon globules or stringer into pore throats. Whenever a pore throat narrower than the globule is encountered, the globule must deform to squeeze into the pore (Figure 5.10). The smaller the pore throat, the more deformation is required. If the upward force of buoyancy is large enough, the globule will squeeze into the pore throat and continue moving upward. If however, the pore throat is very tiny or if the buoyant force is small, the globule can not enter and becomes stuck until the buoyant force or the capillary entry pressure force changes suitably. This is the simplest model without considering the third force namely hydrodynamic flow which can also modify hydrocarbon movement. If the water (hydrodynamic) flow in the subsurface is in the same

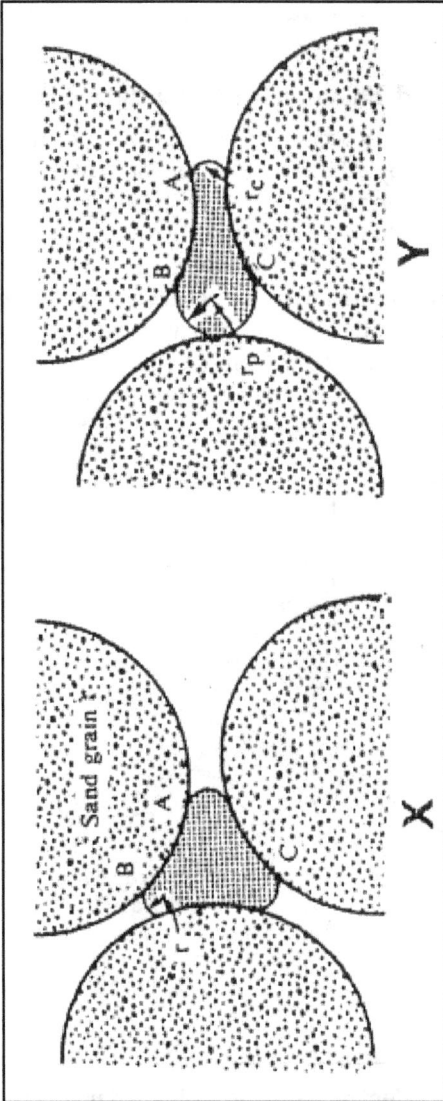

Figure 5.10: Movement of an Oil Droplet through a Small Pore, Showing Deformation Required for Passage through a Pore Constriction (*Source*: Waples, 1985)

direction as the movement of the hydrocarbons by buoyancy, the rate of the latter is increased. On the other hand, if it is reverse, the rate of the hydrocarbon movement will be retarded. In any case, these modifications to the overall migration are however, only minor (Waples, 1985).

5.5.1 Direction and Distance of Secondary Migration

Secondary migration occurs preferentially in the direction that offers the greatest buoyant effect. Thus movement of oil within a confined migration conduit will be updip perpendicular to the structural contours wherever possible (Figure 5.11). This is also facilitated by the movement of hydrodynamic flow in the same direction. In a massive sandstone reservoir rock, secondary migration will occur both laterally and vertically. It means that hydrocarbons which enter sand from an underlying source rock will move toward the top of the sand even as they migrate laterally updip. Vertical migration can also occur across formations. Stacked

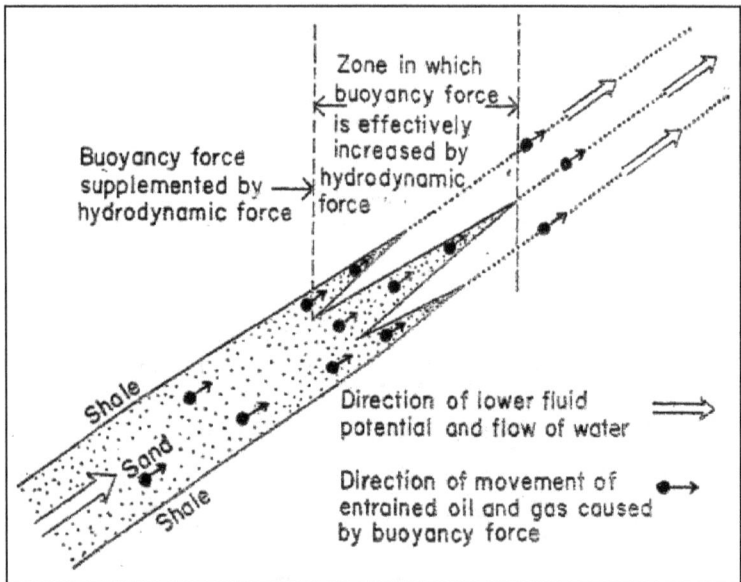

Figure 5.11: The Effect of Water Flow Directed up the Dip and through a Barrier Zone of Relatively Higher Displacement Pressure and Decreased Permeability
(*Source*: Levorsen, 1967)

sands in a paleodelta, for example, can offer possible pathways for vertical migration. Unconformities also can juxtapose migration conduits, thus providing a potentially effective system for combined vertical and lateral migration. Faults may play an important role in vertical migration not only because they often juxtapose carrier beds from different stratigraphic horizons, but also may have high permeability regions adjacent to them (Waples, 1985).

Unlike primary migration which is limited to a small distance, the secondary migration can be a very long distance phenomenon. Indeed, the largest hydrocarbon deposits known including the Athabasca Tar Sands of Western Canada, the heavy oils in the Orinoco Belt of Venezuela and the Saudi Arabian crude oils all must have migrated long distances upto few hundred kilometers to produce the largest hydrocarbon deposits. They occur in extremely stable tectonic settings where major but gentle down warping has deposited huge volumes of matured source rocks, and continuous blankets of carrier (sand) beds. However, such cases are very few. Most basins are brokenup tectonically and have poor lateral continuity of carrier beds as a result of both tectonic disruption and facies change related to tectonic events. In such basins secondary (lateral) migration distances do not exceed few tens of kilometres. Drainage area is one of the important factors influencing the size of hydrocarbon accumulations. The larger the drainage area, longer is the distance of secondary migration leading to larger hydrocarbon accumulations. Vertical migration distances of several thousand metres are commonly known in several basins (Price, 1980).

5.6 Accumulation and its Efficiency

Accumulation of hydrocarbons occurs primarily where their buoyancy driven movement is strongly impeded or stopped. It is well known that caprocks having low permeabilities and high capillary entry pressure are common barriers to migration and therefore act as good traps. Most hydrocarbon traps are either structural or stratigraphic. The seal prevents vertical migration from the reservoir rock into overlying strata, while the structure or lithological change prevents lateral updip migration.

Migrating hydrocarbons often form their own seals. Formation of stratigraphic traps sealed by tar occurs frequently where inspissation from biodegradation and evaporation have occurred.

Tar sealing therefore slows down the loss of oil where reservoir seals have been destroyed. Further, a hydrocarbon accumulation forms when the rate of inmigration exceeds the rate of leakage, and can persist after inmigration ceases if the rate of leakage is only low. In other words, the hydrocarbons are supplied to the trap faster than they can leak away. Rates of leakage considerably vary in different reservoirs. Areas such as Southern California which have major active seeps, are quite leaky, while in areas like the Mid-continent, oil generated and trapped in the Paleozoic persists even today. Tar mats produced by biodegradation can create excellent seals where no other structural or stratigraphic trapping mechanism exists. Despite the rarity of Tar mat seals, they provide largest hydrocarbon accumulations *e.g.*, Athabasca Tar sands and the Orinoco heavy oil belt.

Formation of solid crystalline hydrates with natural gases (particularly methane) offer another efficient trapping mechanism. Methane gas hydrate formed is stable under pressure, temperature regions that occur at depths of few hundred metres below the sea floor in continental shelves or deep sea and in permafrost regions. The methane hydrate forms efficient seal against vertical hydrocarbon migration and thus provide an important trapping mechanism for biogenic methane, particularly in young, unconsolidated sediment (Waples, 1985).

Three critical factors in general, determine the efficiency of migration or accumulation (Waples, 1981). They include:

(*i*) Existence of continuous lateral conduits leading from the centre of the basin, where the most intense oil generation occurs, to the flanks where traps are likely to be found. If lateral continuity is poor, accumulations are likely to be smaller, and much of the oil may never reach the basin flanks. Instead, it is trapped within the basin where subsequent cooking will convert it to gas.

(*ii*) Occurrence of favourable relative timing of structural growth and oil generation for entrapment. If at any time, the structure is too small to handle the amount of oil being poured into it, then it will spill out and migrate onwards, with a correspondingly reduced accumulation efficiency (unless it is trapped by a different trap).

(*iii*) Presence of active faults, such as growth faults during the period of oil migration. These faults may discourage any long distance lateral migration, but on the other hand, they may lead to numerous smaller accumulations that are not associated with structural traps on the flanks of the basin. Migrational efficiency is probably higher in many cases, one typical example being Gulf Coast.

Chapter 6

Geochemical Methods in Petroleum Exploration

As explained earlier, maturation of kerogen occurs in the catagenetic stage (5.2.3). Establishing the level of maturation in source rocks of an area is vital for exploration. When kerogen is immature, no petroleum would be generated. With increasing maturation first oil and then gas are produced and expelled. When the kerogen over- matures neither oil nor gas remains in it. There are several geochemical methods for the measurement of maturity of kerogen. Principles and applications of some of these methods are described in this chapter.

Organic matter in petroliferous rocks can broadly be classified into two categories: (*i*) unstructured/amorphous, insoluble material (kerogen), and (*ii*) soluble organic material (bitumen) that can be extracted into organic solvents commonly called as extractable organic matter (EOM, 5.1.2). Different types of kerogen must be distinguished and identified since they have different hydrocarbon generation potentials and products (5.2.2).

6.1 Isolation of Kerogen and Bitumen

Kerogen is normally isolated from the rock matrix before analysis. The procedure involves removal of carbonates by HCl treatment, followed by successive digestions of silicate minerals by a mixture of HCl and hydrofluoric acid (HF) at low temperature (60° C). This procedure produces a kerogen concentrate which contains other resistant minerals, especially pyrite. Kerogen obtained at this stage can be used for most purposes. If further purification is required, it can be carried out with some success by heavy liquid separation, centrifugation or magnetic separation. Bitumen is generally isolated from the source rock, after the sample has been powdered to about 100-200 mesh and then extracted with an organic solvent like chloroform or a mixture of benzene and methanol. Two types of geochemical techniques are mainly employed for characterising the organic components in sedimentary rocks (*i*) petrographic and (*ii*) geochemical (Brooks, 1981).

6.2 Petrographic Methods

These methods use polished preparations of whole rock and/ or kerogen from rocks to identify organic particles, their content and character. There are three important methods in this category. They include (*i*) microscopic organic analysis (MOA), (*ii*) thermal alteration index (TAI), and (*iii*) vitrinite reflectance (Ro). Though these methods are useful for kerogen maturation studies in source rocks, they have some limitations (Powell *et al.*, 1982).

6.2.1 Microscopic Organic Analysis (MOA)

Microscopic examination of sedimentary organic matter (kerogen) provides valuable information on the type and evolution stage of kerogen. In addition, it is possible to recognize mixing of different organic sources either contemporaneous (*e.g.* marine phytoplankton or terrestrial plants) or reworked from previous sedimentary cycles. MOA is carried out on microscope slides as prepared in palynological studies. It is important, however, that no oxidising or lightening agents be used. The slides can be used to identify the kerogen particles (macerals) based on their visual appearance under transmitted light (Staplin, 1969). Classification of kerogen type is based mostly on the general shape of the organic remnants, because formally identifiable micro-fossils account for only a minor part of the organic matter. Two main types are usually distinguishable:

(*i*) Humic or crystalline material, supposedly herbaceous or woody origin where shapes or structures of the vegetable tissues are still recognisable. It is usually considered to be of a gas source rather than of a oil source rock, and

(*ii*) Amorphous material with a vaporous or cloudy shape and no identifiable structure. It is commonly referred to as sapropelic organic matter, and considered to be a good oil source rock.

When comparing this classification with the types of kerogen based on chemical composition (C, H, O, H/C and O/C atomic ratios, Tissot diagram, Figure 5.2a), there seems to be a broad agreement. Humic material usually belongs to kerogen Type III (vitrinite). Most of the material considered to be sapropelic refers to kerogen Type I (alginite) and Type II (exinite). Visual kerogen analysis is also important for quality control of data. For example, the samples obtained during drilling (conventional cores, sidewall cores, cutting etc.) are normally contaminated with drilling fluids, walnut hulls, rubber and other solid organic debris. Careful microscopic examination, however, can usually identify the problem samples (10.4.1).

6.2.2 Thermal Alteration Index (TAI)

Kerogen colour is widely used as an approximate and inexpensive maturation indicator. The principle involved in this method is that the colour of kerogen macerals observed through a microscope is indicative of their state of preservation of spores and pollen materials. Initially these organic remnants are yellow representing the colour of the living organisms. With increasing evolution, they become successively orange, brown and finally black. Based on these colour changes, a semiquantitative scale of evolution has been proposed ranging from 1 to 5 (Table 6.1) with higher values indicating greater maturity. Since the method is based on colour of spores or pollen grains, this phenomena became known as spore darkening or thermal alteration index (TAI). The TAI scale vary in the range 1-5. The oil generation zone in the source rock approximately falls in the range 2.6 to 3.2. As the technique is subjective, constant quality checks should be performed. Further, it is not applicable to Pre-Devonian rocks because of absence of land plants during that period (North, 1990).

During diagenesis, the kerogen macerals (alginites, exinites and vitrinites) are initially yellow, representing the colour of living organism as stated above. During catagenesis, the colour becomes progressively darker passing through orange, brown and black (Haseldonckx, 1979). Because of their extensive oxidation, inertinites are generally quite dark regardless of the stage of their maturation. Very few kerogens consist of a single maceral type. Most of them are complex mixtures containing at least as many or all maceral types to a small extent. Interpretation of total kerogen colour can be difficult in many samples because of different maturation rates of different species. Consequently, colour interpretation is limited to individual kerogen species. Sophisticated techniques use microscopes with optics capable of measuring the transmitted light through individual spores or pollen grains relative to a standard. Some microscopic systems are setup with ultraviolet (light) excitation to measure the fluorescence of kerogen macerals which also undergo changes with maturation (10.8.1).

Table 6.1: Organic Maturation Indices

Approxi- mate Temp., ^{o}C	Maturation Index*	Kerogen Color	Maturity Level	Petroleum Prospects
30	1 Unaltered	Light yellow	Immature	Dry gas
50	1+ Slightly altered	Yellow		Dry gas, heavy oil
100	2 Moderately altered	Orange	Mature	Oil, wet gas
150	3 Strongly altered	Brown		Condensate, wet gas
175	4 Severely altered	Brownish black		Dry gas
> 200	5 Metamor- phosed	Black	Metamorphosed	Dry gas to barren

*Also referred to as thermal alteration Index (TAI).

Source: Hunt, 1996.

6.2.3 Vitrinite Reflectance (R_o)

The use of reflectance measurements on the kerogen macerals, particularly vitrinite have long been known for evolution of coals. It has been subsequently adapted to disseminated organic matter in source rocks (Hood *et al.*, 1975). Vitrinite reflectance is certainly the most widely used technique for characterising kerogen evolution and it is probably the best facility offered in respect of microscopic examination. Reflectance of light on a polished surface of vitrinite increases with maturation because of a change in its molecular structure. Vitrinite is composed of clusters of condensed aromatic rings linked with chains and stacked on top of each other. On increasing maturity, the clusters fuse eventually forming sheets of condensed rings that assume an orderly structure. The increase in the size of these sheets and their orientation is responsible for increase in their reflectivity (Hunt, 1996).

The technique involves first separation of kerogen from organic matter by digestion with HCl and HF acids. It is then freeze dried, mounted on epoxy resin and polished. Ro values are determined with a reflecting microscope using oil immersion objectives. About 50-100 reflectance measurements of individual vitrinite macerals are made on a single sample and plotted as histogram (Figure 6.1). A good histogram (Figure 6.2A) has a single mode and small standard deviation. Poor histograms (Figure 6.2 B&C) are broad, bimodal or diffuse. Much of the scatter in such histograms may be due to the

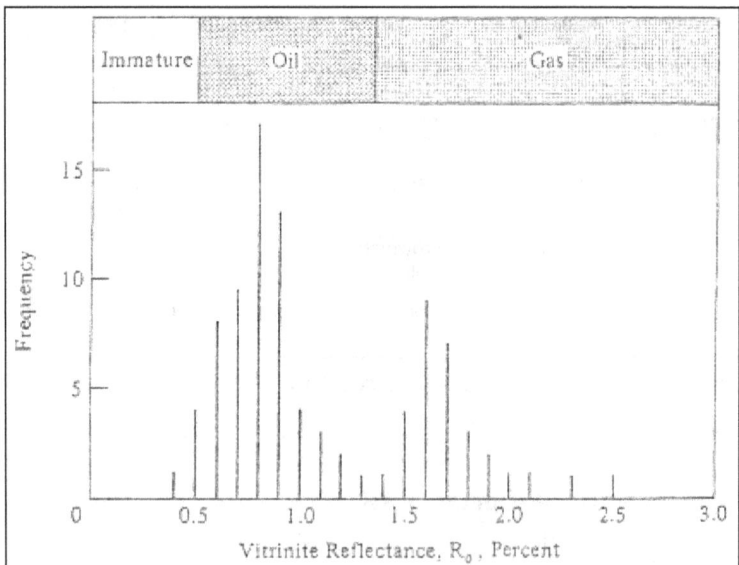

Figure 6.1: Histogram of Reflectance of Vitrinite Particles from a Single Sediment Sample
(*Source*: Hunt, 1996)

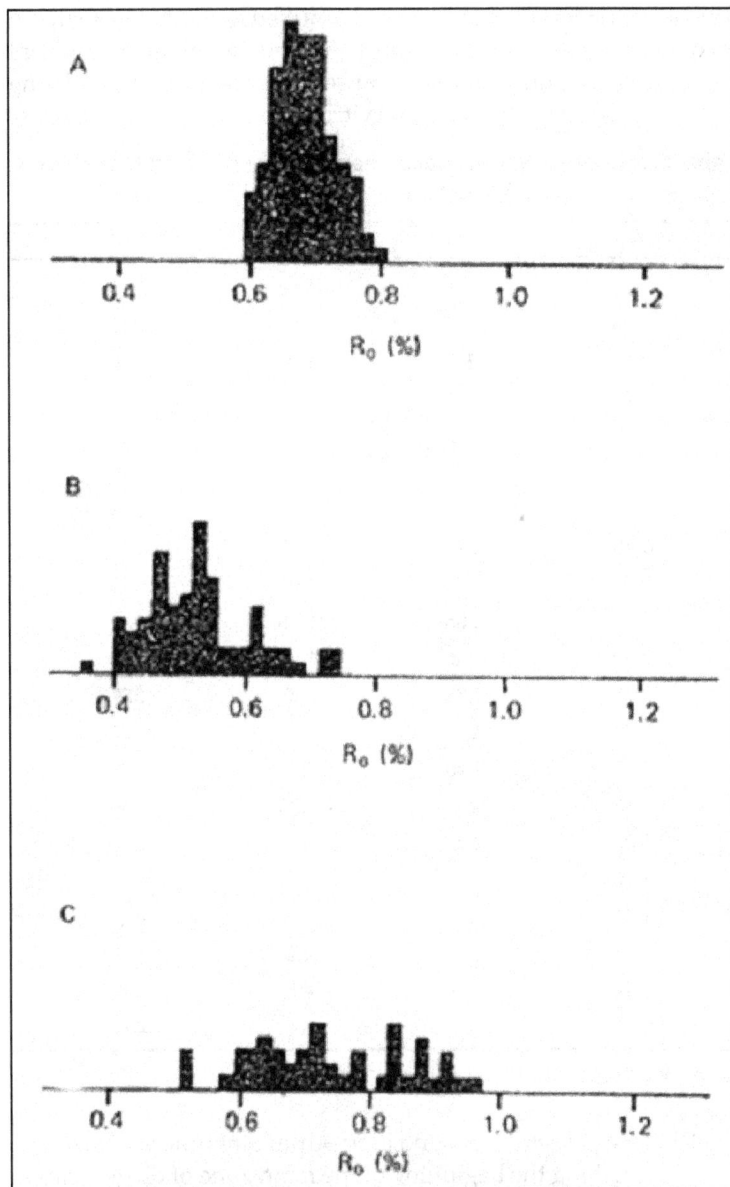

**Figure 6.2: Vitrinite Reflectance Histograms for
Three Samples: A–Good histogram, B and C–Poor histograms
(*Source:* Waples, 1981)**

presence of reworked material, misidentified non-vitrinite material or contamination. Ro values approximately less than 0.5 per cent are subjected to uncertainty because the organic material being analyzed has not yet become true vitrinite.

Table 6.2: Conversion Between Thermal Alteration Index (TAI) and Vitrinite Reflectance (R_o)

R_o	TAI	R_o	TAI
0.30	2.0	1.26	3.15
0.34	2.1	1.30	3.2
0.38	2.2	1.33	3.25
0.40	2.25	1.36	3.3
0.42	2.3	1.39	3.35
0.44	2.35	1.42	3.4
0.46	2.4	1.46	3.45
0.48	2.45	1.50	2.5
0.50	2.5	1.62	3.55
0.55	2.55	1.75	3.6
0.60	2.6	1.87	3.65
0.65	2.65	2.0	3.7
0.70	2.7	2.25	3.75
0.77	2.75	2.5	3.8
0.85	2.8	2.75	3.85
0.93	2.85	3.0	3.9
1.00	2.9	3.25	3.95
1.07	2.95	3.5	4.0
1.15	3.0	4.0	4.0
1.19	3.05	4.5	4.0
1.22	3.1	5.0	4.0

Source: Waples, 1981.

Vitrinite particles are found in 80-90 per cent of all well cuttings. Broadly speaking the beginning of the main zone of oil generation may start in the range of Ro values 0.5 to 0.7 per cent and terminate at 1.3 per cent or some what earlier according to the kerogen type. Wet gas and condensate may extend upto about 2 per cent ; beyond that limit (upto 3.5 per cent) is dry gas (methane) zone (Dow, 1977; Vassoevich et. al., 1974). Table 6.2 shows a relation between Ro and TAI. The principal zone of oil generation lies between 0.6-1.2 per cent and it corresponds to the range of 2.6-3.2., in terms of TAI.

It should be recognized that vitrinite reflectance is only a maturation indicator; it can not predict the presence of oil or gas. Other geochemical data would be needed to determine whether the organic matter in the source rock is capable of generating oil or gas. A technique such as Rock-Eval (6.5.0) pyrolysis or Gas chromatography (10.6.0) must be used with vitrinite reflectance to define the most productive zones of oil or gas. Vitrinite reflectance profiles are also used to interpret some of the sedimentary and tectonic history of basins such as major erosional unconformity, igneous intrusion, change in the geothermal gradient etc. Borehole logs that plot Ro against depth indicate the intervals in which oil or gas may have been generated. Abrupt shifts in the values of Ro with depth may indicate faults or unconformities. An abrupt increase in Ro with depth followed by return to the previous gradient may be caused by igneous intrusions (Dow, 1977). Vitrinite reflectance has the advantage over other maturation techniques in that it covers almost the entire temperature range from early diagenesis through catagenesis into metamorphism. It is in fact used as one of the best paleothermometers in source rock maturation studies. Some limitations of this technique include (*i*) an experienced microscopist is required to recognize vitrinite particles and to distinguish them from other kerogen macerals, (*ii*) atleast 50-100 vitrinite particles from each sample have to be identified and subjected to analysis, and (*iii*) the technique as in the case of TAI, is not applicable to Pre-Devonian rocks (North, 1990).

6.3 Geochemical Methods

Geochemical methods are employed to determine the type and the amount of organic matter in a rock and its potential to generate hydrocarbons (Brooks, 1981). There are five important methods in this category. They are: (*i*) combustion methods, (*ii*) stable isotope analysis, (*iii*) pyrolysis (Rock–Eval), (*iv*) time–temperature index (TTI) and (*v*) level of organic metamorphism (LOM). The combustion methods are again divided into two types based on (*i*) carbon ratio (fixed carbon), and (*ii*) total organic carbon (TOC).

6.3.1 Carbon Ratio Method

One of the oldest and the most fundamental method of maturation indices is the carbon ratio which is also called as fixed carbon method (White, 1915). It is based on the theory that as organic carbon undergoes maturation, it loses first CO_2 during decarboxylation and later during bitumen/petroleum generation. The net result of this is a gradual decrease in the proportion of volatile (movable) carbon and a relative increase in the non-volatile (fixed) carbon. This phenomena, namely increase of non-volatile or fixed carbon with increasing temperature and hence increasing burial depth is known as Hilts law. This law has been confirmed in all coal basins except where local magmatic heating has disrupted the normal geothermal gradient. The fixed carbon or carbon ratio is obtained by taking the ratio of non-volatile (fixed) carbon C_R and the total carbon C_T, as shown in the equation

$$\text{Carbon Ratio} = (C_R/C_T) \times 100 \qquad (6.1)$$

This theory first proposed by petroleum geologists was successfully applied to the study of metamorphism of coals. The coals in general, have a fixed (non-volatile) carbon content in the range 40-50 per cent, gas accumulations were found in the range 60-65 per cent and there was no oil or gas above 70 per cent. Isocarb maps are made by joining lines of equal carbon ratio which are useful in identifying prospective areas of gas or oil in coal fields (Thom, 1934). Although the basic concept is correct, the method failed because of insurmountable sampling problems. Suitable coal samples were rarely available from drill cuttings and hence there is no way to make an accurate evaluation of carbon ratio down a drill hole. This problem was to some extent, surmounted by initially extracting the sample with an organic solvent to remove the soluble bitumen and than treating with HCl to remove carbonate carbon. One aliquot of the treated sample was burned in air at about 900°C to obtain the total organic carbon (C_T). Another aliquot was burned in an inert atmosphere (nitrogen or helium) at the same temperature to obtain the volatile matter. The difference between total organic carbon and the volatile organic carbon gives the non-volatile organic carbon (C_R). This technique enabled carbon ratios to be determined on the source rock kerogen at different depths in any drill hole (Gransch and Eisma, 1970). However, the carbon ratio method did not prove to be successful to oil source rocks because different types of kerogen macerals (alginite, exinite, vitrinite etc.) yield different non-volatile fractions as they mature unlike coal where only one maceral (vitrinite) predominates. In fact the carbon ratio method works well if it is possible to analyze only one maceral type at different depths of a drill- hole.

6.3.2 Total Organic Carbon (TOC)

The method is based on the measurement of CO_2 produced on heating a source rock or kerogen sample. An aliquot of the crushed sample is subjected to heating in a furnace to a temperature of about 900°C and the CO_2 evolved is measured. The amount of CO_2 formed depends directly on the total carbon content both organic and inorganic. Another aliquot of the same crushed sample is digested with HCl to remove carbonates by the reaction (Eq. 6.2) and the amount of CO_2 evolved is measured.

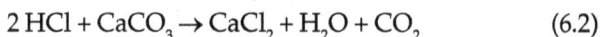

$$2\,HCl + CaCO_3 \rightarrow CaCl_2 + H_2O + CO_2 \qquad (6.2)$$

This quantity is related to the inorganic carbon content. Finally the organic carbon content is calculated by subtracting inorganic carbon from total carbon. This method is satisfactory for all rock samples except carbonates which contain large amount of inorganic carbon than organic carbon, thus leading to erroneous results because of very small difference between the two fractions (Waples, 1981).

6.3.3 Elemental Analysis

Elemental (carbon, hydrogen and nitrogen) analysis of kerogen or organic matter in source rocks can be conveniently carried out using commercially available automatic C, H and N analyzers. The range of elements has subsequently been extended to sulphur and oxygen (S and O) in recent years. The principle involved in this method consists of conversion of organic matter in the sample into volatile gases on heating to a preset temperature and analysing them by appropriate methods. For example, a small quantity of finely ground source rock or kerogen sample is burned and the amounts of CO_2, H_2O, N_2, SO_2 and O_2 gases evolved are measured. Close monitoring, careful standardisation with a reference (standard) material and carrying duplicate runs, are essential for reproduceable and accurate results (Waples, 1981). Changes in the distribution of elements (C, H, and O) and their atomic ratios (H/C and O/C) in kerogen with depth are extensively used to interpret the type of kerogen and the oil or gas generating potential of source rocks (5.2.2).

TOC measurements described above are combined with a Rock-Eval (pyrolysis) analyzer (6.5.0) in a single instrument known as "Oil Show Analyzer". It is designed for rapid well site estimation of source-rock potential where the results can be used in an on going drilling programme.

6.4 Stable Isotope Analysis

Atoms of an element whose nuclei contain the same number of protons but different number of neutrons and thus different atomic masses, are called isotopes. For example carbon has 6 protons but three different number of neutrons (6, 7 and 8) giving rise to three isotopes expressed as ^{12}C, ^{13}C and ^{14}C with atomic masses 12, 13 and 14 respectively. Similarly hydrogen has three isotopes ^1H, ^2H and ^3H and sulphur has four isotopes ^{32}S, ^{33}S, ^{34}S and ^{36}S. All of them except ^{14}C are stable. Stable isotopes have been extensively used in petroleum exploration since the early investigations of Silverman and Epstein (1958). Among them, carbon isotopic ratios ($^{13}C/^{12}C$) have extensive applications in organic geochemistry (Schoell, 1984). They are determined by completely oxidising (burning) the substance (source rock or kerogen or bitumen) to CO_2 and measuring the relative amounts of $^{13}CO_2$ (mass 45 amu) and $^{12}CO_2$ (mass: 44 amu) with a special mass spectrometer. The measured values of the sample are compared to a standard substance, and the ratio represented as del^{13}C ($\delta^{13}C$) by the equation,

$$\delta^{13}C \text{ (per mil)} = \left[\frac{(^{13}C/^{12}C)\text{sample}}{(^{13}C/^{12}C)\text{standard}} - 1 \right] \times 1000 \tag{6.3}$$

The most commonly used standard for carbon is a carbonate belemnite fossil from the Pee-Dee formation of South Carolina, abbreviated as PDB. A few groups report values relative to a standard, National Bureau of Standard Oil NBS–22. δ^2H and $\delta^{34}S$ are calculated in a similar manner using the above equation, by measuring the ratios ($^2H/^1H$) and ($^{34}S/^{32}S$) of sample and standard. They are converted by burning the samples to H_2O and SO_2 respectively and measuring their relative amounts. The standards usually employed are standard Mean Ocean Water (SMOW) for hydrogen and Triolite mineral (FeS) from the Canyon Diablo meteorite for sulphur.

Because the variation in $^{13}C/^{12}C$ ratio is quite small in many cases, it should be measured with great accuracy if one has to derive any meaningful data from them. The usual precision attainable in $\delta^{13}C$ measurements is ± 0.1 to ± 0.02, but if greater precision is required it can be obtained by using more sophisticated mass-spectrometers. Carbon isotopic measurements can be applied to any material that contain carbon *e.g.*, natural gas, petroleum, bitumen, kerogen, coal etc., (Waples, 1981). They are extremely useful in identifying the origin of hydrocarbon gases (particularly methane) namely biogenic (produced during diagenesis) or thermogenic produced during (catagenesis/metagenesis) (Bernard *et al.*, 1976; Sackett, 1977). For example, biogenic methane is depleted with $\delta^{13}C$ (values ranging from –55 to –90‰ relative to PDB) while thermogenic methane is relatively enriched with $\delta^{13}C$ (values ranging –30 to –50‰ relative to PDB). Carbon isotopes in some cases are useful in determining the nature of organic matter in source rocks (Figure 6.3). Most land plants (particularly C_3 type) have $\delta^{13}C$ values in the range –24 to –34‰ (versus PDB), while marine plants (phytoplankton) fall in the range –6 to –19‰ (versus PDB).

Apart from studies on maturation and origin of petroleum, following are some of the major applications of stable carbon isotopes in petroleum geochemistry (Brooks and Bindra, 1977).

(*i*) Comparison between the oils of the same reservoir or between different oils in a particular basin based on isotopic similarity,

(*ii*) Comparison between a crude oil and the organic matter of its source rock based on isotopic composition,

(*iii*) Use of isotopic composition of kerogen as possible paleoenvironmental indicator,

(*iv*) Variation of isotopic composition as indicator for migrational and thermal alteration effects of bitumen or oil, and

(*v*) Variation of isotopic ratios of various crude oil components (n-paraffins, isoprenoids, aromatics, asphaltenes etc.,) as an index of the degree of maturation.

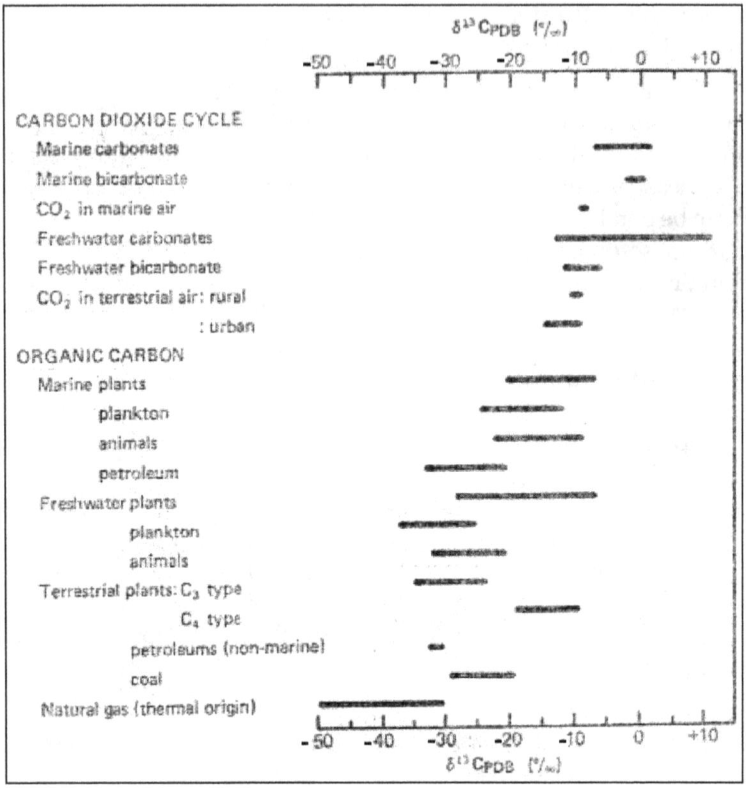

Figure 6.3: δ^{13}C Ranges for Various Sources of Organic and Inorganic Carbon Versus PDB Standard.
(*Source*: Waples, 1981)

δ^{34}S, has been used in maturation of kerogen, biodegradation, sulphurisation–desulphurisation of petroleum. Oils of the Willston Basin were distinguished by the δ^{34}S values of their sulphur containing compounds in the bulk oil (Thode, 1981). Three major types of crude oils were found with δ^{34}S values 5.8 ± 1.2‰, 2.8 ± 0.8‰ and −4 ± 0.7‰ respectively. Long distance sedimentary migration over some 150 km resulted in little or no charge in δ^{34}S values. However, thermal maturation, water washing and biodegradation may affect the δ^{34}S values.

6.5 Rock-Eval Pyrolysis Method

The term pyrolysis refers to heating of a finely powdered sample at a controlled temperature in the absence of air and analysis of the gaseous products produced. A procedure and an apparatus (Rock-Eval Analyzer) have been development for rapid source rock characterization based on pyrolysis (Espitalie *et al.*, 1977; Clementz *et al.*, 1979). Later developments in this technique such as pyrolysis-GC and pyrolysis-GC-MS (10.7.0) have been now extensively used to characterize source rocks, level of maturation and composition of generated hydrocarbons (Horsefield, 1984).

All the organic petrographic methods described earlier (6.2.0) require preliminary separation of kerogen from the mineral matter which is rather time consuming and cannot always be achieved due to high residual content of pyrite. On the other hand, pyrolysis assay is fast (completed in 20 minutes) and can be directly carried out with powdered rock samples of cores or cuttings. It can be performed on large number of samples for routine work even at the well site and hence useful in geological logging.

Rock-Eval pyrolysis method uses a special device (furnace) which heats the powdered rock sample (100 mg) in an atmosphere of helium according to pre-selected temperature programme. The gaseous hydrocarbons and carbondioxide generated are expelled from the furnace, fed to a gas chromatograph (GC) attached with a flame ionisation detector, and their concentrations are quantitatively measured. The following important parameters are obtained from Rock-Eval analysis in a 20 minute cycle.

(*i*) Hydrocarbons already present in the sample which are volatilised by heating to a moderate (upto 250° C) temperature. No kerogen decomposition occurs at this stage. Figure 6.4 represents them by the area S_1.

(*ii*) The Hydrocarbon type compounds generated during pyrolysis of kerogen present in the sample by heating upto 400° C. This process is analogous to natural kerogen catagenesis. They are represented by the area S_2 (Figure 6.4).

(*iii*) Oxygen containing volatiles, particularly CO_2 generated during pyrolysis in the temperature range 600-700°C. Care should be taken to ensure that no decomposition of inorganic carbonates occurs at this stage by keeping the temperature around 650°C. The amount of CO_2 is shown by the area S_3 (Figure 6.4).

(iv) Temperature corresponding to the peak of hydrocarbon generation during pyrolysis (S_2) shown as T°C (Figure 6.4). This is related to the thermal maturity of kerogen.

From these four measurements, it is possible to compute the kerogen type, petroleum potential and the evolution stage of the source rock. In earlier pyrolysis method the values of S_2 and S_3 are normalised with organic carbon content determined by a separate experiment as detailed in (6.3.2) in order to make them independent of organic matter. The ratios thus obtained *e.g.*, S_2/org.C and S_3/org.C are called the hydrogen index and oxygen index respectively. However, in the present (Rock-Eval pyrolysis) method, $S_2/(S_1+S_2)$

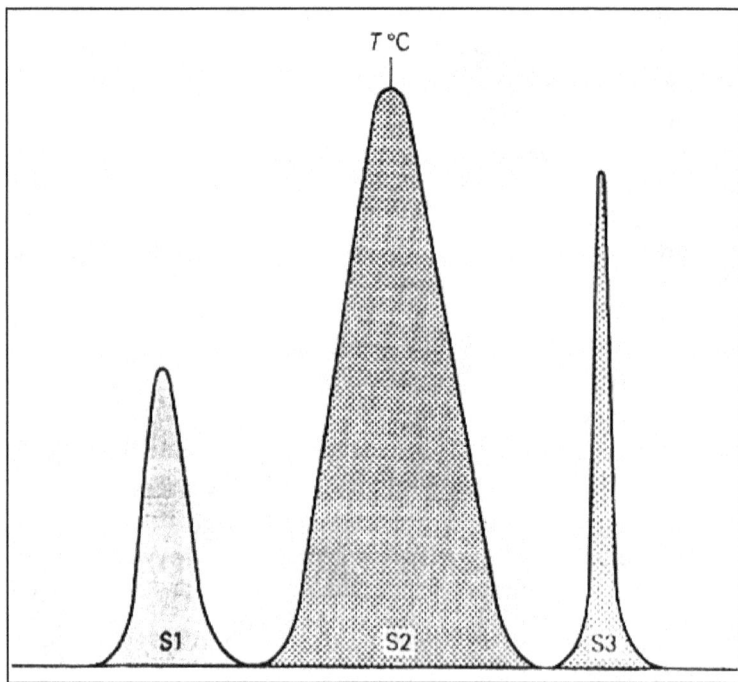

Figure 6.4: Example of Record Obtained by Pyrolysis of a Source Rock (*Source*: North, 1990)

and $S_3/(S_1+S_2)$ are termed as hydrogen index and oxygen index respectively.

Rock-Eval analysis performed on various types of source rocks have demonstrated the existence of a fairly good correlation on one hand between the hydrogen index and H/C atomic ratio, and on the other hand between the oxygen index and the O/C atomic ratio of the kerogen obtained by elemental analysis (6.3.3). The plot of hydrogen index against oxygen index (Figure 6.5) results in the Tissot or van Krevelen type diagram (Figure 5.2a) which differentiates the main types of kerogen macerals (Type I, II and III). Further, a quantitative evaluation of the genetic potential can be made. The quantities S_1 and S_2 represent the respective amounts of petroleum that have already been generated (S_1) and would be generated (S_2) upon further burial and temperature increase. Thus their sum (S_1+S_2) is an evaluation of the genetic potential, expressed in kg of petroleum (oil and gas) per tonne of source rock.

The temperature T, (corresponding to the maximum of the pyrolysis peak, S_2) depends on the type of organic matter (kerogen). It increases with depth (increase of maturation) and becomes progressively same for all types of kerogens. Since kerogen is converted to bitumen during hydrocarbon generation, with increasing maturity (depth), the S_2 peak decreases and S_1 increases (Figure 6.6). Thus T given by the amplitude of S_2 peak and expressed in degrees centigrade (°C) can be used as a source rock maturity indicator. A quantitative evaluation of the Transformation Ratio or otherwise called Production Index may be obtained by the ratio $S_1/(S_1+S_2)$. This in turn, also increases with increasing maturity (depth). These two parameters, T and $S_1/(S_1+S_2)$ can therefore be calibrated against any other maturation parameter described earlier *e.g.* vitrinite reflectance (Ro).

The ratio S_2/S_3 gives a measure of the oil proneness of the source rock sample. For example, if the ratio is greater than 5, it indicates oil prone. On the other hand, if the ratio is 3 or less, it indicates gas prone source rock. If there has been no mobilisation and accumulation of hydrocarbons in the source rock analyzed, the production index increases smoothly with depth. However, accumulation of hydrocarbons interrupt the smooth cruve in the direction of higher index (Figure 6.7). On the other hand, a zone from which hydrocarbons are lost is represented by an interruption

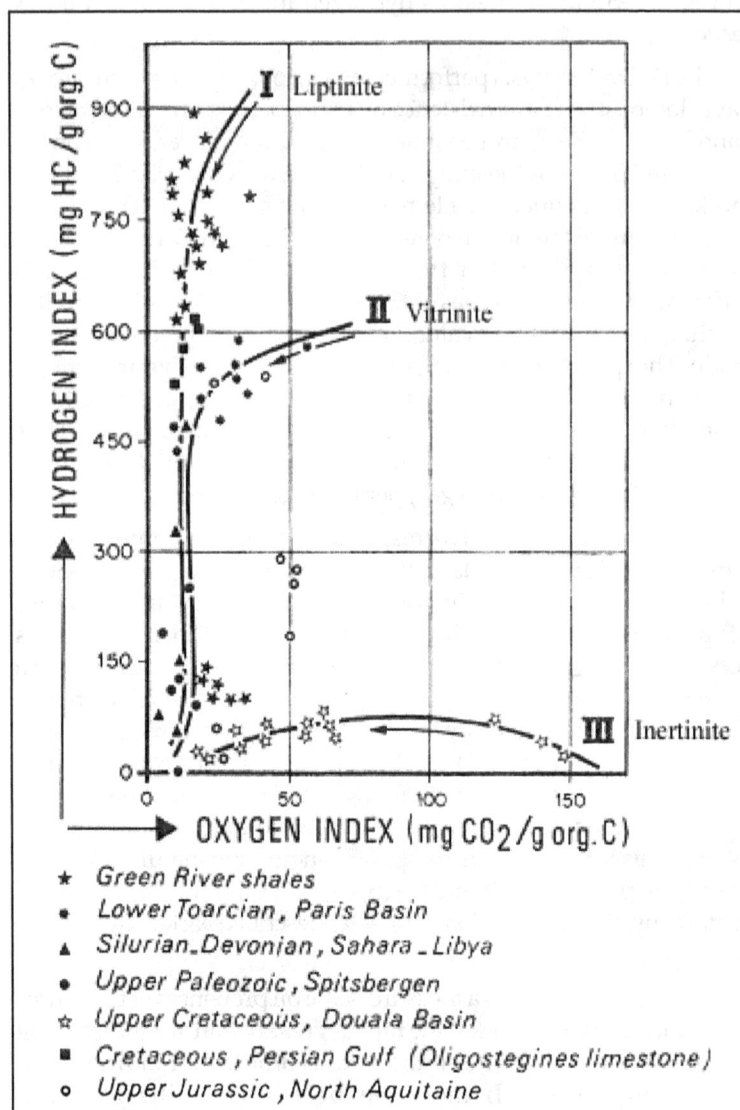

★ *Green River shales*
● *Lower Toarcian, Paris Basin*
▲ *Silurian_Devonian, Sahara _ Libya*
● *Upper Paleozoic, Spitsbergen*
☆ *Upper Cretaceous, Douala Basin*
■ *Cretaceous , Persian Gulf (Oligostegines limestone)*
○ *Upper Jurassic , North Aquitaine*

**Figure 6.5: Classification of Source Rocks,
According to Pyrolysis Indices. The Types I, II and III
correspond to the three main types of kerogen
(*Source*: Tissot, 1977)**

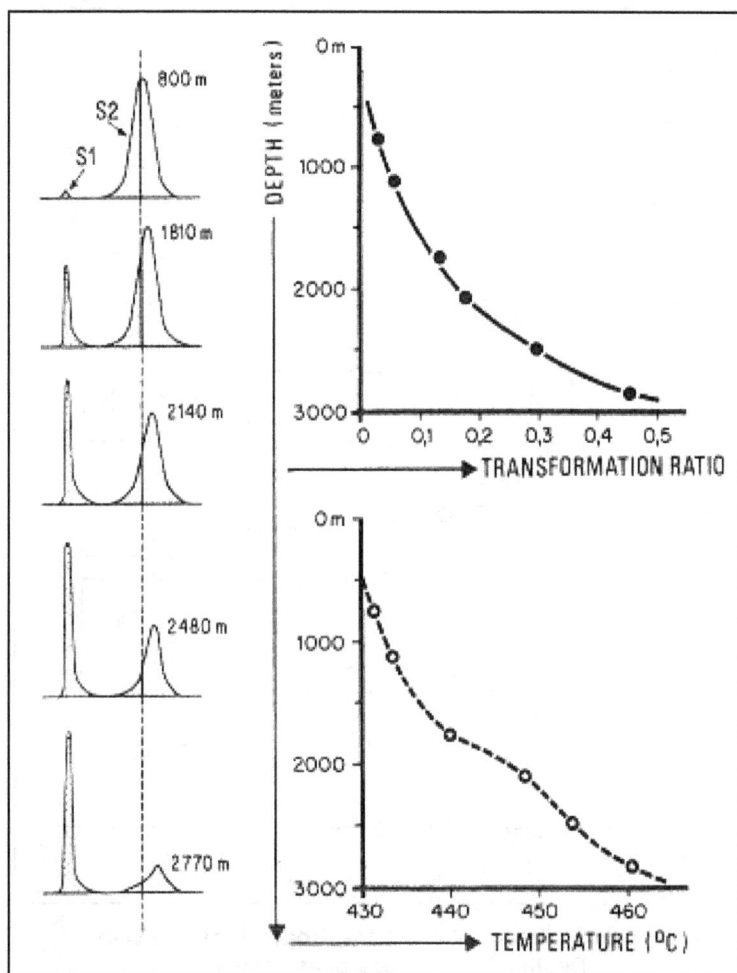

Figure 6.6: Characterization of the Maturation of Source Rocks by Pyrolysis in Sediments of Tertiary Age
Top: Transformation ratio $S_1/(S_1+S_2)$; Bottom: Temperature T corresponding to the peak of hydrocarbon generation during pyrolysis
(*Source*: Tissot, 1977)

in the direction of lower production. If the three indices namely, hydrogen index, oxygen index and Production Index obtained in Rock-Eval pyrolysis analysis are plotted against depth, (Figure 6.8),

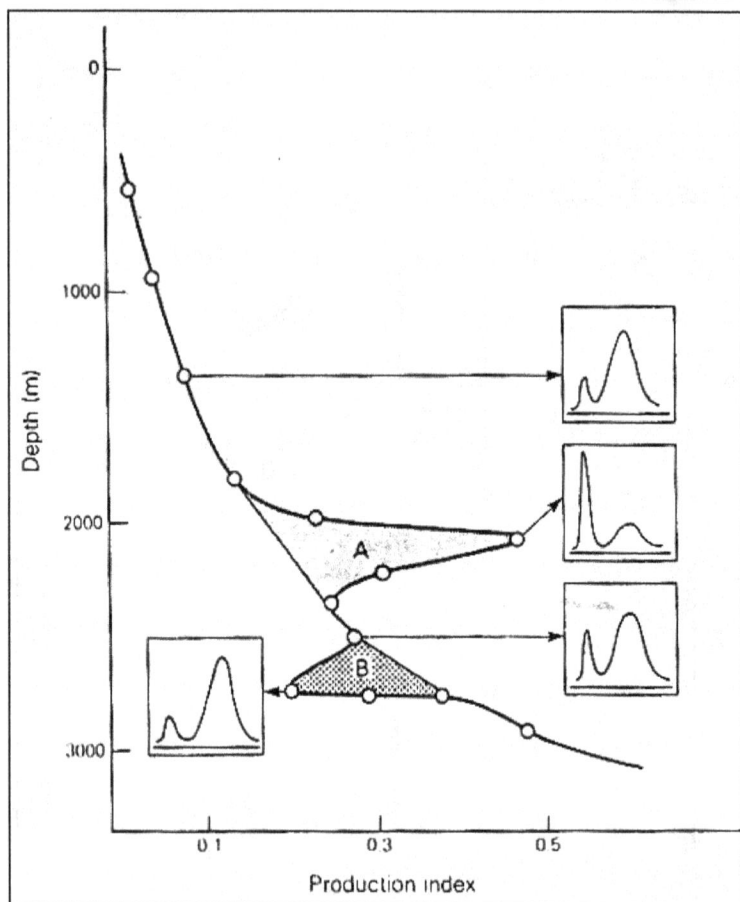

**Figure 6.7: Interruptions in the Production Index with
Depth when Hydrocarbons have been
Effectively Accumulated (A) or Depleted (B)
(*Source*: North, 1990)**

valuable information on the nature and type, quality (immature or
mature) of source rock, oil and gas content, and the zone of oil
accumulation can be obtained. Thus the method is capable of
providing a large amount of data and information about the oil-
source history and future of source rock.

Most pyrolysis techniques do not look at individual
hydrocarbons, instead they measure only total hydrocarbons

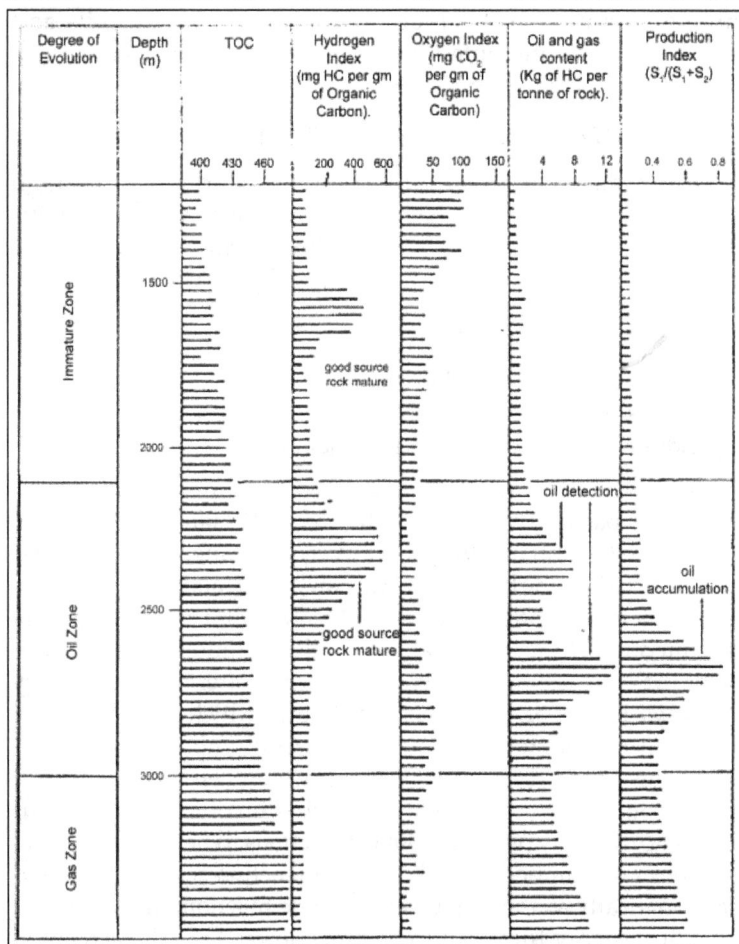

Figure 6.8: Source Rock Maturation Log Derived from Rapid Evaluation of Well Samples by Pyrolysis
(*Source*: North, 1990)

produced. Much valuable information can be obtained if pyrolysis is combined with gas chromatography (GC). In this technique, the hydrocarbons produced during pyrolysis (S_2 peak) are collected until it is complete and then injected into a GC column whereby the individual compounds are separated and identified (10.6.0). Different kerogen types can be identified by their distinct pyrolysis–gas chromatograms. Further more, correlations between oils and kerogens can be attempted by comparing their chromatograms (Schoell, 1984; Waples, 1985).

6.6 Time–Temperature Index (TTI)

It is well known that oil generation is promoted by high subsurface temperature which is in accordance with Arrhenius chemical reaction rate theory. According to this theory, the rate constant (k) of a chemical reaction is given by the equation.

$$k = A\, e^{-Ea/RT} \tag{6.4}$$

Where E_a is the activation energy, T is temperature in (°K), R is the universal gas constant, and A is a constant, the value of which depends on the particular reaction under consideration.

The activation energies calculated for different known oil generating basins are found in the range 11,000-14,000 cal/mol (Tissot, 1969; Connan, 1974). However, this range was far lower than expected for the breaking of carbon-carbon and carbon-oxygen bonds of organic matter (40,000-60,000 cal/mol). The low values are attributed to the mineral catalysis in oil generation since catalysts lower the activation energies, by providing alternate low-energy pathways. However, no catalysts are known which could lower the activation energies for individual kerogen decomposition reactions to such low values (11,000-14,000 cal/mol). Juntgen and Klein (1975) pointed out that hydrocarbon generation involves simultaneous occurrence of many parallel chemical reactions. When the individual reactions are summed, and the overall reaction scheme is treated mathematically as though it were a single reaction, the calculated activation energy turns out to be much lower than the activation energy of any of the individual reactions. It has been generally established beyond doubt, that both time and temperature are important factors in the process of oil generation and in the subsequent cracking of oil to methane. In 1971, Lopatin first described a simple method by which the effect of both time and temperature could be considered in calculating quantitatively the thermal maturity of organic material in sediments using "Time-Temperature Index" (TTI). The method was further developed by Waples (1980, 1981).

6.6.1 Construction of Geologic Model

Implementation of Lopatin's method begins with a reconstruction of the depositional and tectonic history of a geologic section of interest. This is best accomplished by plotting depth of burial versus geologic age (Figure 6.9) as shown in the hypothetical example. In this example, lower Cretaceous sediment was deposited 125 M.Y.B.P. at the sedimentary surface (depth= 0). Since its deposition the sediment has had the time-depth history shown by the solid line (Figure 6.9), moving from left to right. Its history consisted of continual deposition at varying rates until 80 M.Y. B.P. at which time a brief (2 M.Y.) uplift occurred in which the sediment was raised from a depth of 7,000 feet to a depth of 6,000 feet. Uplift was followed by renewed subsidence until a depositional hiatus was reached at 20 M.Y. B.P. The hiatus persisted until 6 M.Y. B.P., when subsidence commenced again. The sediment is at present (time = 0) at a depth of 10,500 feet. The line thus traces the depth-time relation for the sediment. Any shallower strata will have depth-time

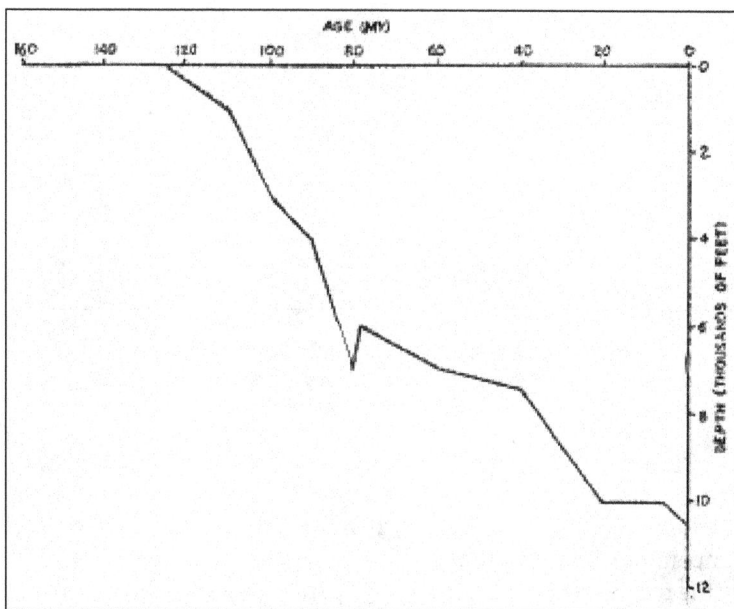

**Figure 6.9: Depositional and Tectonic History
of a Lower Cretaceous Sediment
(*Source*: Waples, 1980)**

lines sub-parallel with the first line, commencing with their deposition. A set of these lines, (Figure 6.10) forms Lopatin's geologic reconstruction. Except in certain situations, the depth-time line segments for the various horizons will always be parallel. The geologic reconstruction is based on the best information available. Some reconstructions will be easy to make with a high level of confidence, particularly where deposition has been continuous. For sediments which have had complex histories, however, the reconstruction may represent only a best guess.

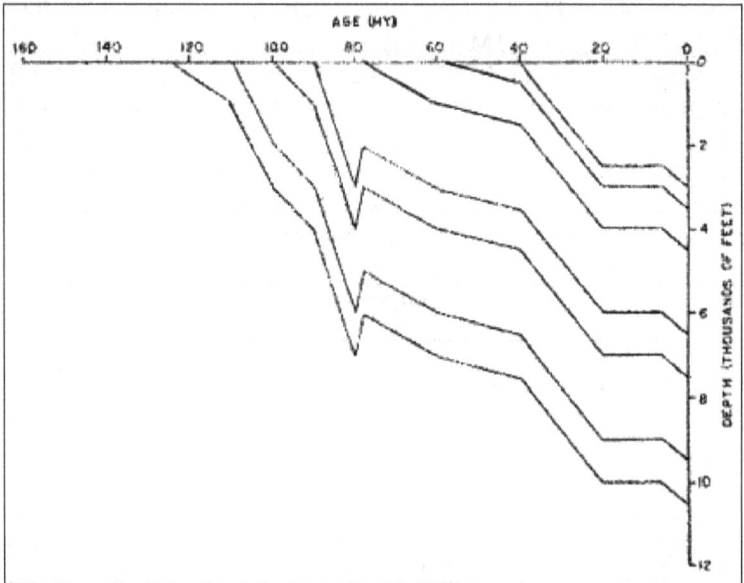

**Figure 6.10: Depositional and Tectonic History
of Several Sedimentary Horizons
(*Source*: Waples, 1980)**

The second aspect of the geologic model is the temperature grid. The simplest way to do this is to compute the present-day geothermal gradient and assume that both the gradient and the surface temperature have been constant throughout the time interval covered by the reconstruction. Hence the temperature grid is simply a series of equally spaced lines of constant depth with a 10°C spacing (Figure 6.11A). On the other hand a complicated situation is shown (Figure 6.11B) in which there is a break in the present-day geothermal

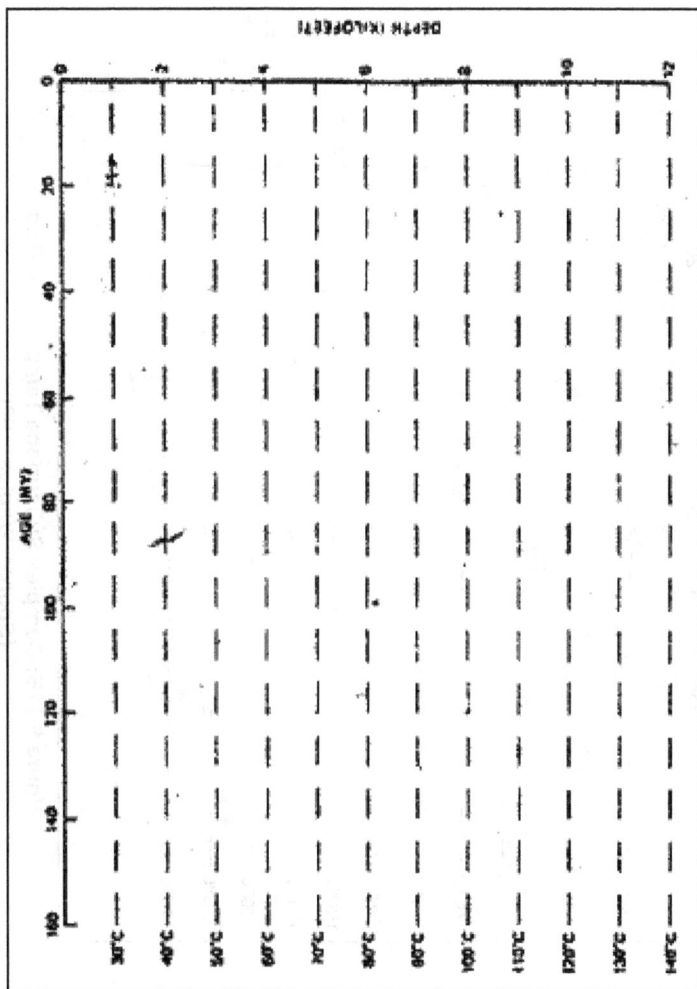

Figure 6.11A: Simple Subsurface Temperature Grid
(*Source:* Waples, 1980)

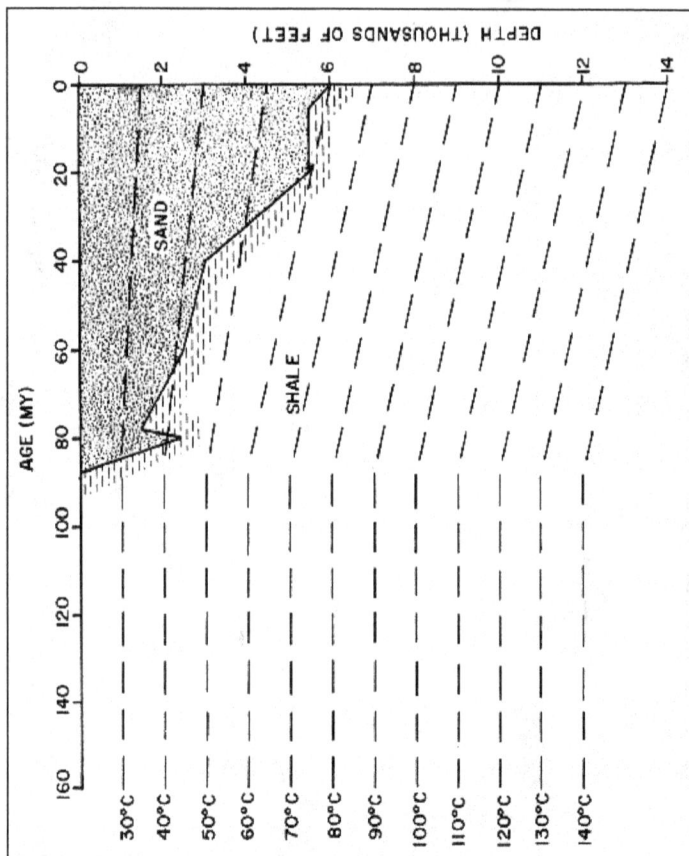

Figure 6.11B: Complex Subsurface Temperature Grid
(*Source:* Waples, 1980)

gradient. The upper part of the section, which is mainly sand, has a lower gradient, but the lower shaly part has a higher gradient. If it is assumed that the geothermal gradient is related to lithology, prior to 88 M.Y. B.P. it must have been high for the entire section, for only shales were present. The low gradient came into existence after 88 M.Y. B.P., when deposition of sand began. The isotherms (dashed lines) thus represent the subsurface temperatures as a function of geologic time.

Although many geologic models can be constructed in a straightforward manner, there are some situations in which caution is advisable, or where special techniques are necessary. When uplift and erosion occur, some section is lost. Thus although the horizon lines remain parallel after such an event, the distance between them will be reduced (Figure 6.12). Another problem can arise when the section under examination is cut by a fault. Such sections above and below the fault may have had different thermal histories. It is thus necessary to make two different geologic reconstructions for the two sides of the fault and combine them to obtain the complete reconstruction for the section.

6.6.2 Theory of Lopatin's Method

Lopatin and many others believed that two factors namely time and temperature, are important in oil generation and destruction. These two factors are interchangeable: a high temperature acting for a short time can have the same maturation affect as a low temperature acting over a long period of time (5.2.3). Lopatin assumed that the dependence of maturity on time is linear-doubling the cooking time at a constant temperature doubles the maturity. Chemical reaction rate theory predicts that the temperature dependence of maturity will be exponential. To take into account this relation between reaction rate and temperature, he divided the temperature profile into $10°C$ intervals, choosing 100 to $110°C$ as the base interval and assigning it to an index value of $n = 0$ and the other intervals were assigned index values as shown in Table 6.3. A factor g was defined which reflects the exponential dependence of maturity on temperature. It is assumed that the rate of maturation increases by a factor r for every $10°C$ rise in reaction temperature. Thus for any temperature interval, the temperature factor is given by $v = r^n$ where n is the appropriate index value given in Table 6.3.

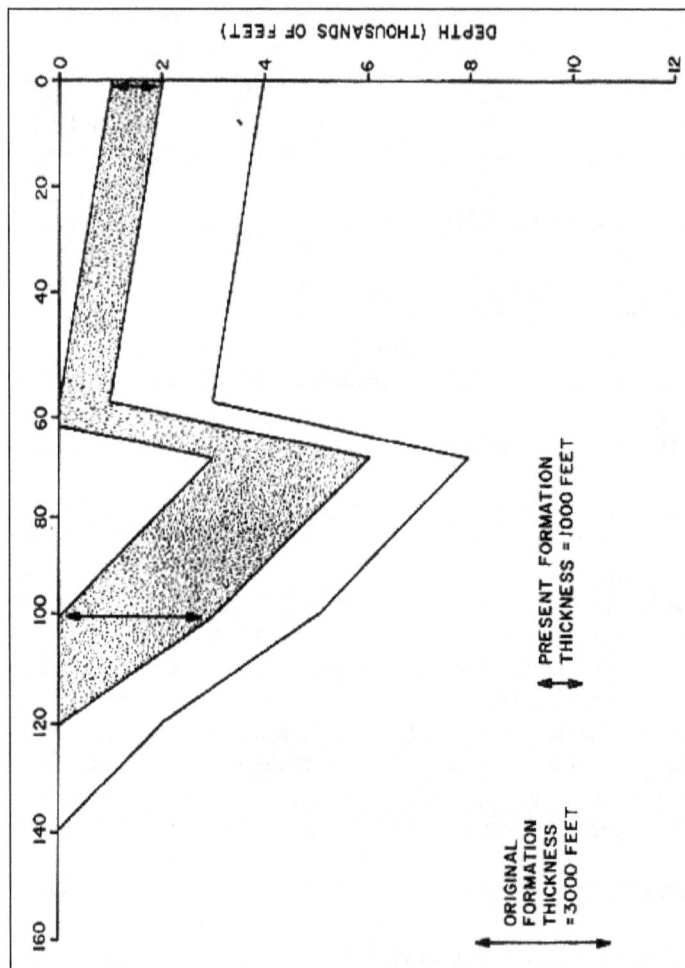

Figure 6.12: Illustration of Section Thinning by Erosion
(*Source:* Waples, 1980)

Table 6.3: Temperature Factors for Different Intervals

Temperature Interval °C	Index Value n	Temperature Factor γ
30–40	–7	r^{-7}
40–50	–6	r^{-6}
50–60	–5	r^{-5}
60–70	–4	r^{-4}
70–80	–3	r^{-3}
80–90	–2	r^{-2}
90–100	–1	r^{-1}
100–110	0	1
110–120	1	r^{1}
120–130	2	r^{2}
130–140	3	r^{3}
140–150	4	r^{4}
150–160	5	r^{5}
–*	m	r^{m}

* Data not available.

Source: Waples, 1980.

For the time factor, Lopatin used the length of time (in M.Y.) that the sediment spent in each temperature interval. The maturity added in any temperature interval i is given by $\text{Maturity}_i = (\Delta T_i)(r^{ni})$, where ΔT_i is the length of time spent by the sediment in the temperature interval i. Because maturation effects on the organic material are additive, the total maturity (TTI) of a given sediment is given by the sum of maturities acquired in each interval. Thus

$$TTI = \sum_{n\,min}^{n\,max}(\Delta T_n)(r^{n})$$

$\qquad\qquad\qquad\qquad\qquad\qquad\qquad\qquad\qquad$ (6.5)

where, n max and n min are n-values of the highest and lowest temperature intervals encountered. A value of 2 was chosen for r as per the Arrhenius equation which states that the rate of a chemical reactions approximately doubles for every 10°C rise in temperature. To test empirically for the most appropriate value of r, TTI was plotted versus Ro (vitrinite reflectance) for various values of r. Ro values ranging from 1.0 to 10.0 were collected from the data for 402 samples from 31 worldwide wells. Correlations between measured and calculated maturities are generally good for values of r between 1.6 to 2.5 as evident from the plot of TTI versus Ro for r = 2, (Figure 6.13). Significant scatter at the top and bottom of the figure, may be due to two main factors: (i) error in TAI or Ro measurements, or (ii) error in the geologic models used or both. Because of the large number of samples, however, the average TTI-Ro line (Figure 6.13) is probably considered as satisfactory.

6.6.3 Calculation and Interpretation of TTI

The principle involved in calculating TTI values can be explained with a specific example. Figure 6.14 shows a geologic model having three sediment horizons (A, B, and C) and a moderately complex temperature grid. The calculation for each horizon is given in Table 6.4. It is possible to calculate TTI value at any time in the past in the same way. Suppose we are interested in the TTI value of horizon A 60 M.Y. ago (represented by point P in Figure 6.14). The calculations are carried out in a manner analogous to that done previously but stop 60 M.Y.B.P.instead of at the present. The calculated TTI value for point P is 5.9.

A scale of TTI correlating with vitrinite reflectance (Ro) has been constructed by comparing measured Ro with TTI values calculated from geologic models (Table 6.5). There is a good correlation between them and other measured geochemical parameters like TAI etc. Table 6.6 shows TTI values together with Ro and TAI for several important stages of oil generation and preservation. The boundaries of the oil generation window are very similar to those proposed by many others. These values effectively define the TTI range in which oil generation occurs (15 to 160); the highest TTI values at which oils of 40° and 50° API gravities will be preserved (approximately 500 and 1,000, respectively); and the highest TTI value at which wet gas can be preserved (1,500). Dry gas is produced in California, Oklahoma, from a horizon having a TTI of about 65,000, but it has not yet been established that this is the maximum possible TTI at which methane is still stable.

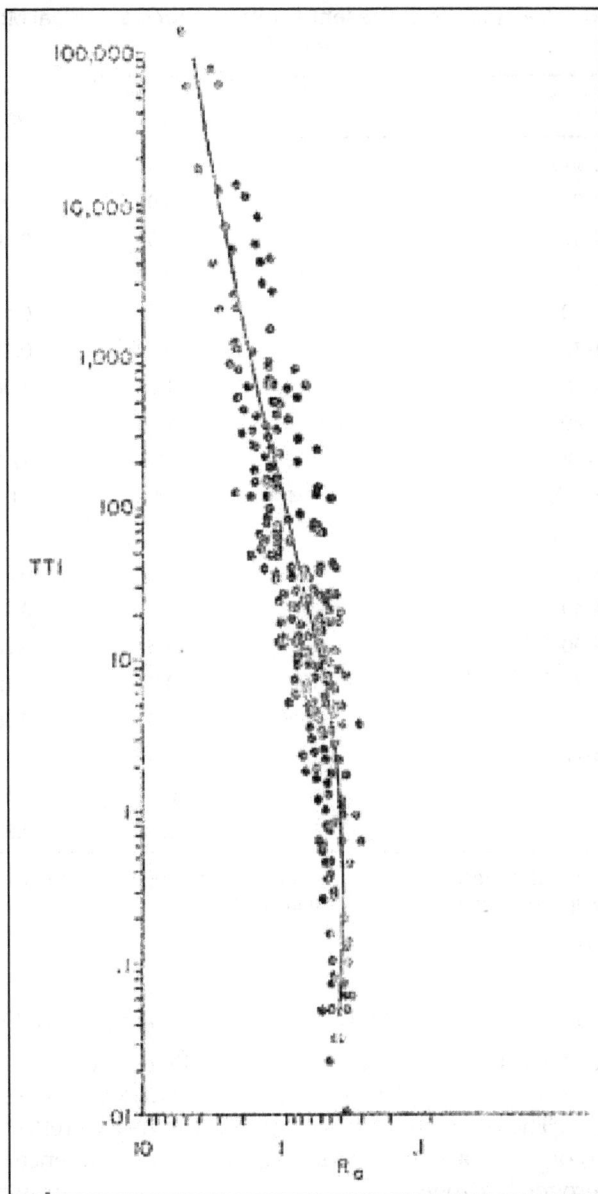

Figure 6.13: Time-temperature Index of Maturity (TTI) versus Vitrinite Reflectance (R_o) for r = 2
(Source: Waples, 1980)

Table 6.4: Calculation of Present TTI Values for Geological Model (Figure 6.14)

Temp. Interval °C	r^n	Δ Time* (m.y.)	Interval TTI	Total TTI
Horizon A				
20–30	2^{-8}	15	0.06	0.06
30–40	2^{-7}	5	0.04	0.10
40–50	2^{-6}	5	0.08	0.18
50–60	2^{-5}	10	0.31	0.49
60–70	2^{-4}	3.5	0.22	0.71
70–80	2^{-3}	(3.5+6.5)	1.25	1.96
80–90	2^{-2}	(4.5+37.5)	10.5	12.5
90–100	2^{-1}	10.5	5.3	17.8
100–110	1	24	24.0	41.8
Horizon B				
20–30	2^{-8}	3.5	0.01	0.01
30–40	2^{-7}	(3.5+2.5)	0.05	0.06
40–50	2^{-6}	(5+38)	0.67	0.73
50–60	2^{-5}	12.5	0.39	1.12
60–70	2^{-4}	24.5	1.53	2.65
Horizon C				
20–30	2^{-8}	10.5	0.17	0.17
30–40	2^{-7}	29.5	0.22	0.39

ΔT for a particular interval is the age at which the sediment enters that interval minus the age at which it enters the next interval.

Source: Waples, 1980.

6.6.4 Correlation of TTI with Other Geochemical Data

Calculated TTI values were compared with measured data from many worldwide samples representing a variety of ages and lithologies. The correlations of TTI with TAI, vitrinite reflectance, bitumen/organic carbon ratio (Bit/C_{org}), carbon preference index (CPI), kerogen hydrogen/carbon (H/C) ratio, and API gravity are satisfactory leading to the conclusion that TTI is a valid measure of thermal maturity of organic material. An example of satisfactory

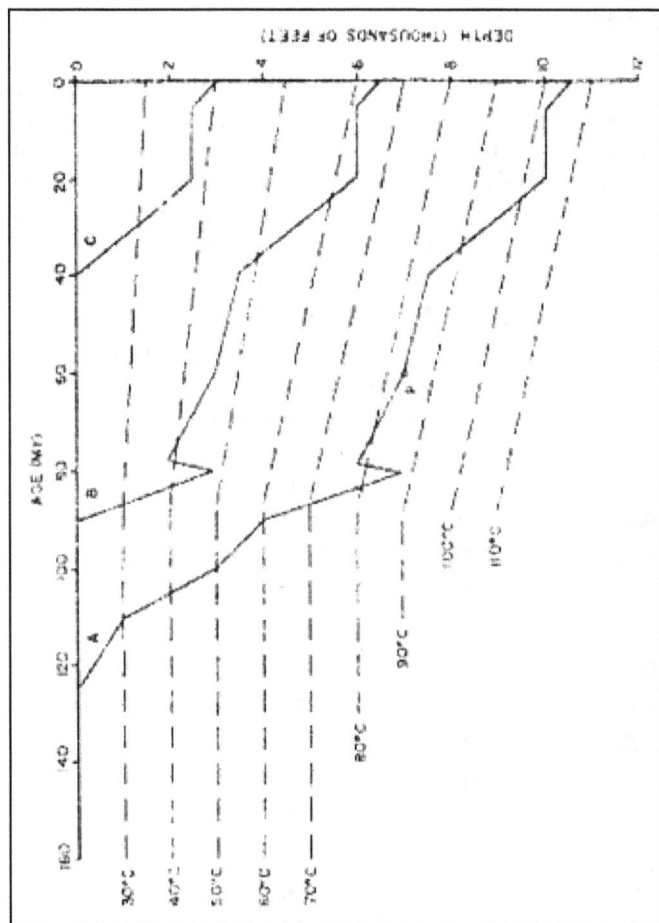

Figure 6.14: Geologic Model for Horizons A, B, and C
(*Source*: Waples, 1980)

correspondence between measured geochemical parameter such as Ro, and calculated TTI can be shown with the data obtained in a well, at Texas, Gulf coast. The geothermal gradient was assumed constant at 1.75°F/100 feet, which is the present day gradient of the bottom half of the well. The surface temperature was taken as 64°F. With this data when plotted, there is in general, a good agreement between Ro and TTI values (Figure 6.15). However, the measured Ro values reach the on set of oil generation (Ro = 0.65 per cent) at a slightly shallow depth (11,700 feet) than do the calculated TTI depth (12,100 feet). Similarly there is slight disagreement about the end of oil generation. A TTI of 160 is reached at a depth of 16,000 feet where a simple extrapolation of Ro data suggests that full maturity is not reached until a depth of 17,400 feet. These minor discrepancies may be due to the assumptions of low geothermal gradient, and lack of Ro data in the maturity region of lower depth.

Table 6.5: Correlation of Time-Temperature Index (TTI) with Vitrinite Reflectance (R_o)

R_o	TTI	R_o	TTI
0.30	<1	1.36	180
0.40	<1	1.39	200
0.50	3	1.46	260
0.55	7	1.50	300
0.60	10	1.62	370
0.65	15	1.75	500
0.70	20	1.87	650
0.77	30	2.00	900
0.85	40	2.25	1,600
0.93	56	2.50	2,700
1.00	75	2.75	4,000
1.07	92	3.00	6,000
1.15	110	3.25	9,000
1.19	120	3.50	12,000
1.22	130	4.00	23,000
1.26	140	4.50	42,000
1.30	160	5.00	85,000

Source: Waples, 1980.

Table 6.6: Correlation of TTI with Important Stages of Oil Generation and Preservation

Stage	TTI	R_o	TAI
Onset of oil generation	15	0.65	2.65
Peak oil generation	75	1.00	2.9
End of oil generation	160	1.30	3.2
Upper TTI limit for Occurrence of oil with API gravity <40°	500	1.75	3.6
Upper TTI limit for Occurrence of oil with API gravity <50°	1,000	2.0	3.6
Upper TTI limit for occurrence of wet gas	1,500	2.2	3.75
Last known occurrence of dry gas	65,000	4.8	>4.0
Liquid sulfur in lone star Baden 1 (below dry gas limit)	972,000	>5.0	>4.0

Source: Waples, 1980.

6.6.5 Application of TTI Data to Exploration

TTI values obtained by application of Lopatin's method can be useful in several ways for oil exploration. If we are concerned with how deep we can expect to find preserved accumulations of oil, wet gas, or dry gas, we need only to calculate the present-day TTI values of the suspected reservoir and find the TTI regime into which they fall. For example, if it is expected that a certain reservoir rock will be encountered at 3,758 m in a proposed well, we can predict the occurrence of oil or gas, and if oil, of what API gravity? Suppose that we obtain by calculation a TTI of 1,200 for the reservoir formation. This means that the reservoir has a higher TTI value than that at which a 50° API gravity oil can be preserved (1,000 from Table 6.6). We would predict from the TTI calculations that the reservoir lies beyond the oil deadline, and could therefore contain only wet gas.

A second way in which TTI data can assist in oil exploration is in answering the question of whether or not the thermal maturity necessary for hydrocarbon generation has occurred in a region. For example, organically rich shale has been found in a basin, and we want to know whether this shale has reached thermal maturity. By making time-depth reconstructions for several points in the basin,

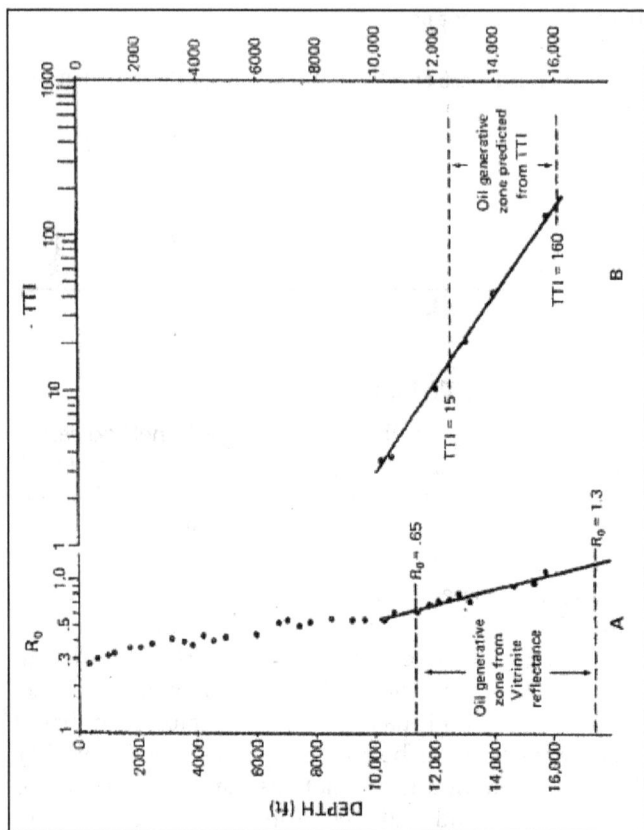

Figure 6.15: Comparison of Oil-Generative Zones for C.O.S.T. #1 Well. Texas Gulf Coast A. from measured vitrinite reflectance values; B. from calculated TTI values. (*Source:* Waples, 1980)

we can calculate present-day TTI values for the shale at these points, as shown in the hypothetical example (Figure 6.16). By contouring the TTI values we can get an idea of the areal extent of rich shale which has entered the generative window. In this example, the generative area (within the TTI > 15 contour) represents only a small part of the total basin; hence only a small fraction of the rich shale could have begun to generate oil. Thus the exploration risk in prospecting this basin would be considerably higher than if the whole basin had already reached thermal maturity

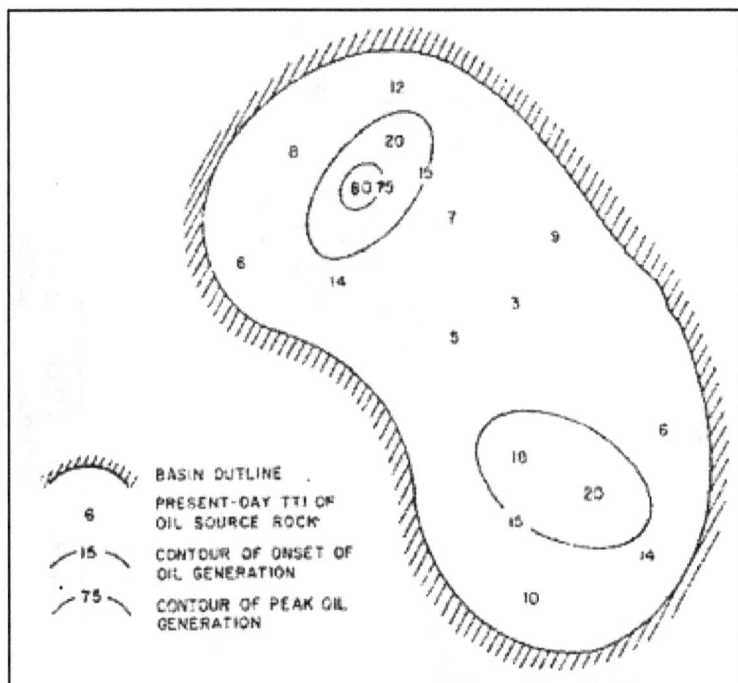

Figure 6.16: Present-day TTI Values of Organically Rich Shale in a Hypothetical Basin (*Source*: Waples, 1980)

A third application of TTI data in exploration is in answering the questions about timing of oil generation (Zieglar and Spotts, 1978). Figure 6.17 shows a geologic model in which TTI values of 15 and 160 have been located on each of several horizons. If we contour iso–TTI values of this model we have two shaded lines which delimit

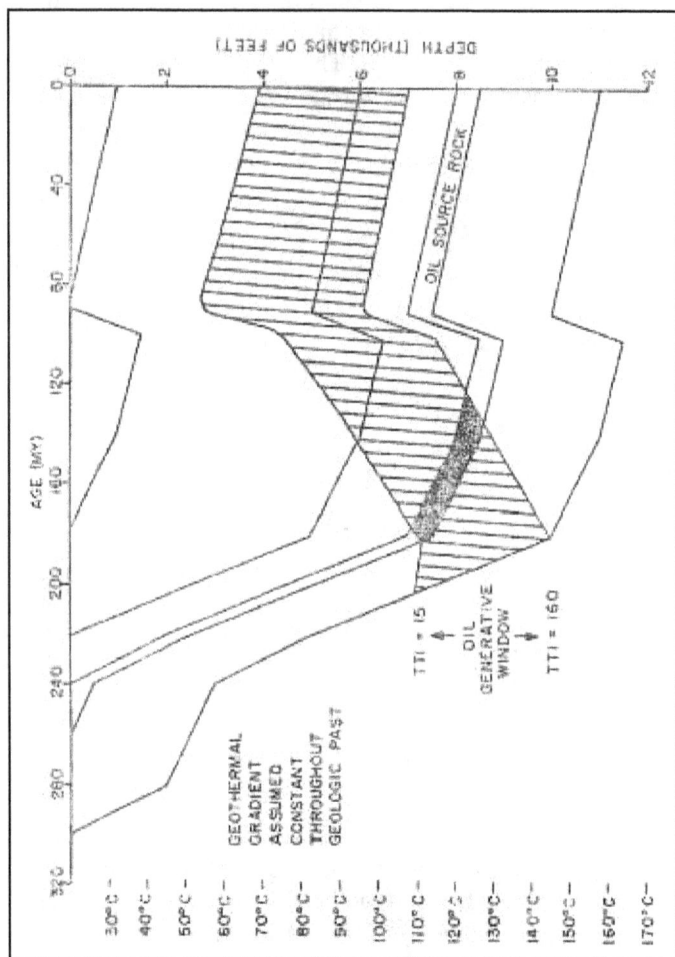

Figure 6.17: Iso–TTI Lines on Geologic Model
(*Source:* Waples, 1980)

the oil-generative window for the entire section throughout the geologic past. Let us suppose that one particular formation, indicated as "oil source rock" (black strip) is the only plausible oil source rock (OSR) for this region (Figure 6.17). We can determine when in the geologic past, the OSR generated oil by inspection of this figure. The OSR entered the generative window 181 M.Y. B.P. and ceased generating oil 120 M.Y. B.P. One can begin to answer important questions about the timing of oil generation and trap formation. Suppose that in this case the only structural traps in this region were created during the uplift lasting from 100 to 90 M.Y.B.P. Because trap formation occurred at least 20 M.Y. after the end of oil generation, the oil had already migrated out, since there existed no barrier to its movement.

Lopatin's method, though simple and versatile, is semi-empirical. It is mostly valid in areas with a simple burial-time relationship. In view of this, a mathematical model was developed (Tissot, 1977) for quantative evaluation of oil and gas generation as a function of time. The simulation of oil and gas formation takes into account the nature of kerogen, the history of burial, and the geothermal gradient. The nature of kerogen is represented by the distribution of activation energies used in the model. They are adjusted on the basis of observations (petroleum compounds extracted with solvents from well samples) and data obtained from laboratory tests (pyrolysis). They may also be adjusted some times on the basis of pyrolysis data alone. The history of burial is reconstructed by means of isobath and isopach maps based on well logs. By using regular grid, a curve is drawn giving the depth of source rock as a function of time, for each point of the network. The problem of reconstructing the eroded thickness is resolved, in general, by extrapolation from the deepest zones of the basin.

Determining the geothermal gradient is more difficult. In general, high gradients are observed in the areas of rifting, and in the zones where folding and magmatic events are young. In areas where these phenomena are old, the gradients seem to be stabilized. For example, one can reasonably use the present-day gradients for the Jurasssic, Cretaceous and Tertiary provinces where the last Orogeny was in the Paleozoic or earlier. At other locations (*e.g.*, basins on continental margins bordering the Atlantic Ocean) the ancient geothermal gradient can be calculated. A parameter sensitive to thermal evolution, *e.g.*, vitrinite reflectance, is measured on a certain number of samples from a well. Simulation is then used to calculate the same values of reflectance, and the geothermal gradient is adjusted by successive iterations until the difference between the calculated and the measured values is minimised.

The main reason for using the mathematical model, rather than the graphic plot, is to handle more complex situations, *e.g.*, changing geothermal gradient or successive cycles of sedimentation separated by folding and erosion. An example of this last situation is the Hassi Messaoud area in Algeria (the northern Sahara Basin) which shows a two cycle of sedimentation. The Palaeozoic system contains sandstones and shales of Cambrian to Carboniferous age, among which the Lower Silurian shales are the source rock and the Cambrian and Ordovician sandstones are the reservoirs of the oils. Moderate Hercynian folding was followed by major erosion which left only Cambrian or Ordovician reservoirs on the anticlines. Sedimentation was resumed in Triassic time and resulted in important Mesozoic deposits, including a thick Triassic-Jurassic salt, which actually seals the Cambro-Ordovician oil reservoirs (Figure 6.18). This situation poses the problem of the timing of oil accumulation. If the petroleum was generated and accumulated in the Hassi-Messaoud dome during the Palaeozoic, it was exposed to the atmosphere because of erosion for millions of years at the end of the Carboniferous and during the Permian. Oil should have been either destroyed or converted to heavy oils and tars. If, on the contrary, the accumulation was formed after the Triassic, the oil could have been trapped and protected by the salt beds, which constitute an effective cover.

Figure 6.19 shows the burial of the Lower Silurian source rocks, as a function of depth in one of the synclines bordering the Hassi-Messaoud dome. The formation of oil and gas is shown in the bottom, as calculated by the mathematical model. A vident from the figure, little oil was generated in the course of the first 300 million years, and the principal stage of oil formation was not reached until the Cretaceous. Thus, accumulation of oil in the reservoirs occurred after the domes have been sealed by the salt cover.

6.7 Level of Organic Metamorphism (LOM)

Geochemists define mineralogical changes that can be reasonably attributed to the action of heat and pressure at depths as

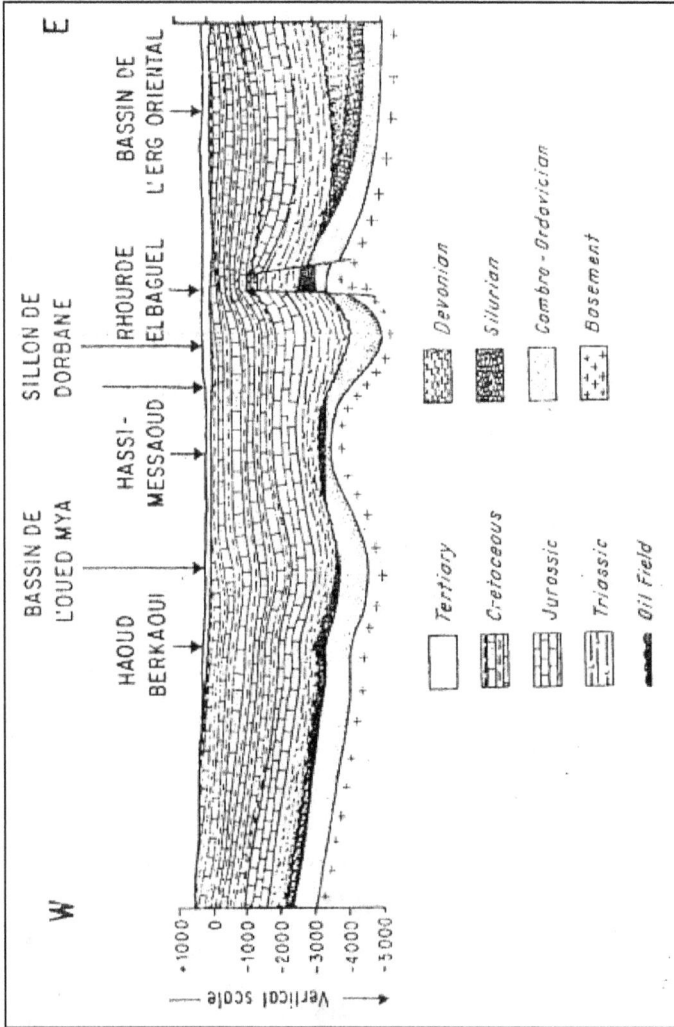

Figure 6.18: Geological Cross-section in the Hassi-Messaoud Area, Algeria
(*Source:* Tissot, 1977)

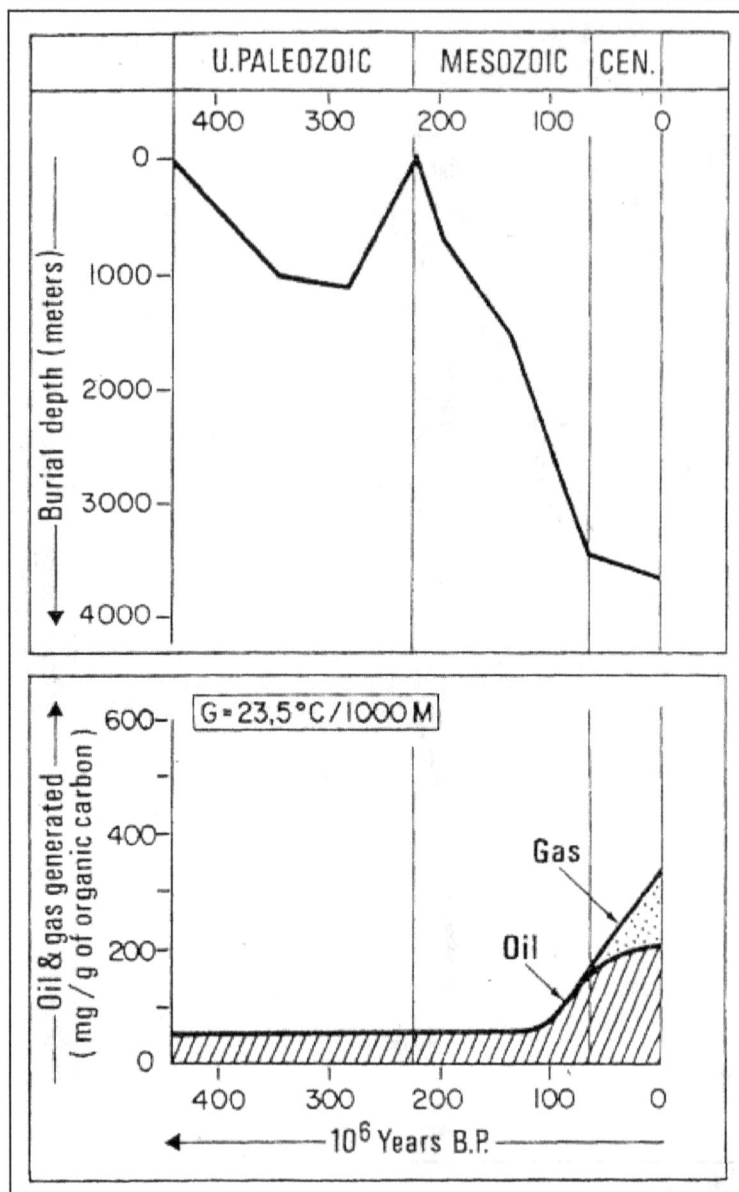

Figure 6.19: Reconstitution of Burial Depth (top) and Mathematical Simulation of Oil Generation (bottom) in the Hassi-Messaoud Area (*Source*: Tissot, 1977)

metamorphic. The low temperature end of the metamorphic scale is considered to be about 200-300° C. This also appears to be the range in which the last trace of methane forms from organic matter in the metamorphic stage. Hence the end process of all thermal alterations of sedimentary organic matter leading to the formation of graphite is called metamorphism. This is also referred by several names such as epimetamorphism, incipient metamorphism or simply maturation.

The coal rank scale has for many years provided the basic frame work for studying coalification process and this was used for sometime for comparing the stages of organic metamorphism in petroleum source rocks. However, none of the coal ranking properties such as volatile matter (or fixed carbon), vitrinite reflectance, calorific value, hydrogen content etc. are satisfactorily applicable over the entire range of interest for petroleum generation. In fact the coal rank scale is not a single numerical scale but, in effect, represents a series of two or more over-lapping scales. In view of this, a new scale called the level of organic metamorphism (LOM) was first proposed by Hood *et al.*, (1975) and developed further by Cohen (1981). Salient features of this scale are:

(i) It is a single numerical scale continuous over the whole range of generation and destruction of organic compounds,

(ii) It can be correlated with available scales of organic metamorphism such as coal rank and major coal rank parameters as indicated above,

(iii) The scale is approximately linear with maximum burial depth in any given geographic location where the sedimentary column exhibits no major time hiatus or temperature gradient anomaly, and

(iv) It reflects both temperature and time effects—thus the temperature history of the sediment.

The scale was originally devised from studies of the Cretaceous-Eocene coal measures of New Zealand as reported by Suggate (1959) with a suitable modification. It was adjusted to become quasilinear, from zero at the surface to 20 at the anthracite-meta anthracite boundary (Figure 6.20). The specific stage of LOM at which oil is generated in a given fine-grained source rock depends to some extent,

| LOM | COAL | | | PRINCIPAL STAGES OF PETROLEUM GENERATION | |
	RANK	BTU x10⁻3	% VM	VASSOYEVICH ET AL.(1970)	MATURITY
0					
2	LIGN.				
4		8		EARLY (DIAGENETIC) METHANE	IMMATURE
6	SUB-BIT. C B	9 10 11			
8	HIGH VOL. BIT. C B A	12 13 14 15	(45) (40) (35)	OIL	ZONE OF INITIAL MATURITY (OIL GENERATION)
10			30 25		
12	MV BIT. LV BIT.		20 15	CONDENSATE & WET GAS	MATURE & POST-MATURE
14	SEMI-ANTH.		10	HIGH-TEMPERATURE (KATAGENETIC) METHANE	
16					
18	ANTH.		5		
20					

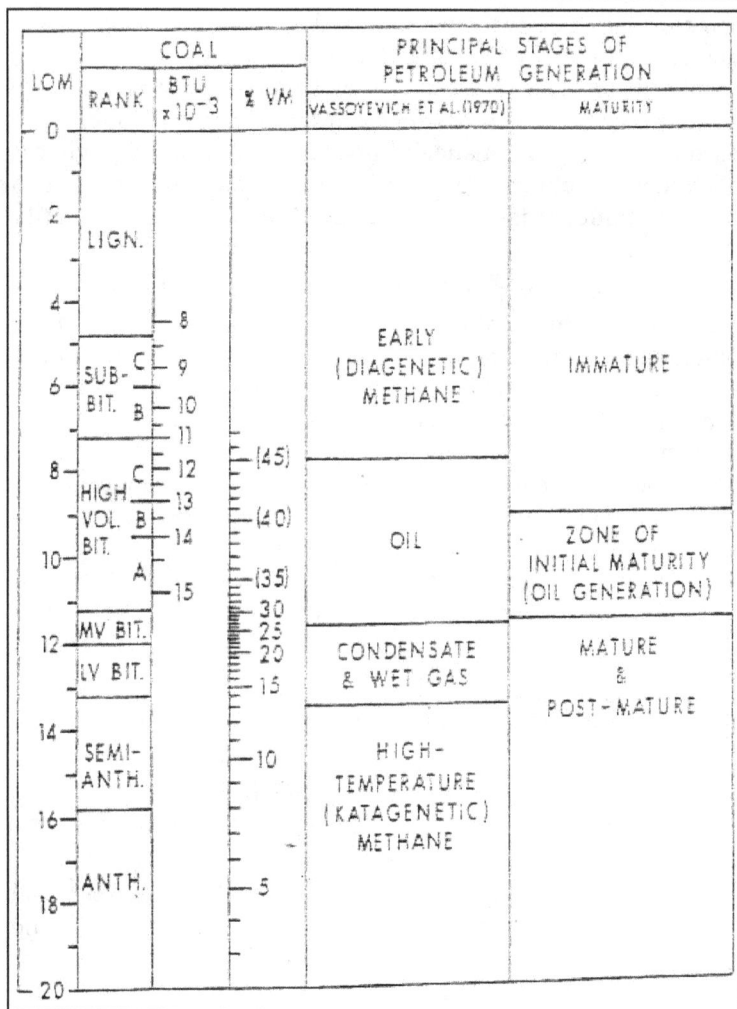

Figure 6.20: Organic Metamorphic Stages of Petroleum Generation
(*Source*: Hood *et al.*, 1975)

on the type of the source rock. However, the principal stages for oil and gas generation interms of coal rank (Vassoevich *et al.*, 1970) and LOM are oil: 9-10; condensate + wet gas: 11.6–13.5; and high temperature (metagenetic) methane: > 13.5.

Figure 6.21 represents the scale of maximum temperature reached (T_{max}) and the effective heating time (t_{eff}). The effective heating

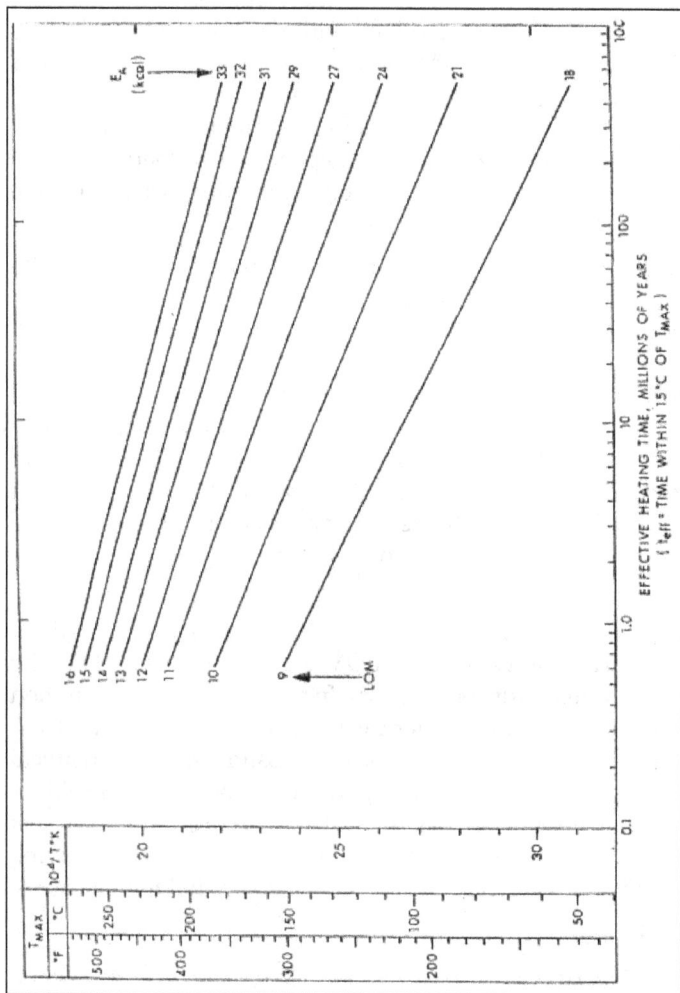

Figure 6.21: Relation of LOM to Maximum Temperature and Effective Heating Time
(Source: Hood et al., 1975)

time (t_{eff}) during which a source rock has been within 15°C interval of its maximum temperature (T_{max}) was arrived using a variety of typical burial histories and a wide range of activation energies (Ea, ranging from 8.4 to 55 kcal/mol). Although approximate, the LOM method is adequate in view of the uncertainities in the geologic data. According to this method, oil window begins at LOM = 7–8 (Ro = 0.5 per cent, T_{max} = 100°C or lower). The oil dead line reaches where LOM = 12 or thereabouts (Ro above 1.35 per cent and T_{max}: about 140°C (equivalent to low volatile bituminous coal). The actual depths of these thresholds, of course, depends on the thermal history which can be indicated by a measurable quantity such as vitrinite reflectance (Ro per cent), electron spin resonance (ESR) etc.

Figure 6.22 shows a plot of LOM determined by vitrinite reflectance against the depth of a deep well in the Anadarko basin, Oklahoma. For comparison, LOM values calculated from T_{max} (using t_{eff} = 260X10⁶ years from Figure 6.21) were also shown. The relationship of T_{max} and effective burial time (t_{eff}) to LOM provides a suitable approximate method for estimating LOM within the range 9–16. Though LOM do not show linear relationship with depth (Figure 6.22), it illustrates the prediction of specific depths at which oil, gas condensate and methane would be generated by source rocks at any given location in the basin.

LOM has an advantage over Lopatin's method in that calculations are simple, since one need not construct in detail the complete burial history of a rock. LOM therefore over simplifies the influence of temperature on maturity. In many cases this assumption is quite acceptable, but in other cases it is not. Both LOM and TTI (6.6.0) must be used with caution. They assume that geochemical gradient has been constant throughout the time, which is seldom likely to be true. Burial curves must be used carefully, not just considering the present day depth of the formation but also allowing compaction, uplift and erosion. It is to be noted that in few basins, subsidence and sedimentation balance each other.

6.8 Correlation Between Geochemical and Petrographic Techniques

We are now in a position to correlate the parameters measured by both geochemical and petrographic techniques and to indicate

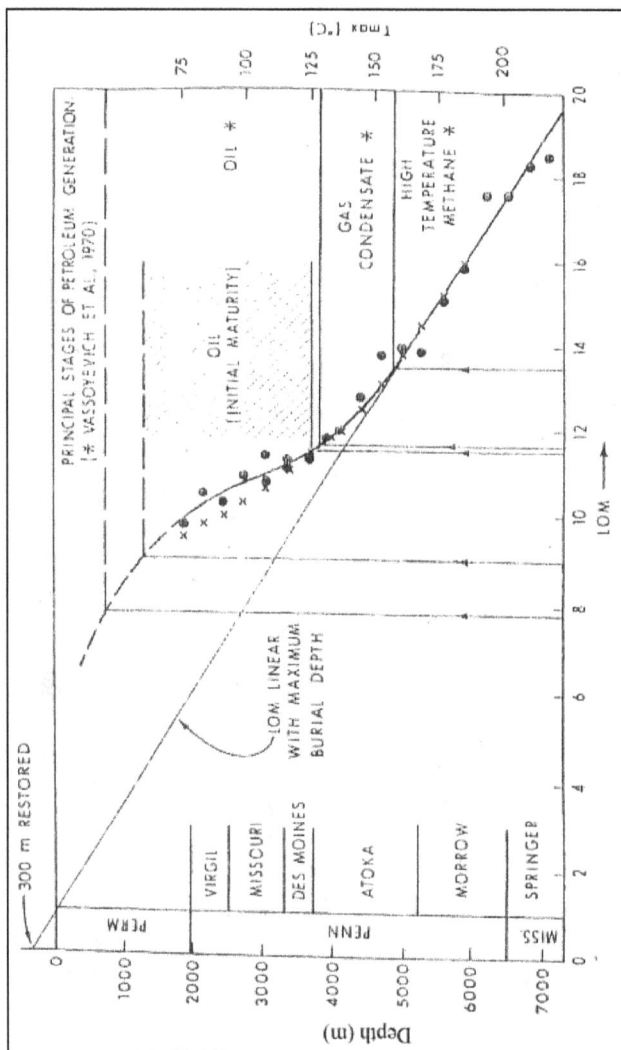

Figure 6.22: Relations of LOM and Petroleum Generation to Depth: Shell Rumberger 5, Beckham County, Oklahoma, Anadarko basin. LOM is based on vitrinite reflectance; for comparison, LOM values based on T_{max} and t_{eff} (see Figure 6.21 using 260 × 10⁶ years for t_{eff}) are shown as x's. (Source: Hood *et al.*, 1975)

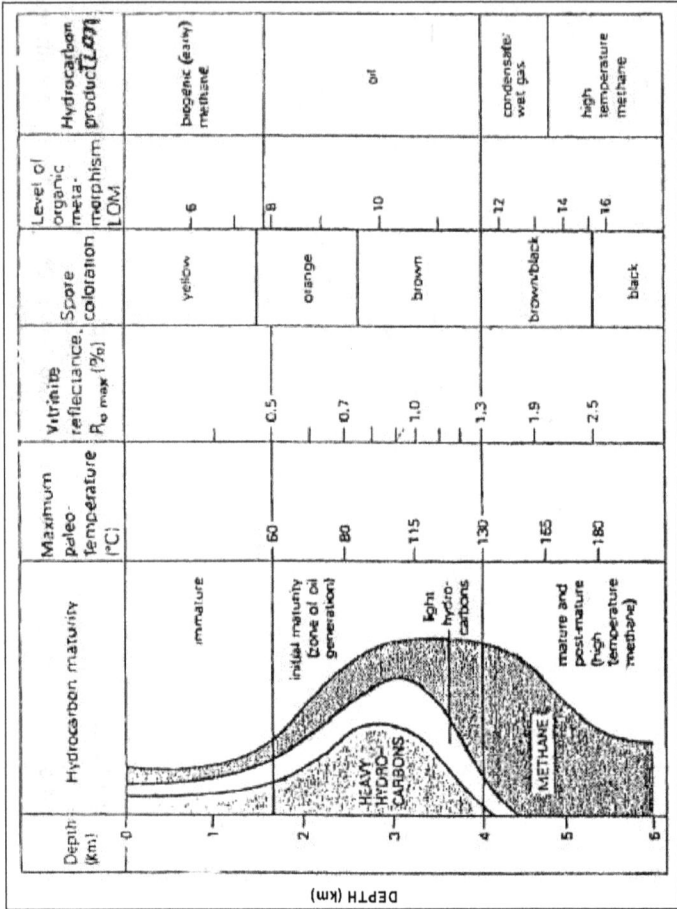

Figure 6.23: General Correlation of Organic Maturation Indices
(*Source:* Kantsler *et al.*, 1978)

what they mean in terms of conversion of organic matter to oil, gas or coal. Figure 6.23 shows general correlation between depth, hydrocarbon maturity, paleotemperature, vitrinite reflectance, spore colouration, LOM and stages of hydrocarbon production. The purpose of all these techniques is four fold (North, 1990).

(*i*) To recognize and evaluate the potential source rocks for oil and gas by measuring their TOC content and degree of thermal maturation,

(*ii*) To correlate oil types with their respective source beds according to the geochemical characteristics, and the optical characteristics of kerogen in the source beds,

(*iii*) To determine the time of generation of hydrocarbons and the time of their migration and accumulation, in relation to the time of formation of traps, and

(*iv*) To enable the geochemists and geologists to estimate the volume of oil and gas initially generated, and thus the potential ultimate reserves or degree of loss of hydrocarbons from the system.

The criteria for a sedimentary rock to be an effective oil source rock is:

(*i*) TOC content should be at least 0.4 per cent.

(*ii*) Elemental carbon should be between 75 and 90 per cent ; values below 75 per cent represent immature, and values above 90 per cent represent super mature,

(*iii*) Ratio of bitumen to TOC should exceed 0.05

(*iv*) Kerogen should be amorphous and oil prone type,

(*v*) Vitrinite reflectance (Ro) should be in the range 0.6-1.3 per cent, and

(*vi*) H/C and O/C atomic ratios should fall in the favourable region of the Tissot diagram (Figure 5.2a). The H/C atomic ratio should be in the range 0.89 to 0.64 for principal phase of oil formation.

In case of conflicting signals between geochemical and petrographic techniques, the former are to be preferred, but wherever possible one geochemical and one petrographic technique should be employed.

Chapter 7
Source Rock Evaluation and Correlations

It is evident that petroleum originates from finely disseminated organic matter buried within sedimentary rocks (Chs. 4 & 5). The criteria for assessment of source rock potential are: (*i*) the amount of organic matter both soluble (bitumen) and insoluble (kerogen), (*ii*) its type (quality), (*iii*) thermal maturity, and (*iv*) expulsion efficiency of source rock (Waples, 1979; Tissot and Welte, 1984).

7.1 Principles of Source Rock Evaluation

Much of petroleum geochemistry depends upon accurate assessment of the hydrocarbon-source capabilities of sedimentary rocks. Although the term source-rock is frequently used generically to describe fine-grained sedimentary rocks, its usage is too broad. It is therefore useful to make the following distinctions among the source rocks:

(*i*) Effective Source Rock

Any sedimentary rock that has already generated and expelled hydrocarbons. It obviously encompases a wide range of generative histories from earliest maturity (premature) to over maturity.

(*ii*) Possible Source Rock

Any fine grained sedimentary rock whose source potential has not yet been evaluated, but which may have generated and expelled hydrocarbons.

(iii) Potential Source Rock

Any immature sedimentary rock known to be capable of generating and expelling hydrocarbons if its level of thermal maturity were higher.

When a source rock sample is analyzed in the laboratory, it gives its present day (remaining or untapped) source capacity, which is represented as G. This quantity is meaningful if the rock's original capacity expressed as G_0 is known. The difference $(G_0 - G)$ between G_0 and G represents the hydrocarbons already generated in the source-rock. It is not possible to measure G_0 directly for a source rock sample that has already began generating hydrocarbons. However, it can be estimated by measuring G for a similar sample that is still immature. It is therefore obvious that G_0 can only be measured directly from immature source rocks, where G and G_0 are identical.

7.1.1 Measurement of Source Rock Capacity

There are two distinct aspects in the assessment of source rock capacity: (*i*) how much total oil could be generated from a given source bed if generation went to completion? This is called "total oil". (*ii*) how much oil has already been generated? This is called "oil already generated". Two types of methods namely (*i*) direct and (*ii*) indirect are commonly used for calculating "total oil" and "oil already generated" in a source-rock.

(*i*) Direct Methods

The most important among direct methods is the Rock-Eval pyrolysis method (6.5.0). The amount of hydrocarbons (S_1 and S_2) after normalization with organic carbon (OC) represents those that are preexisting and remaining at the subsurface respectively. S_2 in fact is a measure of the generative capacity (G). S_3 after normalization with (OC) represent oxygen content of the kerogen. The sum $(S_1 + S_2)$ of S_1 and S_2 represents genetic potential and the ratio $[S_1/(S_1 + S_2)]$ represents Production Index or Transformation Ratio. The Hydrogen Index (HI) can either be computed with S_2 after normalization with OC or taking the ratio. $[S_2/(S_1 + S_2)]$. HI is a direct measure of the kerogen's generative capacity of the source rock. Hence we can estimate the hydrocarbons in place if we obtain the data on total rock volume, its thermal maturity, and of expulsion, migration and accumulation efficiencies. As HI is expressed in milligrammes of hydrocarbons per gramme of OC present in the source rock, it is easy to convert it into more familiar units, barrels per cubic mile or gallons per tonne of source rock. However, one should realise that the hydrocarbons generated by Rock-Eval pyrolysis method contains both gaseous and liquid species. Their relative proportions depends upon the kerogen type. For example, Type II kerogen may probably give as much as 90 per cent oil and 10 per cent gas while Type III kerogen yield 20 per cent oil and 80 per cent gas.

The advantage of Rock-Eval pyrolysis method is it mimics the natural hydrocarbon-generation process occurring in the subsurface. However, there are few limitations to this method. It can be applied directly only when the kerogen being pyrolysed is immature. In case it is not immature, necessary correction should be applied to HI as described in 7.1.3. Another disadvantage is relatively high temperatures normally employed in the pyrolysis. It has been established that the same maturation effects can be produced at much lower temperatures acting over longer times in natural (subsurface) conditions (5.2.3). Two kerogens that behave similarly in the laboratory pyrolysis might act quite differently in the subsurface environments. Furthermore, any effects of mineral catalysts will be much greater in the laboratory than in the subsurface. In such cases where there is a significant catalytic effect by clay minerals, pyrolysis data will give an under-estimate of the true source potential. Under these conditions, clays apparently facilitate conversion of some bitumen to a carbonaceous residue that can never be detected by the Rock-Eval analyzer.

(ii) Indirect Methods

Indirect methods for calculating the quantity G require measurement of two parameters, namely, quantity and the type (quality) of kerogen separately and then combining them. The quantity is invariably estimated by measuring the total organic carbon (TOC) as described earlier (6.3.2, 6.3.3). The samples (cuttings) to be analyzed for TOC more often contain mixtures of different lithologies including caved material and contamination of various kinds. It is, therefore, essential to remove the caving and contamination prior to TOC measurement. When more than one lithology is observed particularly while analysing a core it is advisable to hand pick the sample containing only one particular lithology of interest by MOA (6.2.1).

Elemental (C.H.N) analysis (6.3.3) of isolated kerogen has proved to be very useful and reliable tool for determining its Type. Saxby (1980) developed the following empirical equation relating the atomic H/C and O/C ratios with Hydrogen Index (HI),

$$HI = 667 \, (H/C) - 570 \, (O/C) \qquad (7.1)$$

This equation is satisfactory for predicting oil yields from both coals and oil shales during slow, low temperature pyrolysis. However, it failed to produce satisfactory results for samples containing large amounts of inertinites.

The advantage of TOC and CHN methods are:

(*i*) They give more complete picture of the chemical composition and history of a kerogen, and thus enable us to understand more fully the various geological and geochemical processes that affect source-rock quality.

(*ii*) Normal use of more than one indirect method allows comparison of results from several techniques and thus recognize problem samples.

However, the following are the limitations of indirect methods:

(*i*) Speed and cost of analysis are not as favourable as direct (pyrolysis) method.

(*ii*) Analytical results do not provide direct information about hydrocarbon-generation capacity.

Though the value of G (rock's remaining or present source capacity) is obtained by above factors namely the quantity and type (quality) of kerogen, it is also necessary to know its level of thermal maturity represented by that particular G value. For example, if G is very low, it may represent two different scenarios: In the first one, the rock never had a high initial source capacity (immature). In the second case, the rock had already burned out (over mature) in which case virtually all the initial hydrocarbon source-capacity has already been usedup. The exploration implications of the two scenarios are of course, very different.

Maturity of kerogen can be estimated by several techniques as described earlier (Ch.6). Among them vitrinite reflectance (Ro) method is usually very common (6.2.3). One difficulty with this method as applied to kerogen is that vitrinite is rare or absent in many rocks. What is present is often reworked, its maturity is not related to that of the rock in which it is found. Reworked vitrinite is, in fact, more common in shales than in coals, leading to frequent difficulties in establishing which vitrinite population is indigenous. Despite this and many other weaknesses, it is the most popular technique today for estimating kerogen maturity. Though there are several maturity parameters (Ch. 6), the key to use them effectively lies in evaluating the measured data carefully, and obtaining more than one maturity parameter whenever possible.

7.1.2 Estimation of Original Source Rock Capacity (G_0)

Among the three major methods for determination of kerogen type, namely microscopic (6.2.1), elemental (6.3.3) and Rock-Eval pyrolysis (6.5.0), the first one is relatively unaffected by maturity while the other two methods are strongly affected by it. The most common method for taking maturity effects into account in pyrolysis data is to use a modified van Krevelen diagram (Figure 7.1) to back calculate the original hydrogen index (HI). This diagram in effect represents H/C ratio against Ro instead of conventional O/C ratio (Figure 5.2a). The point P (Figure 7.1) represents such a kerogen. When it was immature this kerogen had the H/C ratio represented by point A. It has already progressed in the maturation pathway to point P and with additional maturation will continue towards B. Once a kerogen has been located on the H/C diagram, its maturation pathway can be interpolated and its immature H/C ratio can be determined. The H/C atomic ratios for an average immature kerogen vary from 0.6 (characteristic of some inertinites) to 1.6 (maximum for pure alginites). This method works fairly well if the kerogen is still within the oil generation window. However, it fails at higher maturation levels.

Like pyrolysis, atomic H/C ratios calculated from elemental analysis represent the present day status of the kerogen rather than its original chemical composition. They must therefore be corrected for the effects of maturation by using the van Krevelen diagram (Figure 7.1) as described above. The immature ratios can then be used to calculate G_0 according to Eq. 7.1.

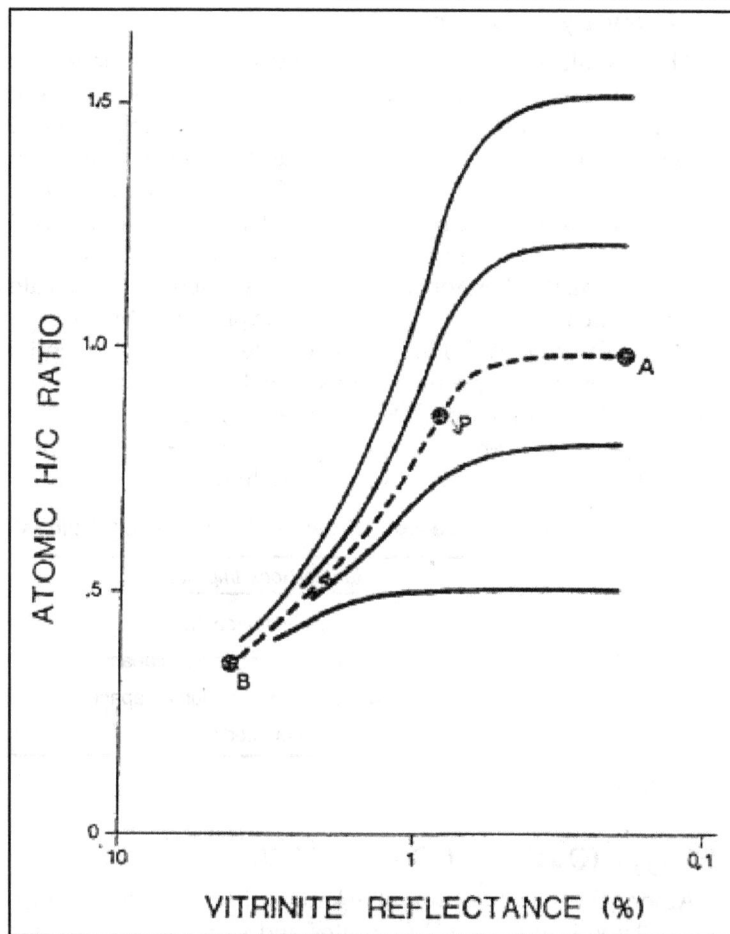

Figure 7.1: Calculations of the Immature Kerogen H/C Ratio from Present Day H/C Ratio and Vitrinite Reflectance Data (*Source*: Waples, 1981)

7.2 Interpretation of Source Rock Data

As already pointed out, the three essential parameters which must be taken into account for interpretation of source rock data are (*i*) quantity of organic matter, (*ii*) its type (quality) and (*iii*) maturity. Some important factors concerned with them are briefly described.

7.2.1 Quantity of Organic Matter

All most all measurements of the amount of organic matter present in a source rock are expressed as TOC values in weight percentage. Source rocks containing 1-2 per cent TOC are associated with depositional environments intermediate between oxidising and reducing, where preservation of lipid rich organic matter occurs, having moderate source capacity (Table 7.1). TOC values above 2 per cent often indicate highly reducing environments with excellent source potential. We therefore use TOC values as screens to indicate which rocks are worthy of consideration for exploration. Many rocks with high TOC values, however, have little oil source potential because the kerogen they contain is woody or highly oxidised (Type III or Type IV). Though high TOC is a necessity, it is not a sufficient criteria for good source-rocks. The kerogen type in them should be of good quality for hydrocarbon source-capability.

Table 7.1: Indices of Source-Rock Potential Based on TOC Values

TOC Value (Weight Per cent)	Source-Rock Implications
< 0.5	Negligible source capacity
0.5–1.0	Possibility of slight source capacity
1.0–2.0	Possibility of modest source capacity
> 2.0	Possibility of good to excellent

Source: Waples, 1985.

7.2.2 Type (Quality) of Organic Matter

Among the macerals present in kerogen, the oil generative are those of Type I and Type II (alginites and exinites) and the gas generating maceral is mainly vitrinite (Type III). Inertinite (Type IV) is generally considered to have no hydrocarbon-source capacity. When HI obtained from pyrolysis is corrected for maturation effects (7.1.2), its interpretation becomes straight forward (Table 7.2). The HI indices below 150 mg/g. of TOC suggests absence of oil generating lipid materials and conforms to the presence of Type III and Type IV kerogen macerals. The HI between 150-300 contain more Type III than Type II kerogen and have marginal to fair potential for liquid hydrocarbons. Kerogens with HI above 300 and up to 600 mg/g. of TOC contain substantial amounts of Type II macerals, and thus are considered to have good source potential for oil. Kerogens with HI above 600 mg/g. of TOC usually consists of mainly pure Type I and Type II with excellent source potential for oil.

Table 7.2: Source Potential of Immature Kerogens Based on Hydrogen Indices

Hydrogen Index (mg HC/g TOC)	Principal Product	Relative Quantity
< 150	Gas	Small
150–300	Oil + Gas	Small
300–450	Oil	Moderate
450–600	Oil	Large
> 600	Oil	Very Large

Source: Waples, 1985.

Table 7.3: Prediction of Hydrogen Indices of Immature Kerogens Based on Atomic H/C and O/C Ratios

Atomic H/C	Atomic O/C	Hydrogen Index*	Product
1.60	0.06	700	Oil
1.50	0.08	622	Oil
1.40	0.09	550	Oil
1.30	0.10	477	Oil
1.20	0.11	405	Oil
1.10	0.12	332	Oil + (gas)
1.00	0.14	254	Oil + gas
0.90	0.16	176	Gas + (oil)
0.80	0.18	98	Gas
0.70	0.20	20	Gas

* Calculated according to equation 7.1.

Source: Waples, 1985.

Atomic H/C ratios of immature kerogens can be correlated with HI obtained from pyrolysis yields or can be interpreted directly (Table 7.3). H/C ratios above 1.2 indicate good to excellent potential for oil generation and they mostly comprises Type I and Type II kerogens. H/C ratios between 0.9 and 1.1 may have modest potential for generating oil along with small amounts of gas. They contain more Type III than Type II kerogens. Gas is likely to be the only product from kerogens with H/C ratios below 0.8.

Typical empirical relationship between TOC and HI expressed as Pyrolysis Yield (Figure 7.2) indicates that samples containing low TOC have relatively lower HI comprising mainly oxidized

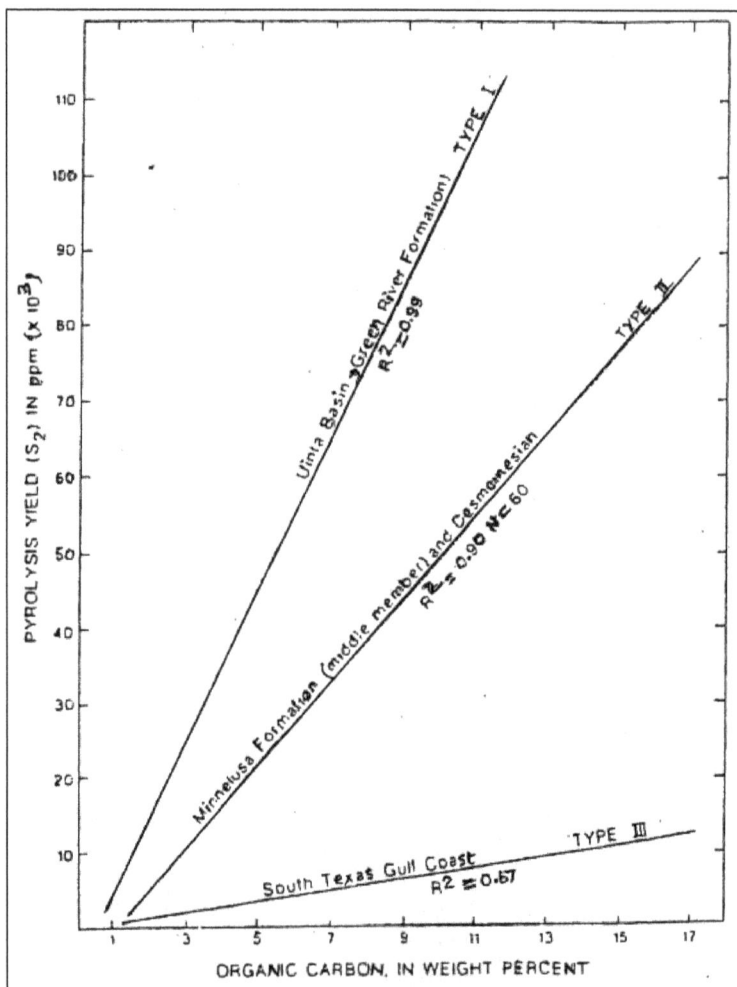

Figure 7.2: Relationship Between TOC and Hydrogen Index for Three Formations Containing Types I, II, and III Kerogens (*Source*: Waples, 1985)

organic matter. In contrast, samples with higher TOC values often contain moderate to large proportions of lipid rich material having good hydrocarbon source potential. However, the correlations between TOC and HI are weak in non-marine or paralic rocks containing large amounts of woody or cellulose material.

7.2.3 Maturity

Determination of oil generation window in a particular section is the objective of most of the maturity analysis performed on possible source rocks. For most kerogens, the limits in terms of Ro are 0.6 per cent, 0.9 per cent and 1.3 per cent for onset of oil generation, peak generation and the end of oil generation respectively. The correlations among maturity parameters such as Ro, TAI and Pyrolysis T_{max} (Table 7.4) have been well established with minor variations from one laboratory to another. Though several other maturity parameters such as CPI, Bit/BFOC (bitumen/bitumen free OC) and some biomarkers (Table 10.2) are available, they have only limited applicability.

Table 7.4: Correlation of Various Kerogen-Maturity Parameters with Vitrinite-Reflectance (R_0) Values

Vitrinite-Reflectance (Per cent R_0)	Thermal Alteration Index (TAI)	Pyrolysis T_{max} (°C)
0.40	2.0	420
0.50	2.3	430
0.60	2.6	440
0.30	2.8	450
1.00	3.0	460
1.20	3.2	465
1.35	3.4	470
1.50	3.5	480
2.00	3.8	500
3.00	4.0	500+
4.00	4.0	500+

Source: Waples, 1985.

7.2.4 Examples of Source Rock Evaluation

It is most useful to display geochemical data of wells in a log format. Many formats for such logs have been developed, Figure 7.3 is one such typical geochemical log (Waples, 1985) which includes data most relevant for source-rock evaluation such as quantity,

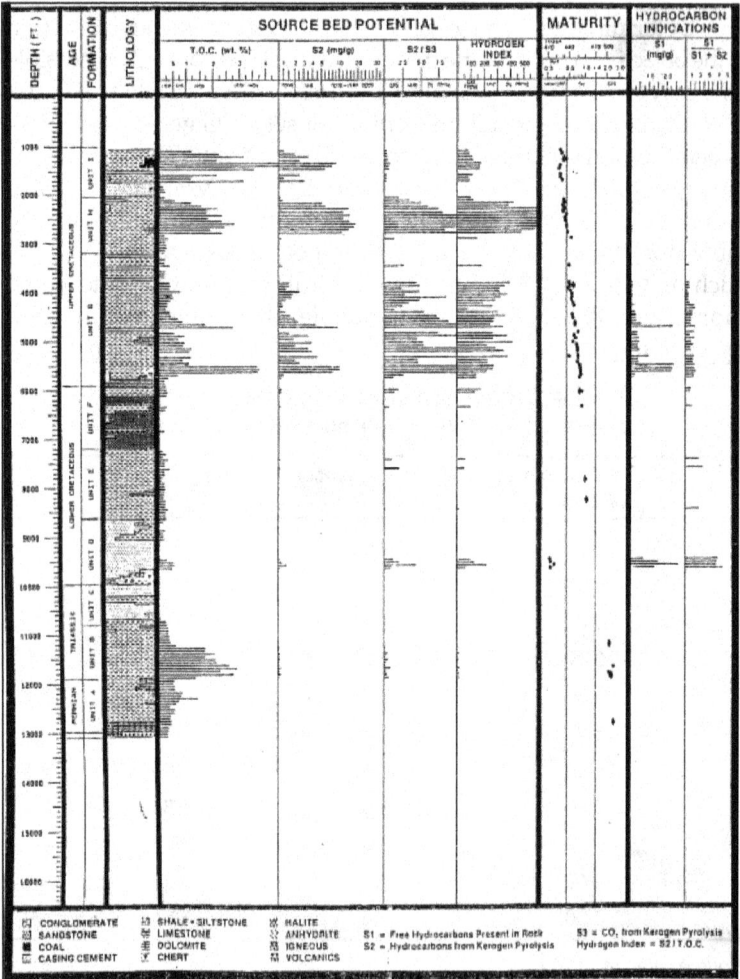

Figure 7.3: Geochemical Log for the Turquoise Well, Showing Results of Source-rock Analysis in a Well-Profile Format (*Source*: Waples, 1985)

quality and maturity of organic matter of Turquoise well. It is apparent from TOC values that four units (I,H,G and B) are worthy of consideration as possible source-rocks. The rest of the section is extremely lean with TOC (<0.5 per cent). Following are some important inferences derived from the geochemical log:

(i) Unit I is the shallowest rock with high TOC values. Lithological description and pyrolysis values, however, indicate the organic matter (OM) of terrestrial (coaly) origin with low HI (100-150) suggesting that the rock is immature with little hydrocarbon generating capacity.

(ii) Upper half of unit H contains a much more oil prone OM with high pyrolysis HI (more than 500) yields. Maturity parameters suggest that it is at or near the top of the oil generation window.

(iii) Unit G includes a few elevated TOC values among many lean samples. Gradual increase of HI downwards together with relatively high production indices and S_1 yields indicate some hydrocarbon generation in the unit. The maturity data suggests that the unit is within the early to peak phase of oil generation.

(iv) Lower part of unit B has high TOC values but pyrolysis yields indicate negligible remaining hydrocarbon-source capacity. The very high maturity levels suggests gas generation window.

Interpretation of geochemical data of the well by another log format (Figure 7.4) reveals two rich zones (based on TOC values) between 3000-4000 ft and between 10,500-13,500 ft. The upper rich zone is of excellent oil resource quality, having H/C ratios near 1.3. However, it is immature with R_0 values around 0.4 per cent. The lower rich zone is much different. Although it is marginally mature for oil generation (Ro near 0.6 per cent), the quality of kerogen is poor as evident by the H/C ratios clustered around 0.8. It is therefore concluded that the lower zone contains coaly material poor in hydrogen, and even if matures, it would generate mainly gas.

Interpretation of source rock data on a basic level is quite simple. With increasing experience one can also learn to derive important information on thermal histories, unconformities and erosional events, and organic facies. In order to achieve this, the measured

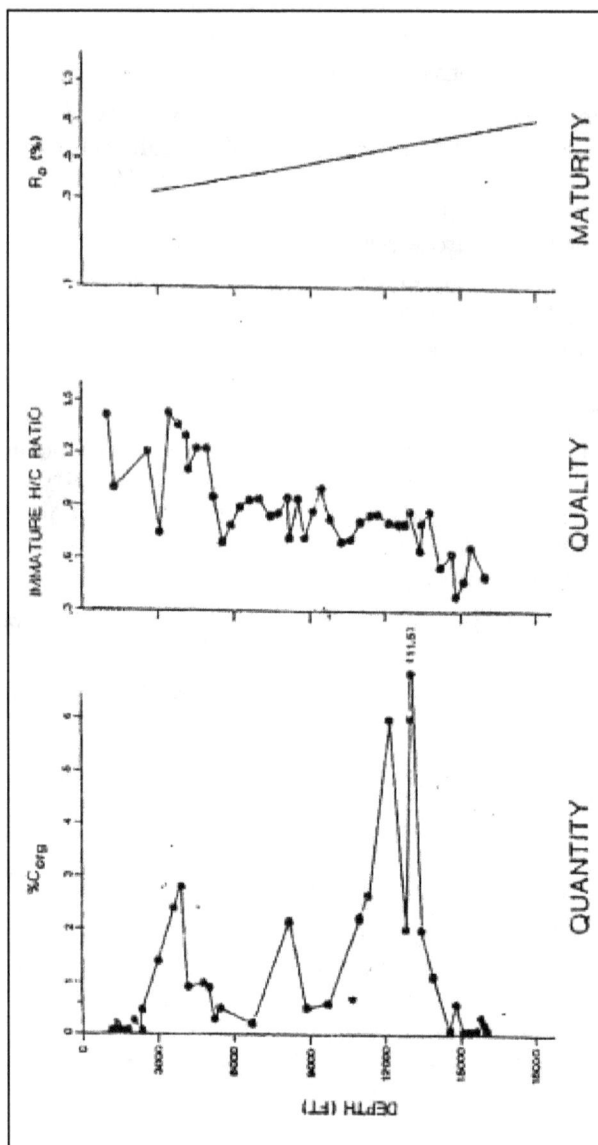

Figure 7.4: Geochemical Log for the C.O.S.T. B-2 Well, Offshore New Jersey (*Source:* Waples, 1985)

data has to be extrapolated to as large an area as possible to develop regional models of organic facies and thermal maturity.

7.3.0 Quantitative Volumetric Estimation of Source Rock Hydrocarbons in a Basin

The convenient way to calculate hydrocarbon volumes in a basin is to break the process of hydrocarbon accumulation into four phases *e.g.*, generation, expulsion, migration and entrapment, and preservation (Moshier and Waples, 1985). The most useful basic equation for hydrocarbon generation utilizes three common parameters of geochemical data (TOC, Rock-Eval pyrolysis yield HI, and maturity factor f). All of them affect the quantity of hydrocarbons that have already been generated. The hydrocarbon volume can be calculated by Eq. 7.2.

$$\text{Volume of HC} = (k)\,(\text{TOC})\,(\text{HI})\,(f) \qquad (7.2)$$

TOC is expressed in weight percentage, HI in mg HC/g TOC, and maturity is expressed as a fraction f between 0 (completely immature) and 1 (fully mature). The value of f is given by $(G_o\text{-}G)/G$. Conversion constant k is governed by the units desired for the hydrocarbon volumes, and by the units for density of source rock and hydrocarbons. If the volume units expressed as millions of barrels for cubic mile of source rock, and if the source rock is a shale with density 2.3 g/cc, and if the hydrocarbons correspond to an oil of 25° API gravity (density = 0.9 g/cc), the value of k is 0.7. If the source rock is a limestone (density = 2.6 g/cc), the value of k is about 0.78. Units of the measured or calculated values of TAI or Ro or TTI must be converted into the unitless scale in which fractional conversion (f) is expressed. Sluijk and Nederlof (1984) have published a useful series of calibrations between Ro and f (Figure 7.5). The calculation taking a typical source rock (shale) as an example, is shown below.

Suppose the measured TOC of the shale sample is 1.2 per cent, whose HI was 120 mg HC/g TOC when the rock was immature, and whose present-day Ro value is 0.9 per cent. From the HI value, the kerogen present in the shale appears to be Type III (7.2.2). Let us assume further that Type III kerogen can generate about 20 per cent oil and 80 per cent gas (7.1.2). The amount of organic carbon capable of generating oil is thus 20 per cent of the 1.2 per cent *i.e.*, 0.24 per cent. The remainder 0.76 per cent, is capable of generating only gas.

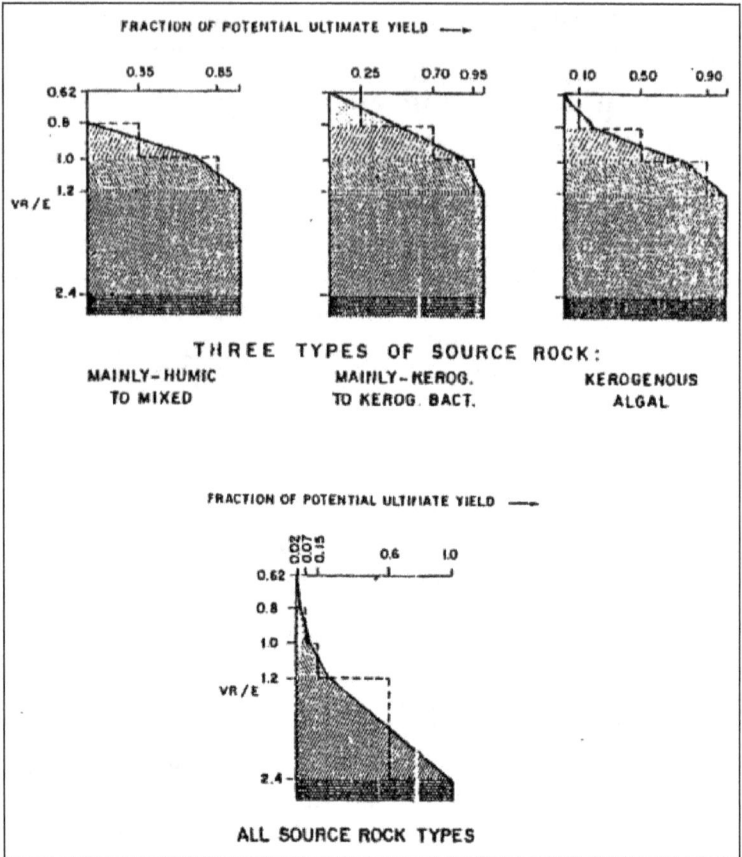

Figure 7.5: Curves Showing the Relationship Between R_o Values and Fractional Conversion (f) of Types III (humic to mixed), II (kerogenous to kerogenous bacterial), and I (kerogenous algal) Kerogens to Oil (top, left to right) and Gas (bottom).
(*Source*: **Sluijk and Nederlof, 1984**)

In order to calculate the volume of oil generated, we can use the relationship between f and Ro of Type III (humic) kerogens (Figure 7.5) which is 0.35. Hence the volume of oil generated in the shale calculated by (Eq.7.2) is given by

Volume of oil = (0.7) (0.24) (120) (0.35)   (7.3)

= 7.056 million barrels oil per cubic mile of source rock.

The volume of gas generated is calculated in a similar manner, except that f for gas generation is 0.07 (Figure 7.5 bottom). Further, if we want the volume of gas in billions of standard cubic feet per cubic mile of source rock, the constant k must be multiplied by 6. Thus the volume of gas generated is given by Eq. 7.4.

$$\text{Volume of gas} = (0.7)\ (6)\ (0.76)\ (120)\ (0.07) \tag{7.4}$$

$$= 26.813 \text{ billion cubic feet of gas per cubic mile of source rock.}$$

In addition to generation, expulsion of oil is critical in order for migration and accumulation to occur. According to Momper (1978), a threshold value of 50 million barrels of hydrocarbons (oil or oil-equivalent) had to be generated in a source rock before any expulsion could occur. Further, once this threshold is reached the expulsion efficiency for oil is about 50 per cent (5.4.3). Let us assume that the expulsion efficiency for gas is higher than oil, about 80 per cent. Based on the above assumptions we can calculate the volume of hydrocarbons expelled from any source rock.

From our example cited above the total volume of hydrocarbons generated is 7 million barrels of oil and another 4.6 million barrels of oil-equivalent in gas (26.813/6). Since the sum of these two quantities is not enough to reach the expulsion threshold (50 million barrels), it is evident that no expulsion has yet occurred.

In contrast, let us assume a rich shale containing 3.8 per cent TOC that had a HI of 500 when immature, with a present Ro 0.8 per cent (f = 0.5). Let us assume that this shale (Type II kerogen) generates 90 per cent oil and 10 per cent gas (7.1.1). Based on this, data, the volumes of hydrocarbons so far been generated and expelled can be calculated by the following equations.

$$\text{Volume of oil generated} = 0.7\ (3.8 \times 0.9)\ (500)\ (0.5) \tag{7.5}$$

$$= 598.5 \text{ million barrels per cubic mile of source rock.}$$

Because this quantity greatly exceeds the threshold, the rock would have expelled about 50 per cent, or 300 million barrels of oil from each cubic mile of source rock.

The volume of gas generated

$$= (0.7)\ (6)\ (3.8 \times 0.1)\ (500)\ (0.02) \tag{7.6}$$

$$= 15.96 \text{ billion cubic feet of gas per cubic} \\ \text{mile of source rock.}$$

If 80 per cent of this gas is expelled *i.e.*, nearly 13 billion cubic feet of gas will accompany each 300 million barrels out of the source rock. This small amount of gas is probably not enough to saturate the oil, and therefore it will move in solution in the oil.

Once the volume of hydrocarbons expelled for unit of source rock has been calculated, we must determine the total volume of source rock available. This step requires that we define the total area of our interest, usually called "drainage area". Then the total volume of hydrocarbons is calculated by Eq. 7.7.

$$\text{Total HC volume} = (\text{HC volume}/\text{cubic mile}) \times (\text{cubic miles of source rock}) \qquad (7.7)$$

Estimation of source rock volume is mainly geological and can be obtained by a variety of geological and geophysical methods through isopach maps. For this purpose, it may be necessary to divide the source rock into packages within which the geochemical properties (maturity, richness, kerogen type) remains relatively constant. If several packages are present, the hydrocarbons contributed by each package are simply summed-up in the final step.

Once the total volume of hydrocarbons expelled from a particular drainage area has been calculated, migration and accumulation efficiencies have to be taken into account. Typical values used by some workers (Barker and Dickey, 1984; Webster, 1984) are in the range of 5 per cent to 10 per cent for rich rocks for the combined efficiencies of expulsion and migration. Sluijk and Nederlof (1984) reported that the migration efficiency probably varies in the range 5 per cent to 30 per cent for good source rocks. Thus the general efficiency of secondary migration and expulsion is probably in the range of 10-20 per cent.

In order to calculate the volume of hydrocarbons that has successfully migrated and accumulated, we multiply the volume expelled (Eq. 7.7), by the proposed efficiency.

$$\text{HC volume trapped} = (\text{HC volume expelled}) \times (\text{migration efficiency}) \qquad (7.8)$$

Finally we have to take into account the preservation of oil in the reservoir. Destruction can occur either by cracking or biodegradation; the widely different thermal regimes required in the two cases prevent both phenomena occurring together in a single sample. If these factors are taken into account, the preserved oil is calculated by Eq.10.9.

Preserved oil = (oil trapped) × (preservation factor)   (7.9)

The preservation factors for oils varies from zero (total destruction) to 1 (no destruction) by the above processes. Total destruction could best be explained by the absence of any producible oil. A preservation factor for gases could also be introduced if oxidation of methane is anticipated.

Volumetric calculations are in principle, simple to carryout, but in practice they offer many complexities. Following are some of the potential difficulties encountered in such calculations.

(*i*) Accurate prediction of oil-mix from a particular source rock.

(*ii*) Certainity about the threshold for expulsion, and the expulsion efficiencies.

(*iii*) Factors influencing migration and trapping mechanisms, and their prediction.

(*iv*) Handling of systems in which geochemical and geological parameters change significantly over the drainage area of interest.

It is evident from the above discussion that the application of geochemistry to exploration should involve full integration of geological and geochemical data. An excellent way to accomplish this objective is to develop integrated models that describe the complete hydrocarbon system in the area of interest. Chapter 11 describes the application of such models in exploration.

7.4.0 Oil-oil and Oil-Source Rock Correlations

In petroleum exploration, identification of various types of oil or gas occurring in a sedimentary basin is of great interest. It helps petroleum geologists to define the number of different targets, which are present in different geological formations. This can be achieved by oil- oil or oil-source rock or gas correlations. Furthermore, the oil-source rock correlations is of special interest for locating precisely the formation responsible for oil generation. This in turn, facilitates the study of the extent of facies change and depth of burial of the source rock formation across the basin (Welte *et al.*, 1975). Parameters which indicate similarities of depositional environments, nature of source organic matter input, thermal history at that time of explusion from source rocks and migration system are investgated in detail for enhancing oil to oil and oil to source rock correlation capcities.

In order to identify source rocks and to determine migration pathways, it is often important to know whether two samples of oil or gas or source rock share a common origin. We therefore frequently attempt to correlate samples by comparing their chemical and physical properties. It is easy to correlate materials of same type, for example, two oils. However, it is somewhat difficult to correlate two dissimilar samples, for example, oil with kerogen or bitumen; bitumen with kerogen. Correlations in general, are carried out by measuring and comparing several characteristics of each sample. These characteristics should reasonably vary from sample to sample but should be affected to a minimum extent by transformations occurring during catagenesis and migration, as well as in reservoirs.

7.4.1 Methods of Characterization

Two fundamentally different types of characteristics, such as bulk parameters and specific parameters are usually measured. Bulk parameter refers to properties of the whole samples such as API gravity, sulphur, saturated hydrocarbons, isotopic ratios etc. Since bulk parameters tend to be affected strongly by processes such as migration, biodegradation and cracking, they have some limitations as correlation tools. Specific parameters such as different types of biomarker ratios, in contrast, measure in detail one characteristic of small fraction of a sample. These parameters are some times also strongly affected by small amounts of contamination, biodegradation or mixing. Bulk parameters are better for detecting transformations that affect the whole sample but generally not sensitive enough to use as positive correlation tools without specific parameters. It is therefore important to utilize both bulk and specific parameters in correlation studies.

Although the term "finger printing" is often applied in correlations its usage is overlaid optimistically. Most samples of oils, for example, are reasonably similar, and no distinguishing parameters as definitive as human finger print is known. In making positive correlations, it is therefore, essential to build up a case of probability based on number of reasonably good positive correlations and no unexplainable negative correlations. This process is analogous to building a case on circumstantial evidence in a court of law. We cannot prove the correlations conclusively but we can make it very plausible. It is thus very important in any correlation study to amass as many different pieces of evidence as possible. Negative correlations are more straight forward than positive correlations. If two samples differ in a single bulk parameter that can not be explained on the basis of transformations, then the samples do not correlate. Disagreement in a single specific parameter is not so serious, because of the possibilities of alteration or contamination. However, a repeated pattern of disagreement on several specific parameters indicates a negative correlation between the samples.

7.4.2 Limitations

There are several factors that make correlations usually difficult. Mixing of oils from different sources can be difficult to detect. Transformations of oils in reservoirs by biodegradation, water washing, cracking or deasphalting also cause severe changes in some of the chemical and physical properties and composition of crude oils (Table 7.5) rendering some correlation parameters completely useless. In such cases, the bulk parameters are often greatly changed, forcing us to focus on those parameters that are not affected (or affected in predictable ways) by the transformations. Another problem can arise in correlating an oil between a sample of immature and mature source rock. Correlation in such circumstances is meaningful by selecting parameters unaffected by maturation which is often difficult.

Despite these few drawbacks, correlation studies have been very useful in the past and will continue to grow in popularity as biomarker (10.1.0) and isotope techniques (6.4.0) are further refined. They have proved to be an aid in establishing the source rocks for some oils as well as in identifying contribution from multiple sources. Correlations involving gases are more difficult than oil and source rocks. However, the former have shown conclusively that vast amounts of gases in many commercial reservoirs are of biogenic origin. Correlation studies should be undertaken wherever a better understanding of the relationships between individual reservoirs would of value in exploration.

Table 7.5: Effects of Alteration Processes on Crude Oil Composition

	Migration	Water Washing	Biodegradation	Gas Deasphalting	Thermal Maturation
API Gravity	↑	↑	↓	↑	↑
Per cent Sulfur	↓	↓	↑	↓	↓
C_{15} + fraction (Per cent of crude)	↓	↓	↑	↓	↓
Asphaltenes (Per cent of crude)	↓	↓	↑	↓	↑ unless deasphalting occurs
Gasoline (C_4–C_7) fraction (Per cent of crude)	↑	↑	↓	↑	↑
Paraffinicity	↑	↑	↓	↑	↑
Porphyrin content	↓	↓	?	↓	↑
n-Paraffins (a) Per cent of crude	↑	↑	↑ generally	↑ or no effect	↑
(b) Maximum in distribution curve	↓	↓ slightly	↑		No effect
(c) CPI	No significant effect		No effect	↓ or no effect	No effect
$\delta^{13}C$	↓	↓	Depends on composition	↑	↑ if gas is lost

Source: Waples, 1981.

7.5 Alteration Processes of Crude Oil Composition

The most important types of alteration processes are migration, water washing, biodegradation, gas deasphalting, and thermal maturation (Table 7.5). During migration, the heavier and more polar components are left behind as in the case of column chromatography (10.5.0). Most of the migrationally induced changes in oil composition probably occur as the bitumen makes its way out of the fine-grained source rocks (primary migration). Secondary migration generally cause smaller changes.

Water washing and biodegradation often go together. Water washing can occur without biodegradation but biodegradation is always accompanied with at least some degree of water washing, the more soluble components of petroleum are simply removed in solution. Lighter hydrocarbons particularly aromatics and the smaller polar molecules become depleted.

Microorganisms are very selective to compounds which they metabolise. Anaerobic bacteria are not considered to be important in biodegradation of hydrocarbons. Compounds containing hetero atoms are often consumed relatively rapidly and n-alkanes are generally severely depleted or totally absent in extremely degraded oils. Occasionally the isoprenoid hydrocarbons are also noticeably depleted. Figure 7.6 shows gas chromatograms of crude oil, in the course of biodegradation simulated in a laboratory experiment.

Gas deasphalting results in the removal of large aromatic molecules called asphaltenes. Since they contain a significant portion of hetero atoms, deasphalting usually lower the sulphur and nitrogen content of an oil. Thermal maturation results in disproportionation of middle molecular weight compounds to light hydrocarbons and asphaltenes, and is often accompanied by gas deasphalting.

Although the information (Table 7.5) is important, it does not reveal the whole information. It is also necessary to know the magnitude of the changes which can be anticipated from the various alteration processes. Unfortunately it is difficult to set rigid limits for most of the processes, because the magnitude of their changes ranges from mild to severe. Furthermore, the magnitude of change will depend to some extent, upon the initial composition of the crude oil. In general, the best way that one could expect the magnitude of

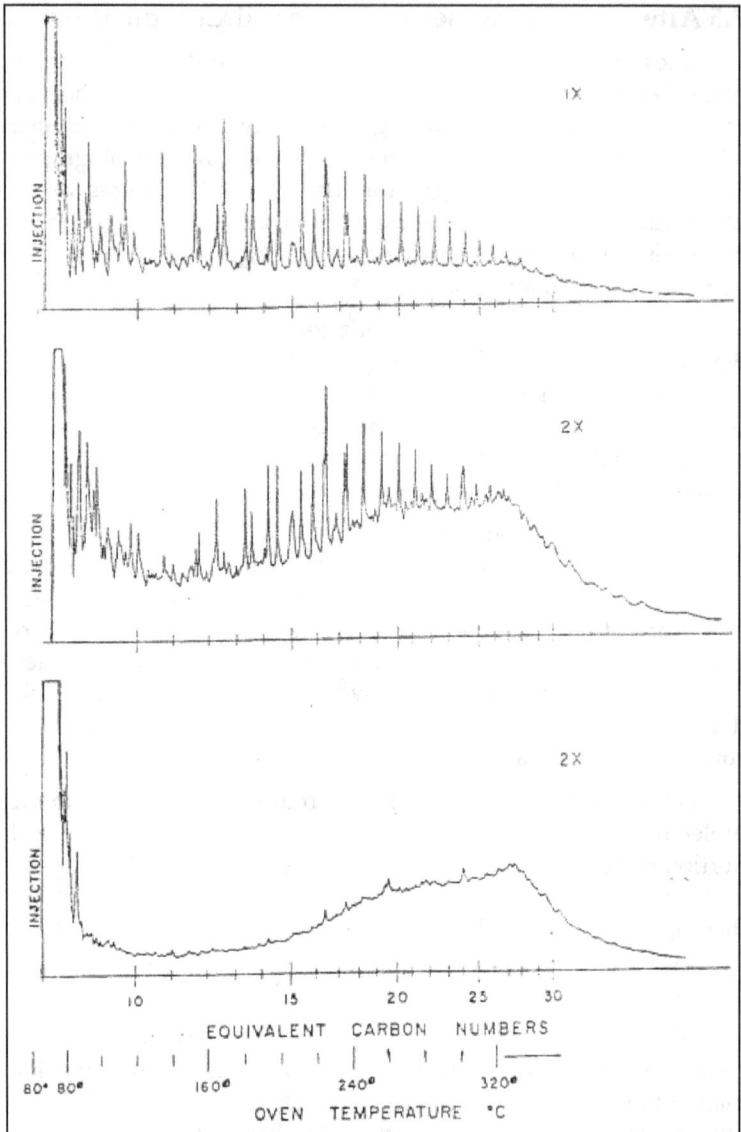

Figure 7.6: Gas Chromatogram of Crude Oil at 0-time (top) and After 6 Hours (center) and 26 Hours (bottom) of Incubation with a Mixed Population of Microorganisms (*Source*: Waples, 1981)

change on the above processes is by making many interpretations. Some of the guidelines on this aspect are given below:

(1) Asphaltenes are very susceptible to loss during migration and to reservoir alteration (deasphalting and disproportionation).

(2) Porphyrin content appears to be quite sensitive to migration. Nickel and vanadyl porphyrins seem to be affected equally so that Ni/V ratio does not change much. Porphyrins are gradually destroyed by high temperature, hence by the refining process.

(3) API gravity may decrease fairly drastically during extensive water washing and biodegradation.

(4) Sulphur content can sometimes be increased significantly by biodegradation.

(5) The hydrocarbon content of petroleum in general, is affected to different extents by biodegradation. However, the effect is much more severe with respect to n-paraffins. It can remove almost all of them during intense biodegradation.

(6) The n-paraffin content may increase markedly with increasing maturity. The maximum in the distribution curve often shifts very significantly towards lower carbon numbers with increasing maturity. Many immature crudes have maxima at C_{25} to C_{31}, at higher degree of maturity the maximum could shift to C_{20}, C_{15} or even lower. CPI gradually approaches 1.0 as oil maturity increases.

(7) The gasoline fraction increases greatly during thermal maturation.

(8) $\delta^{13}C$ values vary by only 1-2 ‰ at the most during these transformations. This is due to two reasons. If the system is closed, the change in a process is merely transfer from one compound to another so that neither ^{13}C nor ^{12}C enters or leaves the system. Further, different fractions (saturates, aromatics, polar, asphaltenes etc.) seldom differ by more than 2‰ so that even migrational loss of all the asphaltenes would not greatly affect the isotopic composition of the oils (Fuex, 1977).

7.6 Oil-oil Correlations

Oil-oil correlations are probably simpler, because one is comparing the same kind of organic material. Two oil samples having a common origin may differ substantially in chemical composition because of changes which occur during migration or storage in the reservoir rock. These changes include loss of heavy, light and polar components; biodegradation and water washing; and thermal disproportionation. In order to attempt oil-oil correlations, it is necessary to know how each of these transformations will affect oil's chemical properties. Milner *et al.*, (1977) reviewed in detail the petroleum transformations in reservoirs.

The general approach followed in attempting a correlation between two oils is to measure several parameters (Table 7.5) and compare the observed differences in the oils with those predicted in the table. It should be noted that some properties of oils are very sensitive to genetic differences. Among them are carbon isotopes, sterane and triterpane contents, and Ni/V porphyrin ratios. Other properties such as CPI and isoprenoid distributions, are somewhat dependent upon the origin of oil, but may be influenced to a greater or lesser degree by such transformation processes as maturation, biodegradation and migration. Still other properties such as API gravity and light hydrocarbon content, mainly influenced by thermal and migrational transformation processes are not very useful in oil-oil correlations. It is therefore important that we should keep in mind which properties provide maximum genetic information while attempting oil-oil correlations. The following examples illustrate the method of approach in this regard.

Let us suppose in one example we suspect that oil A is genetically related to oil B, that oil B has probably undergone some biodegradation. To verify this hypothesis, we analyze both oils for API gravity, porphyrin contents, n-paraffin content and distribution, and $\delta^{13}C$. The data (Table 7.6) indicates that oil B has undergone slight biodegradation as revealed by decrease of lighter n-alkanes and API gravity. Porphyrin ratios and $\delta^{13}C$ values are approximately same. All this is consistant with our hypothesis of their common origin. However, the positive evidence is meagre and by itself would not constitute proof beyond reasonable doubt. Further analysis particularly sensitive to steranes and triterpanes (10.1.0) in the two oils, would greatly strengthen the case for a genetic relationship.

Table 7.6: Analytical Data for Attempted Correlation of Oil A with Oil B

Parameter	Oil A	Oil B
API gravity	31.7°	26.6°
Ni/V porphyrins	1.1	1.2
n-Paraffins (Per cent of total crude)	18.6	12.2
n-Paraffins distribution	See below	See below
δ¹³C (topped oil) (‰ vs PDB)	−27.3	−26.8
	Oil B	

n-Paraffins, carbon numbers

Source: Waples, 1981.

In a second example, let us consider two oils (C and D) obtained from different wells, separated by several kilometres apart and possibly originated from a common source rock. Oil D lies considerably farther away from the proposed source basin than oil C. On the basis of the analytical data (Table 7.7) can it be possible to arrive at their common source?

Table 7.7: Analytical Data for Attempted Correlation of Oil C with Oil D

Parameter	Oil C	Oil D
API Gravity	32.6	36.7
Per cent sulfur	1.08	1.27
C_{15} + (Per cent of total crude)	37.5	34.8
Ni/V porphyrin ratio	0.65	0.81
$\delta^{13}C$ (‰vs PDB)	−26.3	−29.6

Source: Waples, 1981.

The data indicate that API gravity of oil D is more than that of oil C since the former has undergone migration (Table 7.5). Oil D has slightly lower proportion of heavier (C_{15}+) components, a trend which is compatible with the migration hypothesis. Though the sulphur content of the two oils is comparable, their values differ from the expected trend of migration. In fact oil D should have lower value than C, as the former has undergone migration. Porphyrin ratios like sulphur contents are similar but the trend is in the opposite direction to that which would be expected from migration effects. In any case, the difference is fairly small. So all the four parameters mentioned above support a common source. Carbon isotopic ratio ($\delta^{13}C$) however, presents a serious problem. The two oils differ by 3.3‰. No known migrational effect could be responsible for such a large change in $\delta^{13}C$. Therefore, it should be concluded that oils C and D do not have the common source and that the observed similarities are fortuitous. Another possibility is that oil D is actually a mixture of oils with two different source rocks of widely varying isotopic composition.

7.7.0 Oil-Source Rock Correlations

Oil–source rock correlations utilize the same method *i.e.*, comparison of chemical composition of various classes of hydrocarbons and non-hydrocarbons in crude oils and bitumen extracted from source rocks. However, crude oils are severely changed compared to source rock bitumen due to the phenomena of migration. The former are enriched in saturated hydrocarbons and depleted in NSO compounds (Figure 5.5). Furthermore, low molecular weight hydrocarbons are favourable over the heavier molecules. Thus bulk compositional parameters as stated above, though might be useful in oil-oil correlations, but they are useless in oil-source rock correlations. The best tools for this purpose are biomarker compounds (10.1.0) which have characteristic distribution to permit differentiation between individual source beds and the crude oils. However, owing to the possible effects of migration, it is advisable to make use of the distribution within a group of molecules having comparable chemical properties. The most commonly used biomarkers for this purpose are n-alkanes, isoalkanes including isoprenoids, polycyclic naphthenes and aromatics (Seifert and Moldowan, 1978). This order broadly corresponds to increasing complexity of chemical makeup. In addition, some non-biogenic compounds like benzothiophene derivatives present in both oil and source rocks are valuable for high sulphur containing crude oils. The following examples illustrate the method of approach in oil-source rock correlations.

A typical gas chromatogram (Figure 7.7) shows total saturated hydrocarbons of three oils (Trap Spring Field, Eagle Springs Field and Currant Field) and source rock of the Basin and Range Province. The Trap Spring oil, for example, has a very similar chromatogram of a matured sample of Chainman Shale being low in heavy n-alkanes. Pristane/phytane ratios (1.5) and, CPI values (1.0) are close (not shown in the figure). Oils from the Eagle Springs and Currant Field, in contrast, have pristane/phytane ratios below 1.0, display even carbon predominance and contain moderate to large amounts of heavy hydrocarbons (Figure 7.7). However, most powerful arguments for the proposed source rock-oil correlations came from the distribution of steranes and triterpanes. All the three rocks contain large amount of gammacerane indicating their common source (Figure 7.8).

Another typical example of correlations between source rocks and oils can be obtained from the detailed study of four and five-ring naphthenes comprising C_{27} to C_{30} atoms from different basins

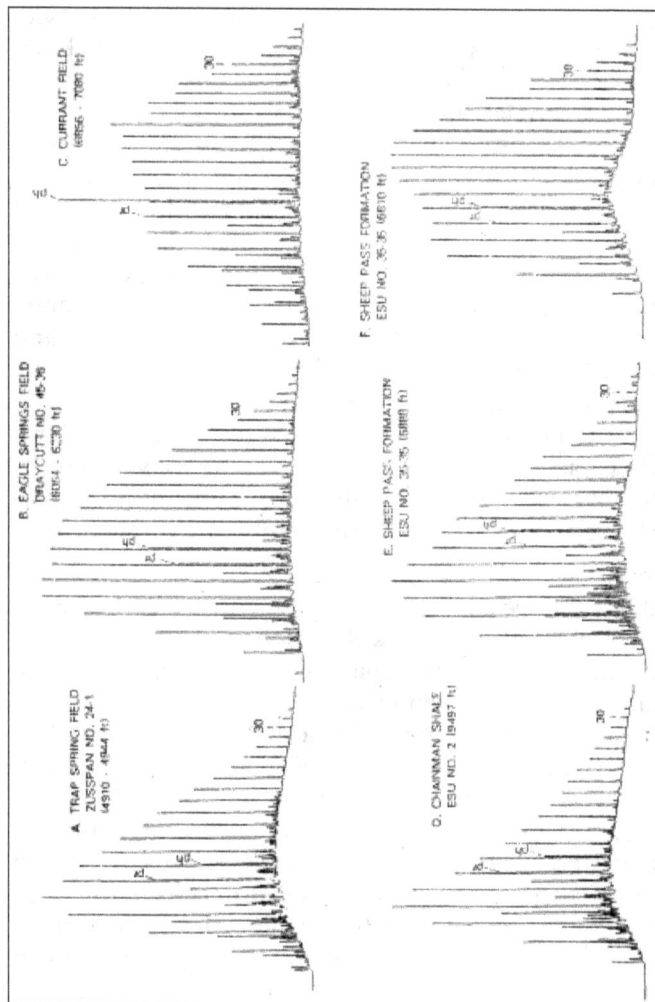

Figure 7.7: Gas Chromatograms of Saturated Hydrocarbons from Three Oils (top) and Three Source-Rock Extracts (bottom) from the Basin and Range Province. Each oil correlates with the extract immediately below it. (*Source:* Waples, 1985)

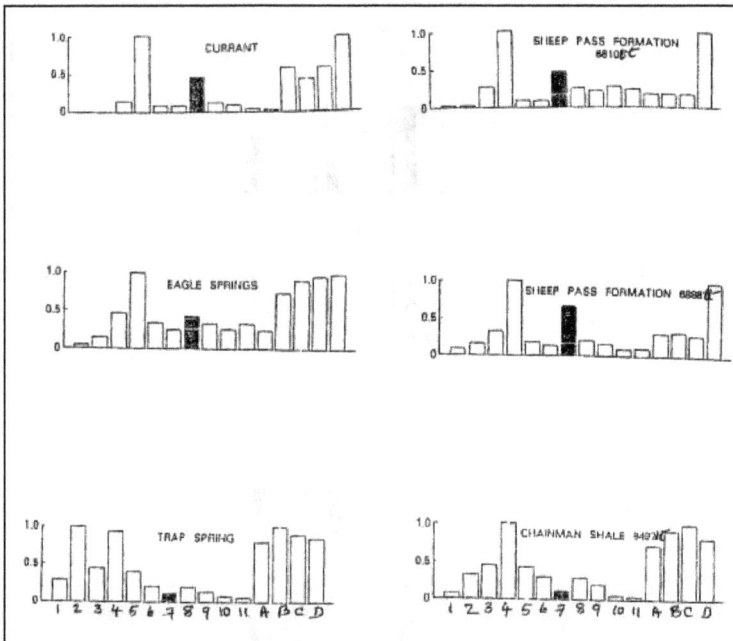

Figure 7.8: Relative Amounts of Selected Terpanes and Steranes for Oils (left) and Rock Extracts (right) from the Basin and Range Province, Peak7 is Gammacerane
(*Source*: Waples, 1985)

(Figure 7.9). The data presented in the figure reveals the following salient features:

(*i*) Good correlations of the Eocene oil and the source rock pairs from the Unita Basin (E-Y and E-I); the Permanian oil and the source rock pairs from the Mid-Land Basin (L-Y and L-I); and between the Oligocene oil and the source rock pairs from the Rhine Graben (D-Y and D-I),

(*ii*) Fair correlation between the Eocene oil (K-Y) and the Middle Jurassic source rock (K-I) from the Rhine Graben, and

(*iii*) No correlation between the Eocene oil (K-Y) and the Lower Jurassic source rocks (K-2 and K-3) of the Rhine Graben.

Welte *et al.*, (1977) while studying the correlations emphasized the distribution of polycyclic naphthenes over the saturated

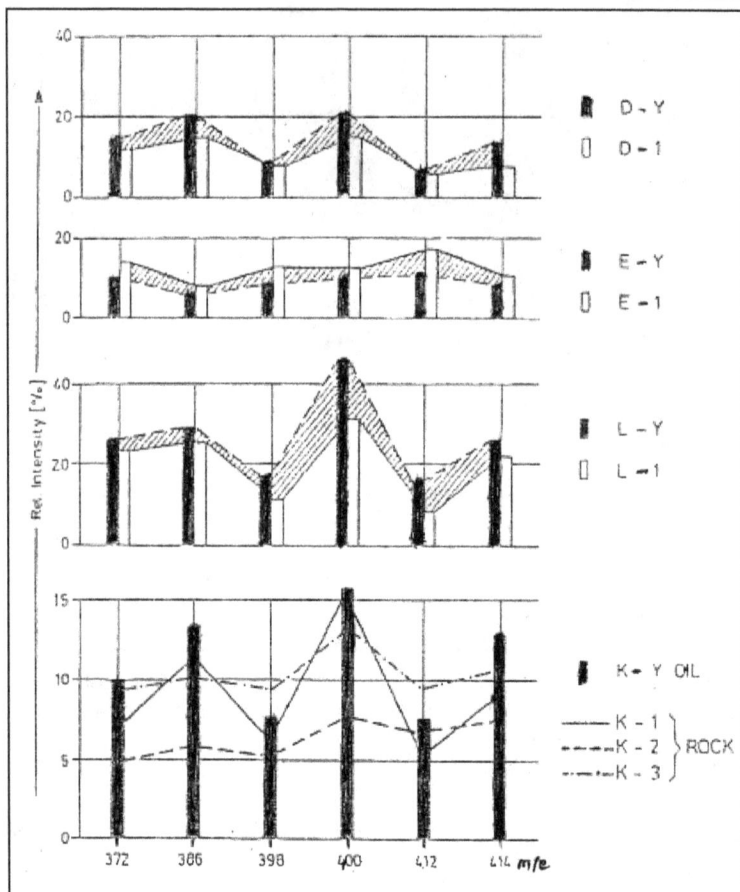

Figure 7.9: Comparison Between Correlative Source Rocks and Oils with Respect to the Distribution of Averaged Molecular Ion Series of C_{27+} Cyclics
(*Source*: Tissot, 1977)

D–Y (oil) D–1 (rock): Oligocene, Rhine Graben, West Germany.

E–Y (oil) E–1 (rock): Eocene, Uinta Basin, Utah, USA.

L–Y (oil) L–1 (rock): Permian, Midland, Texas, USA.

K–Y (oil): Eocene

K–1 (rock): Middle Jurassic

K–2 and K–3 (rocks); Lower Jurassic

Rhine Graben, West Germany.

hydrocarbons for source rock-oil correlations since the former exhibit specific structures and are independent of temperature history (maturation, age and depth). Carbon isotopic distribution has been used as a tool for oil-source rock correlations. It has been observed in many cases that $\delta^{13}C$ of the source rock bitumen is more depleted (more negative) than the related kerogen. On the other hand, most crude oils have slightly depleted or almost similar $\delta^{13}C$ values compared to those of source rock bitumen. Based on these differences in isotopic composition, correlations can be inferred between oil and kerogen or between oil and bitumen of source rocks.

7.8.0 Correlation Parameters for Gases

Gas correlations are much more difficult than oil correlations because of their simplicity and very few distinguishing characteristics. However, carbon isotope ratios have been shown to be useful correlation parameters for grouping gases (Stahl *et al.*, 1977; Schoell, 1980). It is possible to establish the origin of gases in petroleum by measurement of two parameters: wetness and $\delta^{13}C$ content. Natural gases are of two types-wet and dry. Dry gas contains 90-100 per cent methane (sometimes the range may be 95-100 per cent) with very small amounts of higher hydrocarbon gases such as ethane, propane, butane etc. Wet gas contains 5-10 per cent of higher hydrocarbons of the total. Wetness is therefore, defined as the ratio of methane (C_1) to sum of higher hydrocarbons such as ethane (C_2), propane (C_3), butane (C_4) etc. It is expressed as $C_1/(C_2+C_3+$ ....). Dry gases are formed either by bacterial action on kerogen during diagenesis or by high temperature alteration during metagenesis. On the other hand, wet gases are formed by moderate temperature alteration during early stages of catagenesis.

There is no way to distinguish among the various possible origins (biogenic or thermogenic) of either wet or dry gases. However, by combining wetness data with carbon isotopic ($\delta^{13}C$) measurements, we can easily distinguish dry and wet gases. Biogenic gases mainly methane is highly depleted in $\delta^{13}C$ having values in the range –55 to –90‰ relative to PDB (standard). Thermally generated methane in contrast, has $\delta^{13}C$ values below –55‰. It is therefore useful to plot $\delta^{13}C$ values of natural gas against wetness inorder to identify its origin (Figure 7.10). This figure depicts the regions where gases of single origin are found. Biogenic and

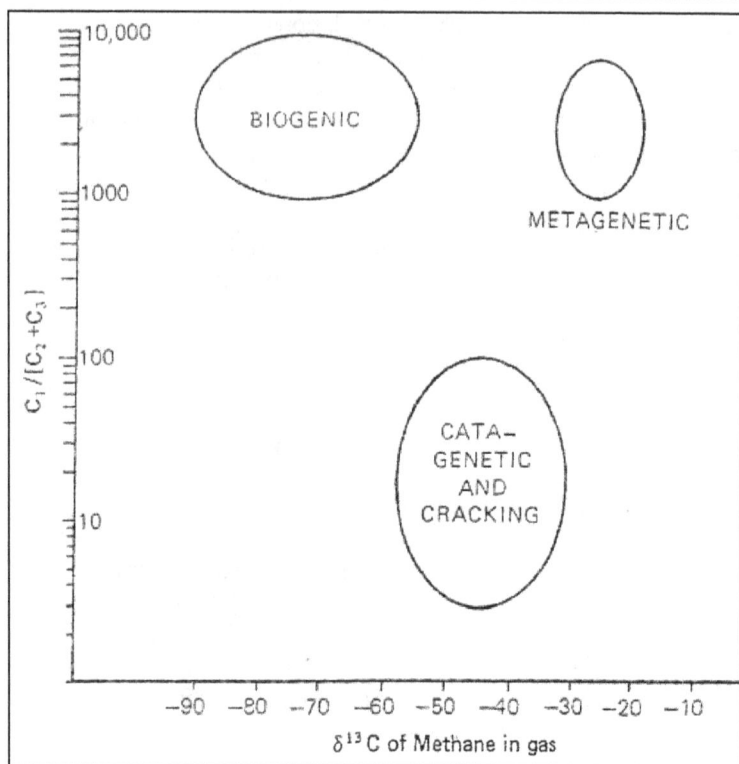

**Figure 7.10: Classification of Natural Gases by
Plotting Wetness Against δ^{13}C. Open areas represent
mixtures of gases from more than one source.
(*Source*: Waples, 1981)**

metagenetic gases are both dry but have very different carbon isotopic
composition. Gases formed by catagenesis of kerogen or by cracking
of petroleum are wet. These three categories are the end member
classes. Gases of other compositions are considered to be the result
of their mixing from more than one source. The wetness parameter is
particularly more sensitive, because only small amounts of wet gas
greatly alters the composition of a biogenic or metagenetic dry gas.

7.9.0 Case Studies

Following are few case studies presented to acquaint the reader
with the problems related to correlations.

7.9.1 Oil-Oil Correlations

Case 1

Figure 7.11 incorporates the data on two types of oils in Lower Tuscaloose Cretaceous reservoir of southern Mississippi and Albama (Central Gulf). We have to identify whether the two types of oils are genetically related. It is evident from the figure that Type I oils contained about twice the n-paraffins and 2‰ $\delta^{13}C$ more in the saturate fraction. They also contain C_{28} steranes dominant than those of C_{27}. Further, the ratio of cyclopentanes/n-paraffins (light hydrocarbon fraction) was much higher in Type I compared to Type II oils. Hence, it can be concluded that the two types are genetically different and belong to two distinct groups (Koons *et al.*, 1974). The authors observed that Type I oils occurred in unfaulted structural and stratigraphic traps, whereas most of the Type II oils occurred in faulted structures. Type I oils were formed in fine-grained shales of terrestrial source adjacent to the producing sands whereas the Type II oils were situated wherein younger and older source rocks had been brought into contact with reservoir sands.

Case 2

Three light oils were obtained from three wells (Blue well, Green well and Aquamarine well) from a Gondwana field which are about 1.7km apart. Based on the data (Table 7.8 and Figure 7.12), we have to find out whether all the three oils belong to a common source rock.

Close similarities among independent parameters (Table 7.8) such as $\delta^{13}C$ of whole oil, CPI and sulphur content of Blue well and Green well indicate that they could have a similar source. Substantially lower aromatic HC content of the Green well oil might indicate slight biodegradation. However, large difference in pristane/phytane ratios can not be attributed to biodegradation, it might instead indicate some mixing. Alternatively, the two oils could have come from different sources, with high levels of maturity being responsible for some of the gross similarities of API gravity, hydrocarbon and n-alkane distribution. Further, similarities in the gas chromatograms of the two oils (Figure 7.12, A and B) indicate their common origin. However, substantial difference in the $\delta^{13}C$ (Table 7.8) of saturated hydrocarbons (−1.6‰) between them suggests a difference in their source. Further evidence of similarities and

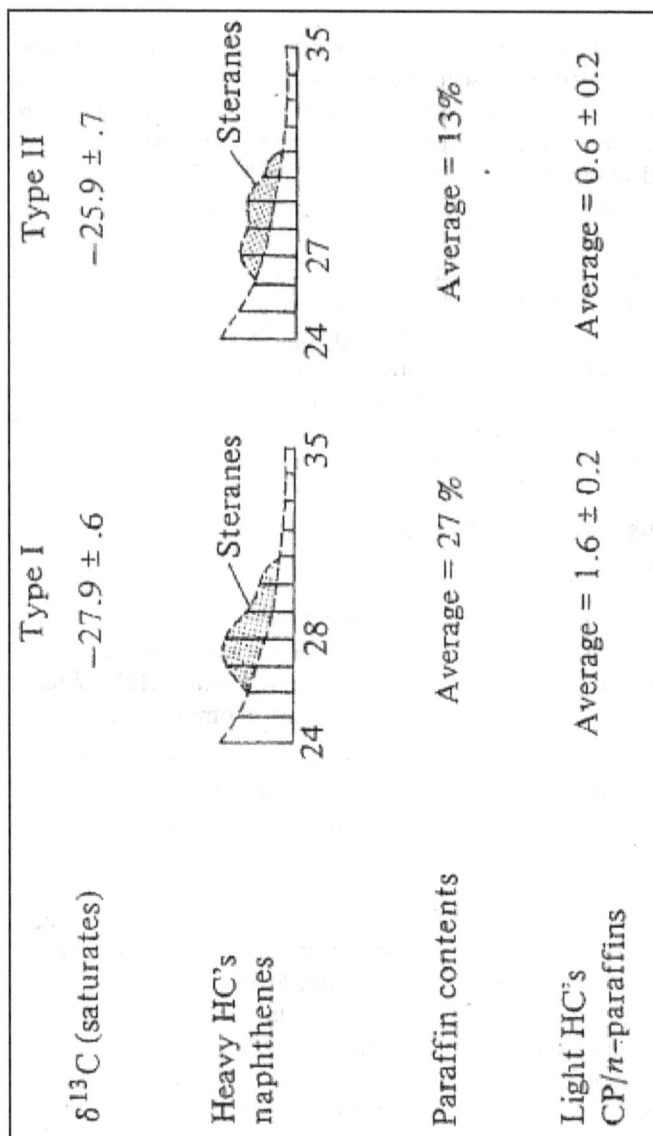

Figure 7.11: Correlation Parameters Used to Classify Lower Tuscaloosa Oils of Southern Mississippi and Alabama into two types. CP: Cyclopentanes *(Source: Koons et al., 1974)*

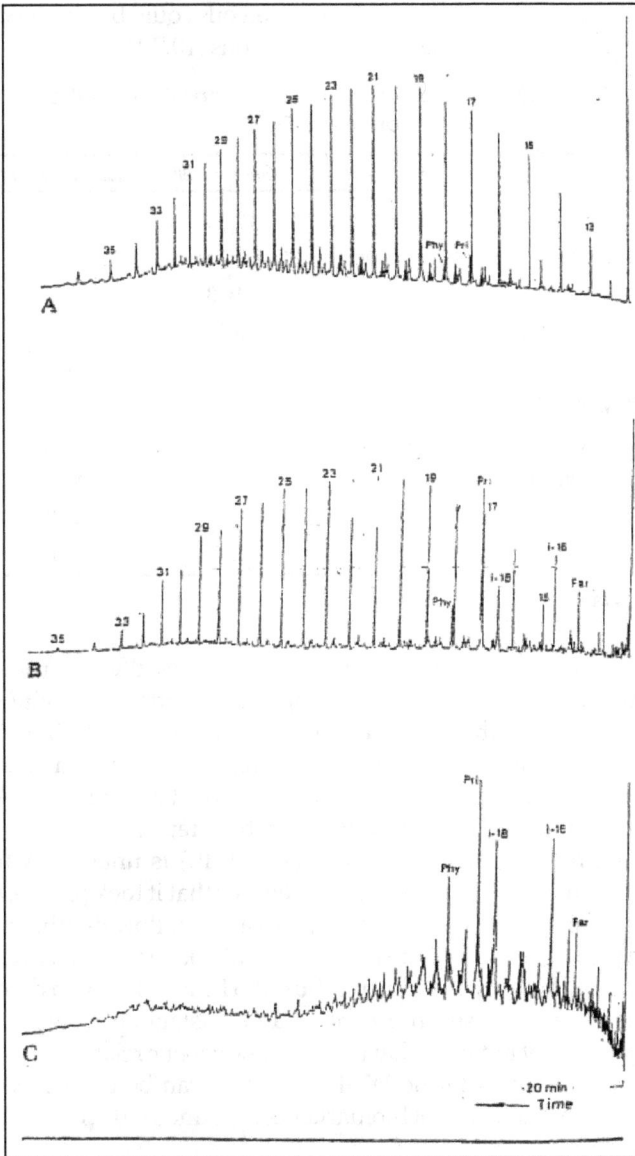

Figure 7.12: Gas Chromatograms of Saturated Hydrocarbons from Oils from the Gondwana Field: (A) Blue Well, (B) Green Well, and (C) Aquamarine Well (*Source*: Waples, 1985)

dissimilarities of the Blue and Green well oils could be obtained by GC-MS analysis of biomarker hydrocarbons (10.7.1).

Table 7.8: Geochemical Data for Three Oils from the Gondwana Field

	Blue Well	Green Well	Aquamarine Well
Depth (ft)	8100	10,200	8700
Reservoir, age	L. Cret.	E. Cret.	Eocene
API gravity (°)	46.1	44.3	32.5
Per cent Sulfur	0.12	0.08	0.21
CPI	1.01	1.02	–
$\delta^{13}C$ whole oil (‰)	–26.1	–27.0	–26.6
$\delta^{13}C$ saturates (‰)	–27.9	–29.5	–28.3
Per cent saturated HC	60.4	72.5	48.3
Per cent aromatic HC	28.5	16.3	21.6
Pristane/phytane	1.5	3.6	2.0

Source: Waples, 1985.

The oil from the Aquamarine well is clearly different from the other two wells. However, all the properties except API gravity and saturated hydrocarbons (Table 7.8) are similar to those of Blue Well oil. Biodegradation could be responsible for the absence of n-alkanes in gas chromatogram and lower API gravity of Aquamarine Well oil. Evidence for biodegradation at the depths (8,700 ft.) of Aquamarine and Green Well oils (10,200 ft.) is unusual, which suggests either low geothermal gradients or that it took place when the reservoir was at lower temperature and shallow depth. In the latter case, migration into the reservoir would have occurred prior to substantial amounts of additional burial. The fact that such diverse parameters as $\delta^{13}C$, sulphur content and pristane/phytane ratios are in agreement indicates that there may be genetic relation between the Blue and Aquamarine Well oils which can be confirmed by further GC-MS analysis of biomarker compounds (10.7.1).

Case 3

The South Hootchickootchi Basin is a prolific oil producing region with many offshore wells. A large oil slick was noticed on one day equidistant from 12 producing (A, B, C, D formations) wells. We have to identity the well responsible for oil leak. For this purpose, fresh samples of oil from each of the producing wells and the least weathered oil slick were collected. Table 7.9 indicates their geochemical data. Before we undertake detailed examination of data, we initially presume that the spilled oil has probably undergone a significant amount of evaporation and biodegradation as a result of exposure to air and sea water.

Data on API gravity and paraffin distribution may not be meaningful because biodegradation removes n-alkanes preferentially. The pristane/phytane ratios however, might not have been affected because branched hydrocarbons are not attacked as readily as straight chain ones. Sulphur content would have probably increased slightly as a result of the loss of light alkanes and light aromatics. Porphyrin contents would have increased as a consequence of the preferential loss of hydrocarbons by evaporation and microbial degradation. However, the Ni/V ratios should not have effected. By taking account into all these facts, we now evaluate the data (Table 7.9).

The pristane/phytane ratio of the spilled oil is 0.2. All those ratios greater than 1.0 (Nos 1, 5, 6, 11) which are sufficiently high can be eliminated. Oil from well number 10 with a ratio of 0.5, also probably is not the source. Further, very high pristane/n–C_{17} and phytane/n–C_{18} are most probably the result of biodegradation after the spill and hence not useful for correlation. As the Ni/V in the spilled oil is 2.7, oils whose ratios not near to this value (Nos 1,2,3,4,5,9,10,11 and 12) can be eliminated. Sulphur content of the spilled oil is higher than in any of the possible source oils. Although it increases in biodegradation, there is a limit for it (about 0.6-0.8). Hence oils with nos. 1,4, and 12 could possibly qualify.

Based on the above logical deductions, we can arrive at the conclusion that only two oils from wells nos. 7 and 8 could be the possible sources of the oil spill. Of these, oil 8 gives a closer fit for all the parameters but there is no clear choice between them. At this point, it would be necessary to carryout some further "finger printing" analysis to identify the spilled oil as one of these two suspects. It would be necessary to look at some parameters which could not have affected by water washing, evaporation or biodegradation. The next analytical step, therefore would be to take

Table 7.9: Analytical Data on Spilled Oil and Twelve Possible Sources for the Spill from the South Hootchiekootchie Basin

Well	Production Depth (ft)	Producing Formation	API Gravity (°)	Per cent Sulfur	$\delta^{13}C$, Topped Oil (vs PDB)	Porphyrins		Pristane	Phytane	Pris		CPI_{23-31}	Maximum n-paraffin (C#)
						Ni	V	$n\text{-}C_{17}$	$n\text{-}C_{18}$	Phyt			
1	8100–8150	C	30.5	0.8	−26.7	1.02	1.55	0.81	0.68	1.2		1.01	17
2	6763–6849	A	27.5	0.3	−29.7	0.00	0.03	0.03	0.43	0.1		1.13	18
3	8460–8610	C	34.5	0.1	−27.9	0.16	0.12	0.23	0.92	0.2		1.02	17
4	7998–8016	C	19.7	0.7	−26.7	1.21	1.02	0.21	0.67	0.3		1.04	16
5	9001–9202	C	31.8	1.3	−28.0	0.92	1.45	1.02	0.88	1.2		0.99	17
6	6887–6990	C	29.7	1.2	−28.1	1.02	0.34	0.92	0.91	1.0		1.07	17
7	7421–7503	C	25.7	1.7	−26.4	0.99	0.29	0.19	0.66	0.3		1.04	17
8	9023–9112	B	24.3	2.2	−26.2	0.15	0.05	0.31	1.72	0.2		0.87	18
9	6810–7020	A	28.1	0.1	−30.1	0.02	0.00	0.11	0.51	0.2		1.17	17
10	9100–9140	D	36.1	0.2	−28.4	0.15	0.13	0.49	0.96	0.5		1.06	18
11	9321–8520	D	24.3	1.7	−27.2	0.88	0.21	0.97	0.79	1.2		1.03	16,
12	8169–8230	C	21.2	0.6	−26.9	1.15	0.96,	0.16	0.81	0.2		0.98	17
Spill Oil	–	–	12.7	3.7	−25.7	0.72	0.27	3.65	16.1	0.2		–	–

Source: Waples, 1981.

up the branched cyclic fraction of saturated hydrocarbons (steranes and triterpanes), and subject to GC-MS analysis of the two oils (nos.7 and 8). Based on these analysis, Seifert and Moldowan (1979) conformed that oil in Well No. 8 was the culprit for the oil slick.

7.9.2 Oil-Source Rock Correlations

Case 4

The Green Well was drilled about 150 km north-west of the Mauve Well. Two depth intervals namely, 2,500–3,000 m and 4,500–4,600 m were considered as possible oil source rocks in the Mauve Well. Based on the data given in Table 7.10, could there be a genetic relationship between the Green Well oil and the source rock strata (bitumen) analyzed in the Mauve Well?

The $\delta^{13}C$ values for the kerogen of Mauve rocks in the 4,500-4,600 m depth interval are 5 to 6‰ higher than those for the Green Well oil. This difference is too large to have been caused by thermal effects or migration. Further more, pristane/phytane ratios, and sterane contents are totally different. The n-paraffin distribution differ substantially-the deep samples of Mauve Well have relatively large amounts of heavy homologs and high CPI values, pointing to a significant contribution from terrestrial waxes. On the basis of this information it can be concluded that the source rocks in the 4,500-4,600 m depth interval can not be the source of Green Well oil.

The 2700 m depth samples, however, show a much closer correlation. $\delta^{13}C$ values are more or less compatible differing by 0.8 and 1.9‰ for bitumen and kerogen respectively of Mauve rocks and Green Well oil. Porphyrin content in 2,700 m sample correlate reasonably well when migrational loss is taken into account. C_{29} steranes/cholestane ratios and n-paraffin distributions are similar. We therefore conclude that the rocks found in the 2,700 m depth interval of the Mauve Bitumen correlate well with the Green Well oil. On the other hand, the 3,000 m rock do not correlate so well even though the differences are not large enough to completely rule out a genetic relationship.

Case 5

Figures 7.13 and 7.14 show the hydrocarbons distributions in the C_4–C_7 fraction, and C_{15}+ n-paraffin of three types of oil, namely Type I, Type II and Type III produced from Winnipeg, Baken, and

Table 7.10: Analytical Data for Green Oil and Mauve Bitumens

Well	Depth (m)	$\delta^{13}C$, ‰ (vs. PDB)	Prophyrins Ni	V	n-paraffin Distribution	Pristane / Phytane	C29 Steranes / Cholestane
Green	1200	−29.3 oil	0.11	0.15	*(histogram, 10 20 30)*	1.15	0.34
Mauve	2700	−28.5 bit. −27.4 ker.	0.42	0.44	*(histogram, 10 20 30)*	1.00	0.41
Mauve	3000	−26.3 bit. −26.1 ker.	0.21	0.60	*(histogram, 10 20 30)*	1.51	0.19
Mauve	4500	−24.7 bit. −23.8 ker.	0.72	0.35	*(histogram, 10 20 30)*	6.41	1.12
Mauve	4600	−25.5 bit. −24.4 ker.	0.41	0.52	*(histogram, 10 20 30)*	5.28	0.87

Source: Waples, 1981.

Tyler respectively and their source rock extracts of Williston Basin. Is it possible to identify any genetic relation between the oils and the source rocks?

Comparison of hydrocarbon type distribution in the C_4–C_7 fraction (Figure 7.13), and the distribution of C_{15}+ n–paraffins (Figure 7.14) of Winnipeg Shale extract all correlate well with Type I oils. Characteristic predominance of C_{15}+ and C_{17} n–paraffins in Type I oils is quite evident in the Winnipeg Shale. The $\delta^{13}C$ values of Bakken

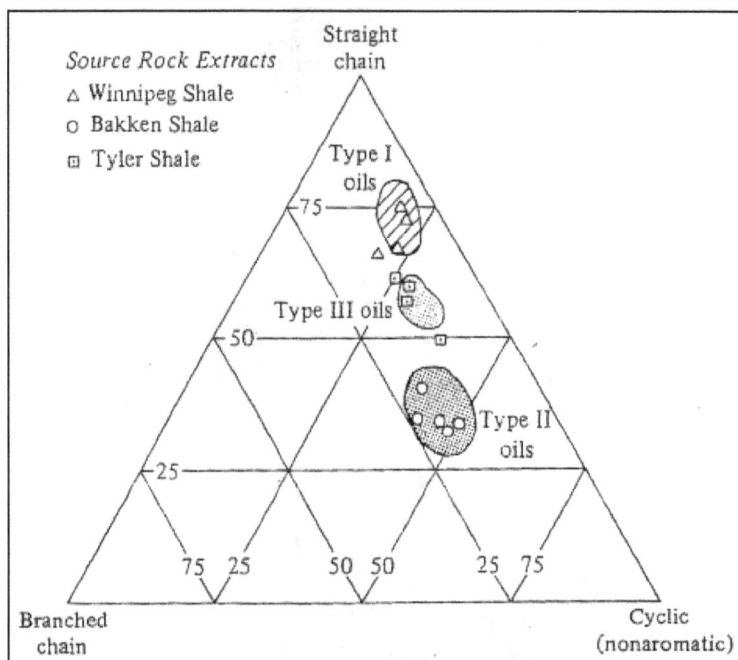

Figure 7.13: Hydrocarbon Type Distribution in
C_4–C_7 Fractions of Source Rock Extracts and Crude Oils
from Williston Basin. Crude oils are inside circled areas.
(*Source*: Williams, 1974)

Shale extract and Type II oils were both in the range –29 to –30‰ indicating similarities between them. Similar correlations were evident between Tyler Shale extract and Type III oils. Based on this information Williams (1974) could categorise the oils of the Williston Basin into three genetic types and relate each type to a specific source sequence.

7.9.3 Gas Correlations

Gases are produced from many fields in the Po Basin of Italy. Reservoir age ranges from Pleistocene to Pre-Miocene and depth of production ranges from 170 to 4,500 m. Maximum reservoir temperatures are about 75°C. Thermal maturity data indicates Ro (max) at about 0.6 per cent at 5000 m depth. A plot of gas wetness

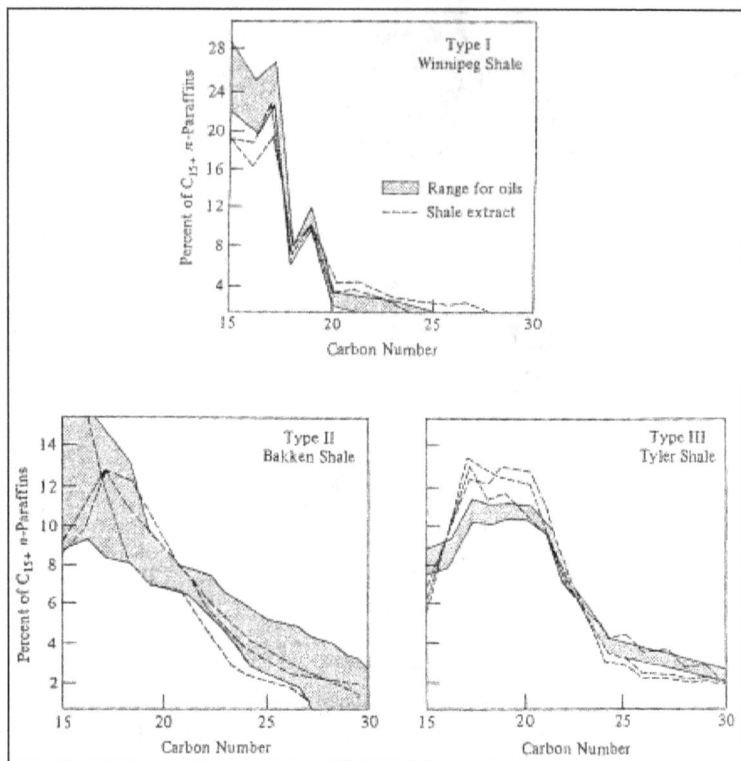

Figure 7.14: Comparison of C15+ *n*-paraffin Distribution of Source Rock Extracts and Crude Oils from Williston Basin (*Source*: Williams, 1974)

versus $\delta^{13}C$ (Figure 7.15) shows a wide range of composition (Mattavelli *et al.*, 1983). From the data can we infer the origin of gases in the Po basin?

Geothermal gradients, subsurface temperatures and maturities are all very low in the Po Basin. Thus if thermally generated gas is present in the reservoirs about 5,000 m, it must have migrated vertically from more mature, older strata at greater depths. Figure 7.15 indicates that most of these gases are predominantly biogenic containing more than 99.8 per cent methane having $\delta^{13}C$ values more negative than 60‰. There is however, a trend that is probably attributed to mixing of thermogenic gas with the biogenic gas; it

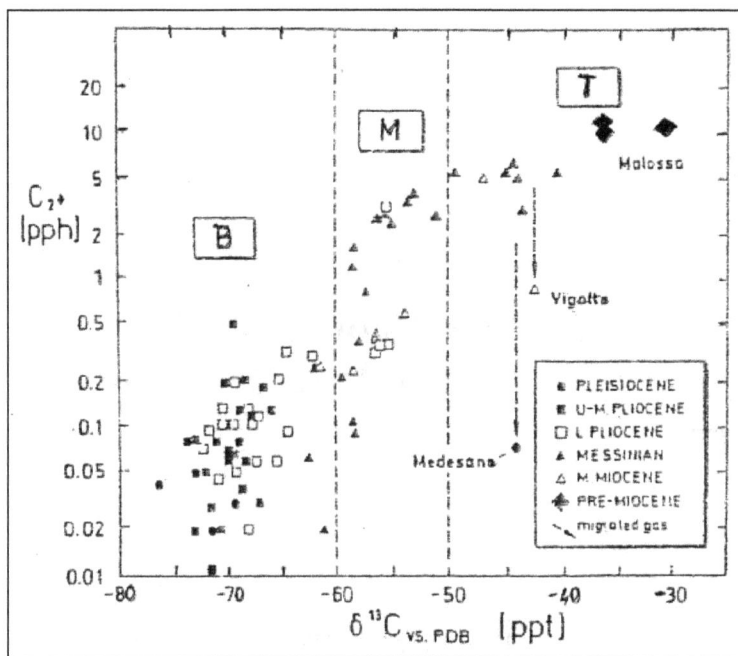

Figure 7.15: Wetness versus Carbon-isotopic Composition for
Natural Gases from the Po Basin, Italy
(*Source*: Mattavelli *et al.*, 1983)

runs from the lower left (B) towards the upper right. Those gases
with $\delta^{13}C$ values less negative than about 50‰ are probably mainly
thermogenic. Most of these gases are from Pre-Pliocene rocks. The
wettest, heaviest (purely thermogenic) gases are those from the Pre-
Miocene.

Chapter 8

Surface Geochemical Prospecting of Hydrocarbons

The term geochemical prospecting refers to methods for finding petroleum, based on the supposition that some hydrocarbons in an oil and gas accumulation migrate vertically to the surface over the reservoir. This is based on the fact that no reservoir seal is prefect and hence hydrocarbon accumulations do leak them to the surface. Although large seepages may be visible to the nacked eye, small and slow micro seepages can only be detected with sensitive instruments.

8.1 Seepages

Visible oil and gas seeps are important in exploration because they indicate the source rocks or reservoir oil or gas. A seep can be defined as visible evidence at the earth's surface of the present or past leakage of oil, gas or bitumens from the sub-surface. This definition does not include micro-seepages. Many large seeps represent tertiary migration, that is migration from an accumulation that has been disturbed by tilting of strata, changes in depth of burial or development in new avenues of escape to the surface such as fracture or fault system. The importance of seeps is reduced to a large extent, in the present exploration because of the use of highly sophisticated instrumentation as well as limited use of ground surveys. Nevertheless, many if not most important oil producing regions of the world were detected or discovered through surface oil and gas seeps (Hunt, 1996). When petroleum leaks to the surface, it undergoes a series of physical and chemical changes that considerably alter its appearance and composition. Some of the changes are:

(i) *Evaporation*: In the first two weeks after an oil comes to the surface, it loses its more volatile hydrocarbons upto about C_{15} equivalent to a boiling point of 250°C. In subsequent months, additional hydrocarbons upto C_{24} are lost.

(ii) *Leaching*: The most water soluble hydrocarbons along with some lighter aromatics, may be leached out by ground waters.

(iii) *Microbial degradation*: The n-paraffins, some isoparaffins and some naphthenes leaking to the surface are subjected to microbial attack and oxidised to various extents depending on the suitability of the environment for microbial activity.

(iv) *Polymerisation*: Some hydrocarbons combine to form bigger aggregates of large complex polymers after elimination of CO_2, H_2O and hydrogen.

(v) *Auto oxidation*: Many constituents of petroleum absorb sunlight and oxygen resulting in the formation of oxidised polymers. Seeps can take up as much as 6 per cent of oxygen on long exposure to air and sunlight.

(vi) *Gelation*: With some types of hydrocarbons, the formation of a rigid gel structure may occur over prolonged time (Dickey and Hunt, 1972).

All these reactions lead to thickening or solidification of the original oil. It is gradually converted to an asphalt, asphaltite and eventually a pyrobitumen. Consequently, unless a seep is supplied by continuous flow of sub-surface fresh oil, it will ultimately harden to a black bitumen deposit.

8.1.1 Classification of Seeps

Seeps are classified for field operations into three groups:

(i) active or live seeps which may be gas, light oil, heavy oil or sticky black asphalt; (ii) inactive or dead seeps, which are generally asphaltites or pyrobitumens not connected to any liquid; (iii) false seeps which are materials that may have the appearance of active or inactive seeps, but actually are in no way related to sub-surface hydrocarbon accumulations (Hunt, 1996).

Active seeps are important for a detailed surface and subsurface studies to determine their relationship to the lithology, stratigraphy and structure. They should be analyzed for their hydrocarbon range, n-paraffin content, H/C ratio and oxygen content (Chs. 6 & 10). Such data will indicate the degree of weathering and the probability of the seep coming from an actual underground reservoir. Detailed gas chromatography–mass spectrometry analysis [10.7.1] will assist in correlating the seep with known crude oils/or source rock extracts in the area.

Inactive seeps in general, leave a residue of asphalt that mature with time into asphaltite such as grahamite or to a pyrobitumen such as impsonite and anthraxolite. The largest known grahamite vein in the world occurs near–Tushkahoma, Oklahoma in a fault zone with shaley sandstone. The shale is 1.6 km long with a thickness upto 7.6 m. This and other Oklahoma grahamite deposits represent ancient oil seeps that have long since become inactive (Hunt, 1996).

8.1.2 Categories of Seeps

Seepages are more prevalent on the margins of basins and in sediments that have been folded, faulted and eroded. Link (1952) categorized seeps into the following five types depending on their origin:

(i) Seeps arising from their homoclinal beds, the ends of which are exposed when they reach the surface.

(ii) Seeps associated with beds and formations in which oil was formed.

(iii) Seeps from large petroleum accumulations that have been bared by erosion or reservoirs ruptured by faulting and folding.

(iv) Seeps at the outcrops of unconformities.

(v) Seeps associated with intrusions such as mud volcanoes, igneous intrusions and piercement salt domes.

Young tertiary sediments in tectonically active areas produce most of the seeps. Seeps are also associated with Cretaceous, Miocene and Eocene sediments. There is a correlation, worldwide, between seeps and earthquake activity, with majority of them being close to plate boundaries where such activity is the highest. For example, the western coast of South America has seeps in Ecuador, Peru and Chile that follows the earthquake belt resulting from the sub-duction of the Nazca plate beneath South America. Other areas where numerous seeps are related to earthquake activity of plate boundaries include: Trinidad, Southern California, Southern Alaska, Philippines, Indonesia and Burma (Hunt, 1996).

Active seepages are often found offshore on the continental shelf from formations that produce oil on the nearby land area. Seepages spreading into the surface takes two dominant forms-vast oil slicks, and floating tar balls. Gas may escape at the surface without forming visible deposit as oil, but can manifest itself in several ways such as fires, clays with appearance of brick red colour, and mud volcanoes. Landes (1973) has summarized offshore seeps including areas such as the Gaspe Peninsula of Quebec, US Gulf Coast, Gulf of Paria, Red Sea, Antarctica, Coasts of Alaska and Canada, and the South China Sea.

8.2 Surface Geochemical Prospecting

Surface geochemical prospecting techniques are most effective as part of an integrated programme that include geophysical (seismic) techniques. Collins *et al.*, (1992) presented an integrated approach that utilize surface geochemistry first as a screening tool and then as a follow up when an anomaly was found by seismic survey. This process can be classified in to two steps: (*i*) petroleum seepage identification (surface geochemical) and (*ii*) trap identification (seismic). The effectiveness of surface geochemical prospecting depends on chosing a technique that will be able to prove or disprove the presence of micro seepages in an area for which the critical factors are survey design and sample density. An insufficient number of samples or unreasonable areal distribution limits proper identification of hydrocarbon prospects. Large data sets will eliminate the need to rely on statistics as the sole means of interpretation. On the other hand, insufficient sample density can cause interpretation meaningless or sometimes mislead identification of an accumulation. Several factors involved in chosing a proper technique include soil conditions (composition, clay type, Eh, pH, moisture content etc), topography and budget considerations.

One complicating factor in the interpretation of surface prospecting data is that seepages are not always located directly above hydrocarbon accumulations. If the seeping hydrocarbons encountered a fault or permeable conduit, their vertical movement may develop a significant lateral component as well. Further, in some cases "hallow or chimney" effect is noted (Figure 8.1) in which the surface anomaly is seen to surround the actual accumulation. The halos have been attributed both to plugging of diffussional pathways immediately above the accumulation and to low reservoir pressure. A more likely explanation is that halos are reflections to topography and surface geology of the area (Tedesco, 1995).

Another complicating factor in interpreting surface prospecting data is deciding which concentrations represent anomalies and which are background values. It is imperative that this decision be made objectively and that the background be determined independently in each area. It is not always correct to assume that the highest values in an area represent anomalies. One way by which anomalies can be identified objectively is plotting the raw data on probability paper. A Guassian distribution of the raw values, which might be expected if all samples were from a single population, gives a straight line when plotted on linear probability paper. If two or more data populations exists, however, a separate line will appear for each population (Figure 8.2). Inflection points occur between the populations and mark the approximate limiting values for each population. The three segments are defined by distinct line segments connected by a transition zone containing an inflection point which approximately defines the limits of each population (Waples, 1985).

8.3.0 Categories of Surface Geochemical Methods

Two types of surface geochemical methods are commonly employed in hydro- carbon prospecting. These include: (*a*) direct and (*b*) indirect methods. The direct methods involve detection and measurement of hydrocarbons (soil–gases) that are expelled from the subsurface into soil substrate. The indirect methods encompass all other forms of byproducts (elements or compounds) of soil–gas reactions with the soil substrate and atmosphere. In general, indirect methods are more likely to detect a solid or a liquid product than gas. A combination of indirect and direct methods is recommended inorder to have checks and balances in prospect evaluation.

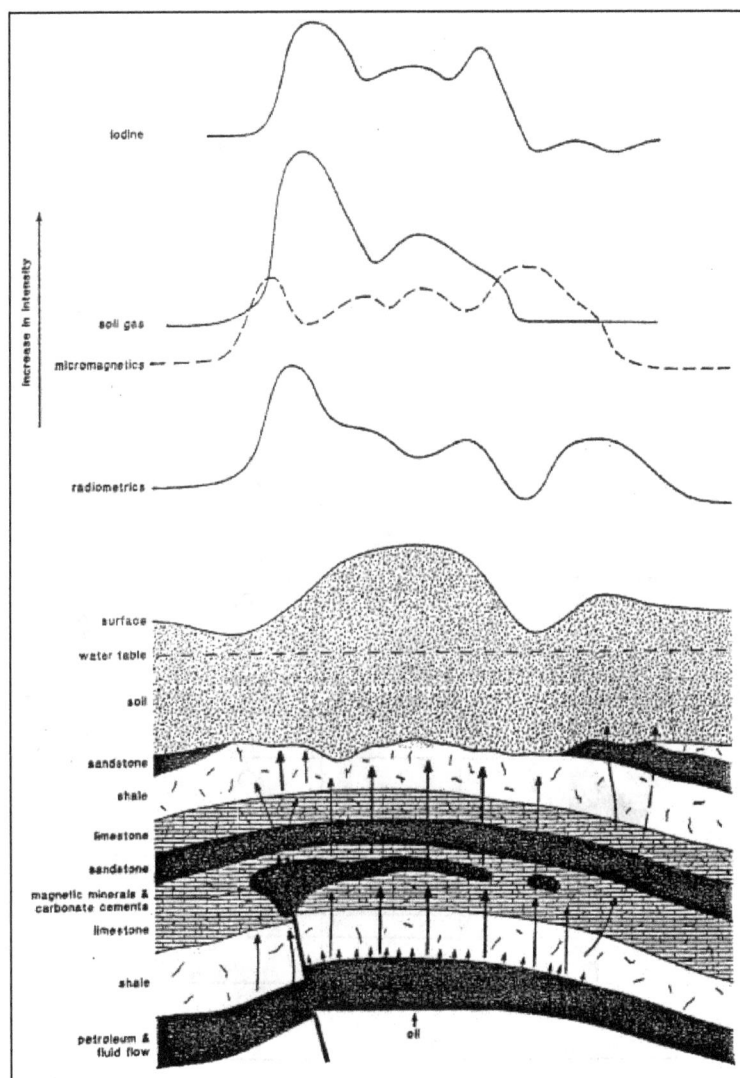

Figure 8.1: A Generalized Cross Section through the Petroleum Accumulation and Chimney Effect Caused by Migrating Hydrocarbons. The surface expressions of various geochemical methods are also indicated (*Source*: Tedesco, 1995)

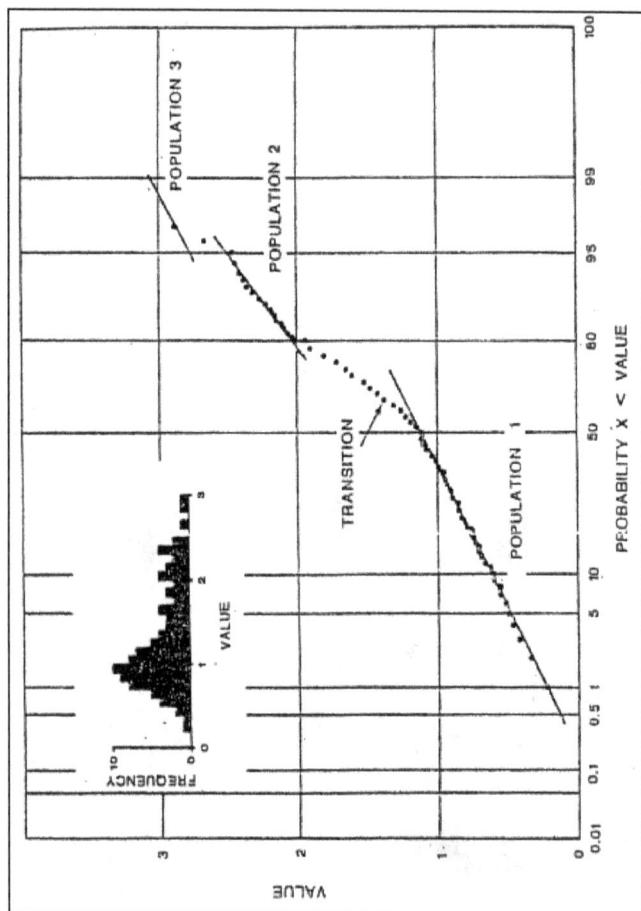

Figure 8.2: Method for Establishing the Existence of Two or more Distinct Populations of Data Values by Plotting Raw Data (histogram, upper left) on Probability Paper (*Source:* Waples, 1985)

Following are some of the methods for detection and measurement of micro-seepages on the surface.

(*i*) Soil gas (light and heavy hydrocarbons)

(*ii*) Radio nuclides (^{40}K, ^{232}Th, ^{238}U and ^{222}Rn)

(*iii*) Halogens (iodine)

(*iv*) Stable isotopes ($\delta^{13}C$, $\delta^{18}O$) and magnetic minerals of iron.

(*v*) Microbial methods

(*vi*) pH–Eh methods

Methods based on stable isotopes, magnetic minerals of iron and pH/Eh methods reflect processes other than seeping hydrocarbons. There are several scenarios in which they would prove inefficient. Hence they are not discussed further in this chapter. Among others, radiometrics and microbial techniques have limited use in some areas but they can be of second priority or followup tools. On the other hand, soil-gas and iodine methods are the best that consistently give good results (Tedesco, 1995).

8.3.1 Soil–gas Methods

Among them free air and head space are more common which directly sample the soil or sediment substrate for hydrocarbons. In free air methods, hydrocarbon gases are extracted from the well connected interstitial pores of the soil. The head space technique measures hydrocarbon gases adsorbed on to, or loosely bound by clay and the organic matter. Both the methods use special types of samplers, and they can be applied to both onland and offshore hydrocarbon prospecting.

8.3.2 Light and Heavy Hydrocarbon Prospecting

Light hydrocarbons (C_1–C_4) have long been among the most popular of surface prospecting tools because of several reasons: (*i*) as hydrocarbons they are closely related to the accumulations being sought, (*ii*) they are the indicators of present day seep activity implying that they represent modern rather than ancient accumulations, and (*iii*) collection and measurement are relatively easy because of their mobility and volatility. However, one difficulty with them is the presence of dominant proportions of methane (particularly biogenic) derived from non-petroleum origin such as sediment and soils. In view of this, higher hydrocarbons such as ethane, propane or butane are preferred under light hydrocarbon measurement. They are usually collected as soil-gas from onland samples using a syringe or by canning of rock or soil and later removing the head space gas. Gas chromatography is commonly used (10.6.0) for quantitative estimation. The data are then plotted on a map showing sample locations and contours containing high concentrations of hydrocarbons. Richers *et al.*, (1982) and, Jones and Drozd (1983) while analyzing light hydrocarbons (propane and butane) on soil–gas and head space gas samples and adsorbed gases, observed largest anomalies along faults immediately above oil accumulations.

Light hydrocarbon sampling in offshore surveys is analogous to onshore sampling. Sediment samples are obtained by piston coring or grab sampling and then they are canned or frozen until gas analysis can be carried out. Since biogenic methane is a much serious problem in sub-aqueous sediments than in most soils, analytical procedures must be able to separate small amounts of higher hydrocarbons from much larger quantities of methane. Another technique for measuring light hydrocarbon concentrations in offshore areas is to sample the bottom water where its composition reflects the rates at which seepages are occurring directly below. Two American companies namely "Inter Ocean System" and "Hydrochemical Surveys International" have developed devices to collect bottom sea water samples (Waples, 1985). In the former, a device (sniffer) towed beneath a seismic vessel pumps water continuously on board the ship, where hydrocarbon gases dissolved in water are analyzed by gas chromatography. The second system is similar to the first except that it analyzes those samples high in methane for carbon isotopic composition ($\delta^{13}C$) in order to distinguish its biogenic or thermogenic origin (6.4.0).

In a technique patented by M/s Petrex, the total quantity of heavy hydrocarbons (having more than ten carbon atoms) adsorbed and their molecular sizes can be analyzed from onshore samples. The technique consists of placing vials just below the surface of the ground for a few days. The vials contain a small wire made up of special metal–alloy, one end of which is coated with activated charcoal. Any hydrocarbons emanating from the earth during the collection period are adsorbed on the charcoal. The samples are then retrieved and analyzed by inserting the wire directly in to the ionizing chamber of a mass spectrometer. When an electric current is passed through the wire it heats up almost instantaneously, the absorbed hydrocarbons are vapourised, and the characteristic mass spectra is recorded. Though the technique is more expensive than light hydrocarbon analysis, it may prove very useful in selecting drilling locations or in chosing several available structures (Klusman and Voohees, 1983).

8.3.3 Iodine

Among the halogens that closely associate with petroleum derived organic compounds, iodine has been specifically used in the search of naturally occurring hydrocarbons because it has a strong affinity for them. Sampling of soil or sediment for iodine is typically done within 2-6 inches of the surface. The sample is dried and the clay fraction (<2μm) is separated for estimation of iodide or I_2. The resulting values presented as molecular iodine (I_2) in ppm (by weight) are plotted and then contoured to determine if an anomaly exists. Typical background values of iodine range from 0.5 to 6.0 ppm with an average of about 2 ppm. Anomalous values are usually 150 to over 1000 per cent higher than the background values.

An iodine survey requires large number of samples to establish the background before identifying anomalous values. It is much simpler to acquire, handle and analyze iodine samples than to collect and analyze soil-gas. Soil conditions as mentioned earlier (8.2.0) do not appear to be a factor as long as the survey is restricted to upper 2 inches of soil. An advantage of using iodine is its reproducibility. Unlike gases or liquids, the organo-iodine compounds (methyl or ethyl iodide) are relatively stable and they are not easily affected by fluctuations of pressure, wetting or drying of soils or changes in the water table. The only disadvantage of this method is the necessity of large number of samples for an accurate interpretation and the need to have a model for comparison. Gallagher (1984) while using this method observed high iodine concentrations (halos) surrounding the hydrocarbon accumulations at several locations.

8.3.4 Radiometrics

Radiometrics detect changes in the total gamma ray count at the surface. The changes result in the radioactive decay of potassium (^{40}K), Thorium (^{232}Th) and Uranium (^{238}U) series. Radiometric anomalies detected over a petroleum accumulation represent a phenomenon that is present only in the upper few millimeters (mm) of the soil. Whether these anomalies extend down to the petroleum accumulation is not known. However, several researchers have documented increased shale radioactivity in the strata overlying petroleum accumulations.

Radon (^{222}Rn) is a short lived decay product derived from either Uranium–238 or Thorium-232 with a half–life of 3.82 days. It is the only gaseous element in the decay chain useful in soil-gas analysis. The short life limits the distance that radon can migrate or be transported and, thus, can be used as an effective tool. Radon surveys have been applied to areas of earthquake activity. However, a few anomalies are also found in association with fault or highly permeable strata. Radiogenic helium (^{4}He) derived from α -decay of naturally occurring radio nuclides such as Uranium-238 and Thorium –232 may be useful as in the case of ^{222}Rn in delineating fracture and fault systems (Sikka, 1959). However, it is not useful in specifically determining the presence or absence of petroleum accumulations.

Radiometric (including radon) methods are viable to a limited extent as exploration tool. The major problem with them is that numerous radioactive sources present must be identified and removed from the data. Corrections must be made for non-radiometric factors such as topography, weather and moisture, and complete success is never achieved. Experience in interpreting the results is essential. Any radiometric anomaly should be verified by other surface geochemical or geophysical methods, regardless of the confidence that all the extraneous sources have been eliminated. As an exploration tool, radiometric methods are best used for reconnaissance rather than identifying exactly a prospect or a oil well site location.

8.3.5 Microbial Methods

Bacteria that digest hydrocarbons have been used to find petroleum accumulations. The basis for microbial detection of petroleum lies in the ability of bacteria to metabolise (oxidise) some portion of the hydrocarbons migrating through and out of the soil substrate. Two types of microbial populations mainly exists in the soils : (i) aerobic bacteria which requires oxygen for their growth, and (ii) anaerobic bacteria which grow in the absence of oxygen.

Among them the first category is important for hydrocarbon prospecting. It is generally believed that most hydrocarbons in the soil have been derived from petroleum migrated from depth. However, methane can be derived from humification process (12.6.1) in which microbes convert plant material to humus in the soil. Oxidation is the most common form of organic degradation in soil resulting in the production of CO_2, water and acetic acid. Apart from methane, generation of ethane, and in some cases propane and butane by bacterial or fermentation processes may itself account for the presence of microbes (Juranek, 1958). Anamolous bacterial accumulations can also be caused by other sources such as hot ground water or hydrothermal fluids that come into contact with organic rich rocks.

Methane oxidisers were the first type of bacteria studied to identify the locations of petroleum hydrocarbons. The presence of methane oxidising bacteria in the soil has been interpreted as methane exhaled from the sub-surface. However, this method has a serious limitation in petroleum exploration in that methane can be generated by microbes in the soil rather than by migrating hydrocarbons. Coleman *et al.*, (1988) indicated that microbial methane can be distinguished from petroleum generated methane by carbon and hydrogen isotope analysis. Subsequently specific bacteria which consume ethane, propone or butane but not methane were developed and used in microbial prospecting (Davis, 1952; Price, 1986).

Sampling depths for surface microbial prospecting vary from 0.2 to 2 m. Several methods are available for determining the presence of anomalous hydrocarbon digesting microbe populations in the soil. All methods require an incubation period of more than two weeks for completion of analysis. One of the earliest methods consisted of placing a sample of soil in a container with hydrocarbons and oxygen. In this environment, the microbes oxidise the hydrocarbons. The possibility of oxygen consuming bacteria interfering with petroleum–consuming bacteria can be taken care of by using control samples that are free of hydrocarbons. In subsequent development, a gas chromatographic detection (flame ionization detector) is used for analysing the change in the hydrocarbon content of the air in the container. However, this method is not effective in detecting small differences in bacterial populations between different samples.

In another method, a cultured soil micro organisms placed in a bottle is mixed with inorganic nutrients and a mixture of gases containing 65 per cent light hydrocarbons, 30 per cent oxygen and 5 per cent carbon dioxide. If petroleum digesting microbes are present in the soil, the gas volume will be reduced. Hydrocarbon consumption is assumed to have occurred if the volume of gas is reduced to less than the volume that can be accounted for by complete utilization of the oxygen and carbon dioxide. Gas chromatography can be used to determine which hydrocarbons were used. A short traverse utilising a device designed for this purpose can be extended across the edge of a field. The data can be plotted to indicate the hydrocarbon consumption over the field.

In addition to the above two methods, there are several patents available on microbial methods. Tedesco (1995) while reviewing the microbial methods was of the opinion that they are effective tools in certain situations. However, problems arise because of several factors such as survey design, sample collection and analytical methods. Application of statistical treatment/techniques are essential to distinguish background and anomalous values which may vary from area to area and through time. Hence the methods are not applicable in all exploration situations or geological environments. Philp and Crisp (1982) reviewed in detail surface prospecting techniques used in hydrocarbon exploration and arrived at similar conclusions on microbial methods.

Surface geochemical prospecting can be useful as a regional tool to indicate the general area of oil or gas accumulation, provided information is available on the geology, the effect of near surface diagenetic hydrocarbons, and the local and regional fluid flow systems in the sediments. Probably the most valuable use of surface prospecting is in identifying low levels of seepage associated with faults, fractures, unconformities and intrusions such as piercement salt domes, mud diapers, and igneous intrusions (Hunt, 1996).

In conclusion, it may be stated that surface prospecting techniques are highly empirical. Successful application of any one technique is likely to be highly dependent upon local factors. One important limitation with all these techniques is that none is capable of specifying whether a leaking accumulation is deep or shallow or whether it will be commercial. The magnitude of modern surface anomalies depend on rates of leakage, not on the size of the accumulation. Any area with multiple pay zones is likely to be difficult to evaluate the surface prospecting, unless one is already committed to drill to the deepest possible reservoir. With these limitations, one should consider the use of surface prospecting wherever it seems appropriate as an auxiliary technique to conventional geological and geophysical exploration methods (Hitchon, 1974).

Chapter 9
Hydro Geochemistry of Oil Field Waters

Water associated with petroleum in a subsurface reservoir is called an oil field water. By this definition any water associated with a petroleum deposit is called an oil field or formation water. The nature and quantity of waters associated with oil or gas bearing regions are vital for exploration and exploitation. Knowledge of their chemical composition and physical properties is important in understanding the origin and evaluation of the interactions of chemical constituents of oil field waters with petroleum and the reservoir rocks. Normal sea water contains about 3.5 per cent (35,000 ppm or mg/litre) of dissolved salts predominantly in the form of sodium chloride. Many oil field waters contain much more dissolved salts than sea water *e.g.*, upto 10 per cent (1,00,000 ppm) or even more. About 70 per cent of the world petroleum reservoirs are associated with waters containing more than 10 per cent of dissolved salts which are classified as brines. Waters associated with the remaining 30 per cent of petroleum reservoirs contain less than 10 per cent of dissolved salts, some of which are almost fresh water (less than 0.1 per cent).

The composition of dissolved salts in oil field waters depends on several factors. Some of them are: (*i*) composition of water in depositional environment of source rocks, (*ii*) subsequent changes by rock-water interactions during sediment compaction, (*iii*) changes by rock-water interactions during water migration (if migration occurs), and (*iv*) changes by mixing with other infiltrating younger waters.

9.1 Types and Definitions of Oil Field Waters

Oil field waters may be divided genetically into three types namely, (*a*) meteoric water, (*b*) juvenile water, and (*c*) connate water (Selley, 1998).

9.1.1 Meteoric Water

It is the water that has fallen as rain water and filled up the pores and permeable shallow rocks or percolated through them along bedding planes, fractures and permeable layers. Such water is relatively fresh containing less than 1,000 ppm of dissolved salts (low salinity), moderate concentration of dissolved oxygen (oxidizing) and acidic (pH<7). Meteoric waters have higher concentration of bicarbonate and sulphate and lower concentrations of calcium and magnesium ions than connate water (9.1.2). Since they often contain both dissolved oxygen and microbes, they can react chemically and biologically with any oil accumulation resulting in its degradation.

9.1.2 Connate Water

The term connate implies born, produced or deposited with the sediment from the beginning. By this definition, it is a fossil water that has been out of contact with the atmosphere for longer part of a geological period since its deposition. It can probably be also considered as an interstitial water of syngenetic type (9.1.4). Other definitions proposed for connate waters include "interstitial waters existing in reservoir rock prior to the disturbance by drilling" (Case, 1956), and " waters which have been buried in a closed hydraulic system and have not formed part of the hydraulic cycle for a long time" (White, 1957). Connate waters contain dominantly chloride with almost no bicarbonate or sulphate. They differ from sea water both in salinity and chemistry. Connate waters contain lower percentage of magnesium and often calcium (possibly caused by the precipitation of anhydrite, dolomite and calcite) and a large percentage of sodium and potassium than does sea water.

9.1.3 Juvenile Water

Juvenile water is water that is in primary magma or derived from it. It is difficult to prove that such hydrothermal water is indeed primary without any contamination from connate water.

The above three definitions lead to a fourth class of sub-surface waters-those of mixed origin. They may be formed by all the three or any two types of waters defined above. Oilfield waters may also be categorised by the manner of their occurrence as (a) interstitial water, (b) diagenetic water, and (c) free water (Collins, 1980).

9.1.4 Interstitial Water

Interstitial or sometimes called as porewater is the water contained in the small pores or spaces between the minute grains or units of rock. It is of two types: (i) syngenetic-formed at the same time as the enclosing rock, and (ii) epigenetic-originated by subsequent infiltration into rock. Interstitial water is present in all petroleum reservoirs, occupying 10-50 per cent or more of the pore space. Its percentage generally increases with decrease of both permeability and porosity.

9.1.5 Diagenetic Water

Diagenetic waters are those that have changed chemically and physically before, during and after sediment consolidation. Some of the reactions that occur in or to diagenetic water include: bacterial, ion-exchange, replacement (dolomitisation), infiltration by permeation and membrane filtration.

9.1.6 Free Water

The water that naturally occurs in the rocks other than meteoric or connate or mixed water, which is free to move in and out of pores in response to pressure differential is called free water. It is considered as a continuous and inter connected body of water in which mineral particles have been deposited. It paves the way along which petroleum moves to concentrate into a pool.

9.2 Classification of Oilfield Waters

Oil field waters are classified according to dominant mineral ions present in them into two categories namely (a) Palmer's and (b) Sulin's classification.

9.2.1 Palmer's Classification

It is the earliest classification (Palmer, 1911) which is out of use now. It is based on groups together with radicals dissolved in water that are either chemically similar or geologically associated. According to this classification, oil field waters are divided into four types based on the combination of the following anions and cations:

Type 1: Primary Salinity (S_1)

Chloride, Sulphate (Cl^-, SO_4^{2-})–Sodium, Potassium (Na^+, K^+)

Type 2: Secondary Salinity (S_2)

Sulphate (SO_4^{2-})–Calcium, Magnesium (Ca^{2+}, Mg^{2+})

Type 3: Primary Alkalinity (A_1)

Carbonate (CO_3^{2-})–Sodium, Potassium (Na^+, K^+)

Type 4: Secondary Alkalinity (A_2)

Carbonate (CO_3^{2-})–Calcium, Magnesium (Ca^{2+}, Mg^{2+})

Waters of secondary alkalinity (A_2) cause hardness, while waters of primary alkalinity (A_1) are termed as soft waters.

9.2.2 Sulin's Classification

It is a widely used classification at present. Sulin (1946) distinguished four types of oil field waters based on the distribution of three dominant anions and cations.

Type a: Sulphate (SO_4^{2-}–Sodium (Na^+)

Type b: Bicarbonate (HCO_3^-)–Sodium (Na^+)

Type c: Chloride (Cl^-)–Magnesium (Mg^{2+})

Type d: Chloride (Cl^-)–Calcium (Ca^{2+})

Types c and d also contain of course, Na^+. Most oil field waters including all typical 'brines' fall into type d. While Na^+ dominates among the cations, both Mg^{2+} and Ca^{2+} are present in addition, the Ca:Mg ratio being about 5:1. Chloride is the sole dominant ion while SO_4^{2-} is notably absent in type c and d waters. Waters of this type characterize deep stagnant basins in continental margins where there is little outcrop to permit invasion of surface waters and create a hydraulic head. Type c waters takes the place of type d in many evaporite bearing sequences. In both types, the concentration of dissolved material tend to increase linearly with depth. Waters of type a and b occur in near surface artesian conditions. Bicarbonate is the dominating anion in meteoric waters. Waters containing it and/or sulphate are oxygen bearing and are of very low salinity, some basins almost containing fresh water.

Because of complex chemical composition of oil field waters, they can best be displayed graphically for comparison. Schemes have been proposed by Tickell (1921), Parker and Southwell (1929), Sulin (1946) and Stiff (1951) in which short bars in the graphs are given in length proportional to concentrations of various cations to the left and anions to the right. Figure 9.1 shows the chemical composition of sea water and typical connate water plotted according to Tickell and Sulin schemes.

9.3 Physical Properties of Oil Field Waters

The most important among them are pH and Eh. The pH of oil field waters usually is controlled by the carbondioxide–carbonate system. Though it is not used for water identification or correlation purposes, it will indicate possible scale–forming or corrosive tendencies of water. The pH may also indicates the presence of drilling–fluid (mud) filtrate or well treatment chemicals. A knowledge of Eh (oxidation–reduction or redox potential) is useful in studies of transport of some compounds containing uranium, iron, sulphur and other minerals in aqueous systems. Solubility of some elements and compounds depends on the redox potential and the pH of their environment.

Physical properties particularly pH and Eh of oil field waters are expressed by a Eh–pH diagram (Figure 9.2). The figure indicates that rain water is oxidising and acidic. It generally contains oxygen, nitrogen and CO_2 in solution, together with ammonium and nitrate after thundershowers. As rain water percolates into the soil, it undergoes several changes as it becomes meteoric water and tends to become reducing since the dissolved oxygen in it oxidises organic matter. If meteoric waters flow deep into the subsurface, they gradually dissolve salts and cause change in pH. Deep connate waters show a wide range of Eh and pH values depending on their history and particularly on the extent to which they have mixed with meteoric waters or contain palaeoaquifiers trapped between unconformities. Oil field brines tend to be alkaline and strongly reducing (Pirson, 1983; Selley, 1998).

TYPICAL CONNATE WATER   SEAWATER

Na + k   Ca + Mg

|_|_|_|_|_|
10% units   ─── CO$_3$
                              + HCO$_3$

Cl   SO$_4$

A

Na 100 ───────────────── Cl 100
Ca 10 ───────────────── HCO$_3$ 10
Mg 10 ───────────────── SO$_4$ 10
Fe 10 ───────────────── CO$_3$ 10

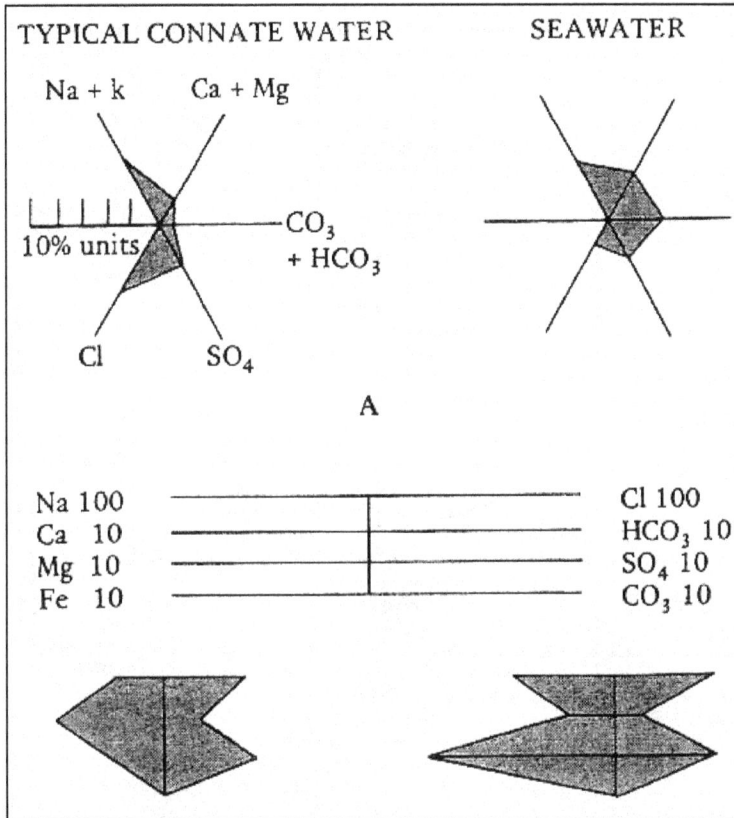

Figure 9.1: Methods of Plotting Water Chemistry.
(A) Tickell Plots and (B) Sulin Plots for a Typical
Connate Water Sample (left) and Seawater (right)
(*Source*: Selley, 1998)

9.4.0 Chemical Composition of Oil Field Waters

Oil field waters contain varying concentrations of inorganic salts together with dissolved gases, traces of organic compounds including hydrocarbons, and stable isotopes (Collins, 1980).

9.4.1 Inorganic Salts

Oil field waters contain number of major and minor (Table 9.1) inorganic constituents (cations and anions). The analytical data

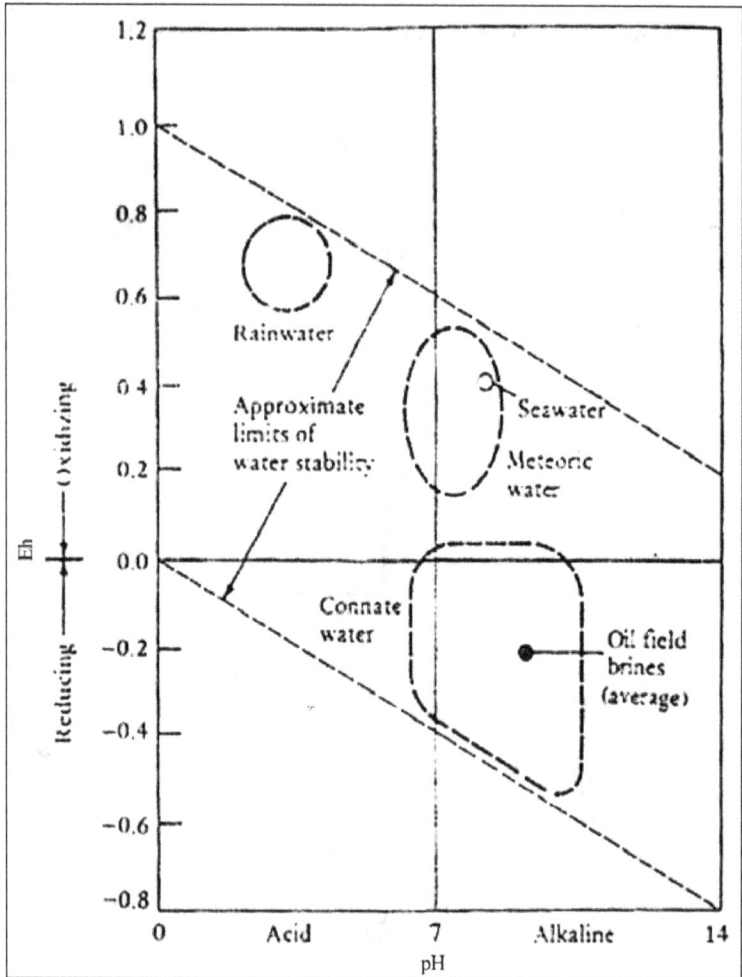

**Figure 9.2: Eh-pH Graph Showing the Approximate
Distribution of the Various Types of Subsurface Fluids
(*Source*: Selley, 1998)**

have several uses including water identification, log evaluation, geochemical exploration etc. (Collins, 1975). For example, a knowledge of iron present in either ferrous (Fe^{2+}) or ferric (Fe^{3+}) ion in oil field brines can be used to calculate Eh (redox potential). Lead

(Pb) concentrations as high as 100 ppm found in some Jurassic oil field brines in Mississippi caused severe oil production problems by scale formation on subsurface equipment (Carpenter *et al.*, 1974). Knowledge of bromide (Br⁻) ion concentration is important in determining the origin of an oil field brine, and it is one of the best geochemical marker constituents. Boron affects electric log deflections if its concentration exceeds 100 ppm. One interesting phenomenon is the enrichment of lithium (Li) in oil field waters upto 2,000 times its concentration in sea water (Collins, 1975).

Table 9.1: Major and Minor Constituents in Oil Field Waters

Concentration	Constituent
Per cent	Na, Cl
> 100 ppm	Ca, Mg, Br, SO_4, K, Sr
1–100 ppm	Al, B, Ba, Fe, Li
ppb (most oil fields)	Cr, Cu, Mn, Ni, Sn, Ti, Zr
ppb (some oil fields)	Be, Co, Ga, Pb, V, W, Zn

Source: Collins, 1980.

9.4.2 Dissolved Gases and Other Organic Constituents

Large quantities of dissolved gases are present in oil field brines. They include hydrocarbons together with others such as CO_2, nitrogen, hydrogen sulphide etc. The chief constituent in hydrocarbon gases is methane along with measurable amounts of ethane, propane and butane (Buckley *et al.*, 1958). Their concentrations in general, increase with depth in a given formation and also basinward with regional and local variations. Waters are enriched in dissolved hydrocarbons in close proximity to oil fields.

In addition to above gases, large number of hydrocarbon constituents in colloidal, ionic and molecular form occur in oil field brines. Knowledge of the dissolved organic constituents is important because of their relation to the origin and/or migration of oil, as well as to the disintegration of an oil accumulation. For example, a knowledge of the concentration of benzene, toluene and other components in oil field brines is used in exploration. Oil field brines also contain amino acids (aspartic, glycine, serine, alanine-B and thiamine) and fatty acids both soluble (C_1-C_8) and insoluble (C_{14}-C_{30}) carbon atoms.

9.4.3 Stable Isotopes

Stable isotopes such as deuterium (H-2), oxygen-18, carbon-13 and strontium-87 are used to make interpretations concerning origin of the brines. Their ratios e. g., δD, $\delta\,^{18}O$, $\delta\,^{13}C$ and $\delta\,^{87}Sr$ are measured with a mass spectrometer (6.4.0) and compared with appropriate oil field water or reservoir rock analysis. With proper analysis and interpretation of δD and $\delta\,^{18}O$ data, a distinction can be made between endogenetic and exogenetic water. In some sub-surface petroleum reservoirs, much of the indigenous connate water or brine may be replaced by infiltrating meteoric water. Conformation of the likely age of water can be made by interpretation of the above isotopic ratios. Similarly, layered ultramaffic rocks contain relatively high $^{87}Sr/^{86}Sr$ ratios, whereas intermediate and silicic rocks show an apparent differentiation. Therefore, the differences in the values are commonly attributed to selective migration of strontium (Sunwall and Pushkar, 1979). Furthermore, it is believed that subsurface fluids play a role in the selective migration.

9.5 Origin and Evolution

The question of origin of oil field brines is difficult to answer in a general manner since the water involved and its dissolved constituents to form the brine may have divergent histories. Subsurface water may either originally present or infiltrated into the subsurface from the surface. If it is originally present, it would be 'endogenetic', whereas if it is infiltrated from the surface and/or penetrated sediment accumulations, it would be 'exogenetic'. Obviously these two types of waters could meet and mix in the subsurface and thus the mixture would contain water of two separate origins. In addition there may be water, which moves upward due to the compaction of sediments.

The chemical composition of an oil field brine is an end product of several variables. These include: (*i*) chemical constituents of water and interactions among them, and (*ii*) interactions of the brine with surrounding rocks and petroleum. Some of these interactions include: (*a*) conversion of calcite to dolomite, which increases Ca, Sr and Ba concentrations in the brine while decreasing Mg and sulphate concentrations, (*b*) bacterial reduction of sulphate which decreases its concentration and increases hydrogen sulphide concentration, (c) formation of potassium and sodium aluminosilicates which decreases the concentration of K and Na in the brine while increasing the concentration of Ca, and (d) ion-exchange reactions between dissolved ions in the brine with those in the rocks.

Studies on formation waters of the western Canada sedimentary basins (Hitchon *et al.*, 1971) indicate that 80-85 per cent of the strata were deposited under marine conditions while 15-20 per cent were deposited under brackish water or fresh water conditions. This is supported by decrease in the deuterium (^2H) concentration due to mixing with infiltrating fresh water and decrease in the concentration of ^{18}O due to exchange with carbonate in the rocks. These observations along with several other basins all over the world led to the assumption that all sedimentary strata originally contained sea water.

9.6 Genetic Relation to Evaporites

The occurrences of petroleum in sediments is mostly associated with evaporites. Evaporative excesses produce dense brines which tend to sink and create density stratification in a water mass, resulting in eutrophication and an anoxic bottom where organic matter could be preserved in sediments that are potential source beds for petroleum. The presence of a thick layer of salt is a prerequisite for the formation of salt domes, which have served as traps of many large petroleum occurrences. Diagenetic processes in the regions of evaporite decomposition may from porous and permeable rocks, such as sucrosic dolomite, which could be excellent hydrocarbon reservoir beds (Schreiber and Hsu, 1980). Petroleum in black shales mostly associates with evaporites in many marine sequences. Further, the lacustrine Green River formation is well known for both its oil shale and evaporite deposits. The common evaporites associated with marine waters are gypsum, anhydrite, celestite, calcite, dolomite, halite etc. Geochemical and geological studies of some very concentrated brines indicate that, in deep quiescent bodies of water, strong bitterns can persist for long periods of time under a layer of near normal sea water. As a result, carbonates can precipitate from the less saline water and fall through the bitterns at the bottom, and as compaction proceeds, the pore spaces remain filled with bitterns.

A common type of origin of many oil field brines may thus probably be the interstitial fluids in the evaporites. These fluids find their way into oil field brines as they are compacted, causing them to move. Another type of brine originates by leaching or dissolution of evaporite minerals by infiltrating waters. These two types can readily be distinguished by adequate interpretation of their bromide concentration. A geochemical model can be built to represent origin and evolution of the type of brine, using relatively simple operations and processes such as (*a*) evaporation, (*b*) precipitation, (*c*) sulphate reduction, (*d*) mineral formation and diagenesis, (*e*) ion-exchange, (*f*) leaching, and (*g*) expulsion of interstitial fluids from evaporites during compaction.

9.7 Patterns and Significance of Salinity Distribution

The chemical composition of subsurface oil field brines changes both laterally and vertically relative to the total volumetric area of subsurface water-bearing horizons. Salinity or total dissolved salts (TDS) in sub-surface (oil field) waters tend to increase from the margins of a basin towards its center in many instances (Case, 1945; Youngs, 1975). This is due to the fact that the basin margins are more susceptible to circulating meteoric water than the basin center, where flow is negligible or coming from below. Thus regional isohaline maps can be useful exploration tool, indicating areas of anomalously high salinity. These areas are presumably stagnant regions unaffected by meteoric flow, where oil and gas accumulations may have been preserved. This is illustrated with the salinity distribution of Lower Wilcox formation (Figure 9.3). Based on salinity distribution, the formation can be divided into five zones: (*a*) fresh water zone at the outcrop, (*b*) saline water zone with less than 5,600 mg/l (ppm) as NaCl, (*c*) salt water zone with 5,600 to 70,000 mg/l, (*d*) salt water zone with 70,000 to 1,70,000 mg/l, and (*e*) salt water zone with greater than 1,70, 000 mg/l as NaCl. The heavy brine zones (*d* and *e*) occur in the deeper areas (Collins, 1980).

Detailed mapping of sub-surface water characteristics provides information concerning sand fingering, diagenetic changes that affect reservoir, source rocks and stratigraphic traps. Petroleum accumulations occur in the areas of salinity transition zones, where it varies from 50,000 to 1,00,000 mg/l NaCl. Mapping of subsurface water characteristics (sulphate, chloride, calcium, magnesium and TDS) of Bellcanyon formation (Eastern Delware Basin, Taxas) showed

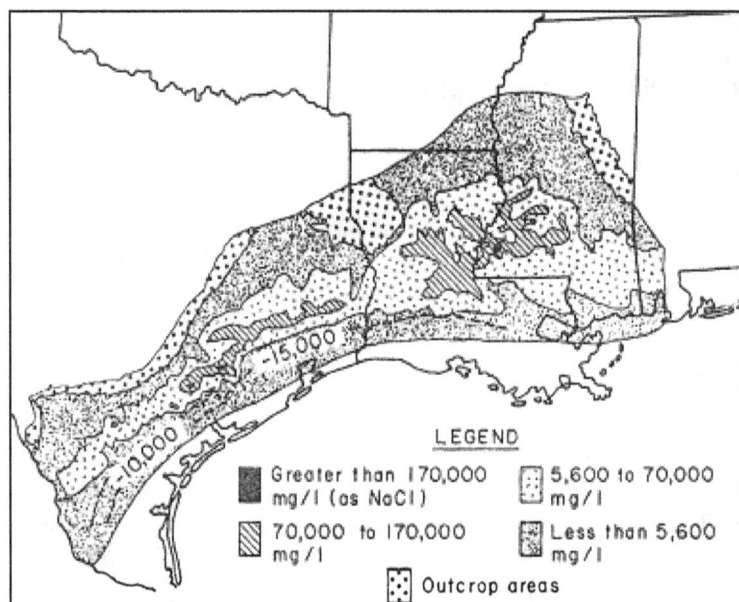

Figure 9.3: Salinity Distribution in Water from the Top of the Lower Wilcox Formation in Portions of Texas, Louisiana, Arkansas, Mississippi and Alabama (*Source*: Collins, 1980)

systematic variations in their distribution which are related to proximity to the outcrop and the degree of transmissibility of the basin (Collins, 1975). Rapid changes in the distribution of magnesium and TDS in waters is related to permeability changes of the sediment. In areas of low permeability fine-grained argillaceous rocks, the relative concentration of magnesium decreases in waters of high TDS. This type of relationship between the composition of formation waters and permeability of sediments can be of use as the basis for exploration technique since the latter in turn, is related to the producibility of reservoir rocks.

Sulin (1946) based his water classification on hydrogeochemical zones. He distinguished three zones namely (*i*) free exchange zone, (*ii*) zone of difficult exchange, and (*iii*) zone of little or no exchange (stagnant zone). The first zone is the SO_4–Na and HCO_3–Na types of water (*a* and *b* respectively), the second zone contains HCO_3–Na and Cl–Mg types of water (*b* and *c* respectively), and the third zone is of Cl–Ca type (*d*) water. Petroleum is usually associated with type *d* water which is often found in a hydrostatic or stagnant environment. Useful maps can be constructed of these water types after application of the classification to water analysis data.

Primary mechanisms of origin, migration, trapping, accumulation of petroleum and its subsequent disintegration involve water. It is therefore obvious that knowledge of certain water characteristics is useful in exploration and production. Figure 9.4 illustrates some characteristics related to or found in oil field brines that are indicative of a petroleum accumulation, and some that are indicative of a dry reservoir. Many of these characteristics can be mapped for use in exploration efforts.

9.8   Role of Oil Field Waters in Enhanced Oil Recovery (EOR) Operations

Oil field waters play an important role in enhanced oil recovery (EOR) operations. The main objective of EOR is to recover oil that cannot be recovered by primary and/or secondary recovery operations in a reservoir. This can be achieved by injecting liquids such as formation water, sea water or fresh water, whereby the original reservoir pressure is maintained or reestablished. However, careful analysis and treatment of injected water is essential to prevent unwanted chemical changes that may likely to take place, damaging the reservoir rocks. These include:

(*i*) *Plugging*: Any type of solid material suspended in or precipitated from the injected water affects the permeability of reservoir rocks.

(*ii*) *Clay sensitivity*: When fresh water is used for injection, clays such as smectite and illites are sensitive for swelling which results in the reduction of permeability of the formation (Mungan, 1965). This is however, prevented by the addition of requisite quantities of common salt (NaCl) to the fresh water.

(*iii*) *Corrosion*: Prolonged use of sea water corrode injection pumps and pipe lines. Interaction of major cations (Ca^{2+} and Mg^{2+}) in sea water with reservoir fluids leads to plugging problems. Bacteria and other organisms such as copepods, diatoms and dinoflaggelates degrade oil in the reservoir resulting in increase of its viscosity and decrease of its API gravity.

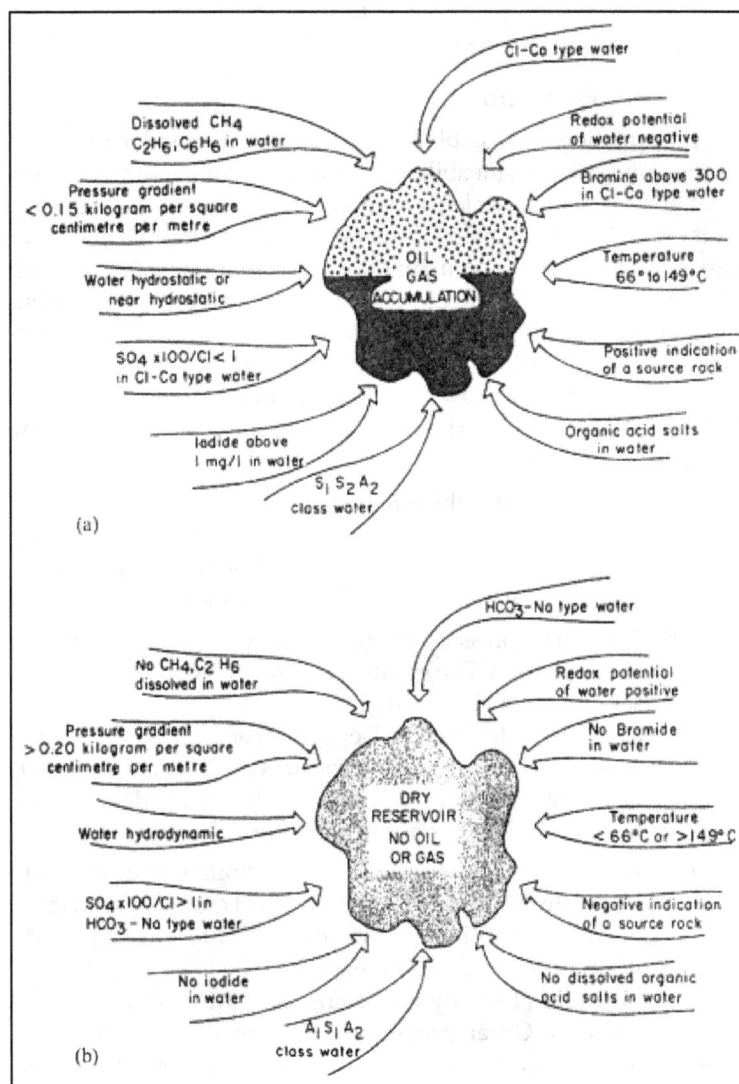

Figure 9.4: Genetic Indicators Associated with an
Oil and Gas Accumulation (*a*) Compared to indicators
associated with a dry reservoir (*b*)
(*Source*; Collins, 1980)

It is therefore imperative to conduct rigorous water quality tests and treatment if necessary prior to their use in EOR operations (Collins and Wright, 1985).

9.8.1 Scale Formation

Precipitation of insoluble compounds from formation (oil field) waters reduce the permeability of a porous producing petroleum formation and cause scale formation. This results in plugging of injection wells in water flooding systems, water pumps and flow lines. Some commonly occurring ions in formation waters such as $Ca^{2+}, Sr^{2+}, Ba^{2+}, Fe^{2+}, Fe^{3+}, HCO_3^-, CO_3^{2-}$ cause precipitation of insoluble compounds such as $CaSO_4, BaSO_4, SrSO_4, CaCO_3, FeS, Fe_2S_3$ leading to the scale formation. Deposition of scale not only restricts production, but also cause inefficiency and equipment feature (Collins and Wright, 1985). Following are some of the important variables in scaling:

(i) Temperature of the formation in relation to possible scale formers in the fluids passing through it. For example $CaSO_4$ and $SrSO_4$ becomes less soluble with increasing temperature whereas $BaSO_4$ becomes more soluble.

(ii) Subsurface pressure changes while the fluid flows through the formation. The greatest pressure charge occurs at the sand face of the producing well where the solubility changes are the largest. Deposition of scale at this point is most damaging to the oil production. For example $CaSO_4$ is known to decrease its solubility with decrease in pressure at NaCl concentration upto 10 per cent.

(iii) Brine concentration exclusive of precipitating compounds, also influences scale formation. Most electrolytes in ionic from cause an increase in the solubility of compounds which form scale. This is explained by lowering of $BaSO_4$ saturation level by increasing amounts Ca^{2+} ions in the solution. Other properties of oil field brine known to influence the solubility levels of scale formers are gases in solution, pH, ion-pairs and dissolved organic complexes (Davis and Collins, 1971).

9.8.2 Corrosion

Electrochemical corrosion is the usual type of corrosion that occurs in oil field waters (Ostroff, 1979). This is because the brines are good conductors of electricity and their conductivity increases with increase of salt content. Presence of dissolved gases such as O_2 H_2S and CO_2 in brines cause severe corrosion problems of steel equipment.

Among the gases dissolved oxygen presents the worst corrosion problem at concentrations less than one ppm. Its solubility in water is a function of temperature, pressure and concentration of dissolved salts. It is less soluble in salt water than in fresh water. Dissolved CO_2 in oil field waters increases corrosion by lowering the pH due to formation of carbonic acid (H_2CO_3) and establishing CO_2– bicarbonate–carbonate equilibria. On the other hand, increase of pH results in scale formation due to precipitation of carbonates. The solubility of CO_2 gas is influenced by temperature, pressure and dissolved salts. H_2S gas is commonly present in many oil field waters and its origin is some times attributed to sulphate reducing bacteria. The gas is very much soluble in water and its solubility is also a function of temperature, pressure and dissolved salts. The corrosivity of oil field waters increases with increase in the concentration of H_2S.

9.9 Oil Degradation in the Reservoir

The degradation processes that occur in a petroleum reservoir reduce the economic value of oil by destroying the paraffins, lighter ends and remaining fractions whereas thermal maturation (2.6.0) improves the commercial value of oil accumulation by increasing the lighter fraction and paraffinicity, and reducing the asphaltene and sulphur contents. Degradation decreases API gravity and increases viscosity. The degradation processes can be broadly classified into three categories, (*a*) physical, (*b*) chemical, and (*c*) biochemical.

9.9.1 Physical Processes

Major changes in petroleum composition can be produced by contact with flowing water (Milner *et al.*, 1977). As the water moves past the oil in the reservoir, it removes the most soluble (lighter) components, thus enriching insoluble (heavier) paraffin components resulting in the formation of a tar layer nearest to the oil-water contact. This type of disintegration begins with formation of slope of the petroleum (oil-water or gas-water) interface (Figure 9.5). The water flows from F towards G in the reservoir rock since the potentiometric

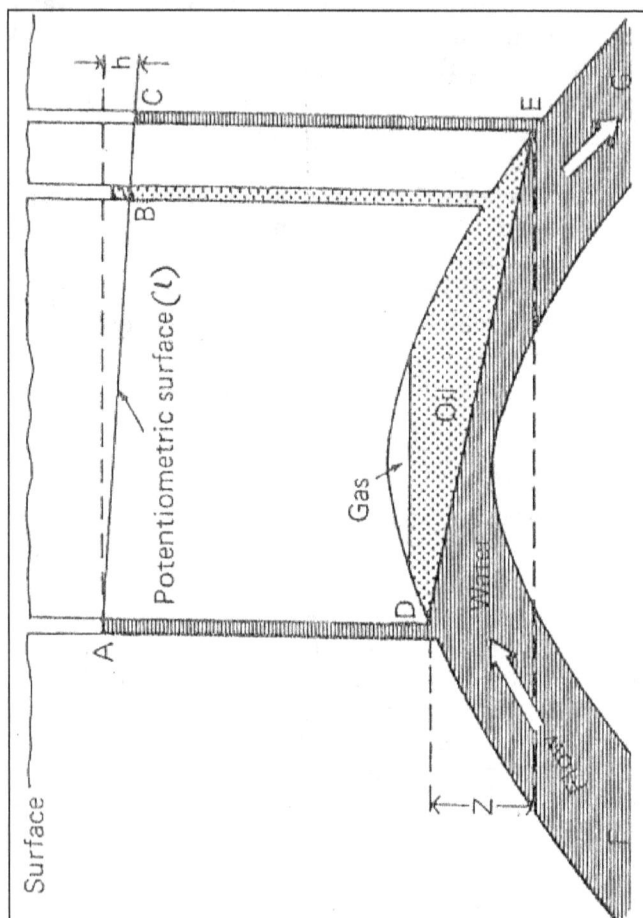

Figure 9.5: Entrapment of Buoyant Oil and Gas at the Tops of Reservoirs. Hydrodynamic flow creates a slope in the oil-water contact. (*Source:* Waples, 1981)

surface and slope (dh/dl) at G is lower than at F. Thus the oil accumulation is swept away by the moving water while the former is in a state of equilibrium. The slope of the petroleum-water contact is a function of the piezometric subsurface, and density of both petroleum and water. Tectonic disturbances can also cause the hydraulic potential associated with a hydrocarbon accumulation to become dynamic. The resulting hydrodynamic situation can cause severe alteration of hydrocarbon accumulation, in addition to water washing. For example, any tilting of the sub-surface strata and reservoir trap can cause the oil to spill and move. This can cause migration of the oil to another trap accompanied by additional water washing. This type of phenomenon can be observed in the Rocky Mountain area of USA, where the first row of anticlines at the edge of a basin are often barren of any hydrocarbon accumulation (Collins, 1980).

When water comes in contact with oil in the subsurface, some of the more water-soluble components of oil will be removed and the oil becomes enriched with heavy components such as asphaltenes. In short, the residual oil is degraded with a decrease in its API gravity. In general, a combination of water washing and biodegradation (9.9.3) accomplish such a change. Water washing occurs when fresh water infiltrates through an outcrop and flows through the zone containing a structurally trapped hydrocarbon accumulation. When water washing is accompanied by chemical degradation such as oxidation with sulphate or dissolved oxygen, an asphaltic mat is created at the oil-water interface. Such asphaltic mats of 30-80 m in thickness were noticed in the Permian oil fields of the Urals at the oil-water contact of the pools (Yarullin, 1961). This was attributed to the oxidation of oils by the formation waters which were high in sulphate ion concentration. The Burgan fields in Kuwait and the Hawkins field in east Texas also contain asphalt mats at the oil-water interface.

9.9.2 Chemical Processes

Chemical oxidation of reservoir oil is caused by dissolved molecular oxygen and sulphate anions associated with water. Dissolved oxygen was found in some subsurface waters at depths of about 500 m in quantities as high as 10 mg/l. However, this type of water is an infiltrating but not indigenous water. It occurs only when subsurface waters are in hydrodynamic contact with the surface and are actively receiving surface recharge waters. Oxidation of reservoir oil by sulphate may be more common than dissolved oxygen because of its presence in many subsurface waters and brines at a higher concentration than the latter. This reaction is represented as:

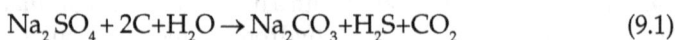

$$Na_2SO_4 + 2C + H_2O \rightarrow Na_2CO_3 + H_2S + CO_2 \qquad (9.1)$$

Significance of this reaction is the formation of H_2S. Liquid (lighter) hydrocarbons are more likely degraded than gaseous hydrocarbons by this oxidation. The size of the hydrocarbon pool is a function of the degradation rate. As a general rule, the oxidation increases with increase of surface area of the petroleum-water contact.

9.9.3 Biochemical (Microbial) Processes

Sediments, soils and waters contain a wide variety of micro organisms that can utilize hydrocarbons as a sole source of energy in their metabolism. Paraffins, naphthenes and aromatics, including gases, liquids and solids are all susceptible to microbial decomposition. Microbial populations comprising many species of bacteria, fungi and yeast, are highly adaptable and can alter their metabolic processes depending on the hydrocarbons available. Though all kinds of hydrocarbons appears to be susceptible to bacterial action, the process is selective. In general, aliphatic hydrocarbons and aliphatic side chains on cyclic compounds are most susceptible to attack, while naphthenes and aromatics are least affected (7.5.0). Because bacterially altered oils contain less paraffin hydrocarbons than unaltered oils of the same type, they have higher correlation index (CPI) values and lower API gravities (Collins, 1980). Three types of microbial activity can occur in the oil field waters: (*i*) aerobic bacteria–growth depends upon oxygen, (*ii*) anaerobic bacteria-growth maximum in the absence of oxygen, and (*iii*) facultative bacteria-growth independent of oxygen. Among them types (*i*) and (*ii*) are more important in degradation of oil in petroleum reservoirs.

(*i*) Aerobic Bacterial Degradation

The principal agent of degradation (oxidation) of this type is aerobic bacteria (but not dissolved oxygen) generally carried into the reservoirs by meteoric waters. Biodegradation of a crude oil by aerobic bacteria requires the following three conditions:

1. Circulating waters with about 8 mg/l of dissolved oxygen, to support its life.
2. Formation temperature of 20-50°C to permit its growth.
3. Hydrocarbons essentially free from H_2S (which are poisonous to aerobes) to serve as food.

The aerobic bacteria preferentially consume normal alkanes so that the first indication of incipient biodegradation is a reduced paraffin content. With progressive degradation, all the paraffins are removed, whereas multibranched compounds such as the isoprenoids remain, which are also ultimately consumed (Barker, 1985). This sequence is clearly seen in the gas chromatograms of hydrocarbons (Figure 9.6). The Athabasca heavy oil deposits of Eastern Alberta, Canada, represent the largest degraded oil accumulation in the world that have undergone extensive biodegradation and water washing (Hunt, 1976).

Aerobic bacteria is most effective for degradation of heavy oils and tars common to surface seepages and reservoirs (Connan, 1984). However, in petroleum reservoirs the presence of free oxygen which is essential rapidly decreases with depth. Dissolved oxygen concentration even less than 1 mg/l is sufficient to sustain aerobic bacteria, but the rate of degradation of petroleum in the reservoir becomes too slow that it may require thousands or even millions of years to achieve a perceptible change.

(ii) Anearobic Bacterial Degradation

As stated earlier, anaerobic bacteria strives in waters that contain no oxygen. They take their oxygen from sulphate dissolved in water by the general reaction

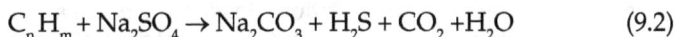

$$C_nH_m + Na_2SO_4 \rightarrow Na_2CO_3 + H_2S + CO_2 + H_2O \qquad (9.2)$$

Since H_2S and CO_2 are fairly soluble in water, their concentrations will be high in subsurface oil field brines. For example, natural gas produced from the Smackover formation has been found to contain 2 to 78 per cent by volume of H_2S, 1.6 to 55 per cent methane, and 9 to 20 per cent CO_2. Since these gases are in contact with

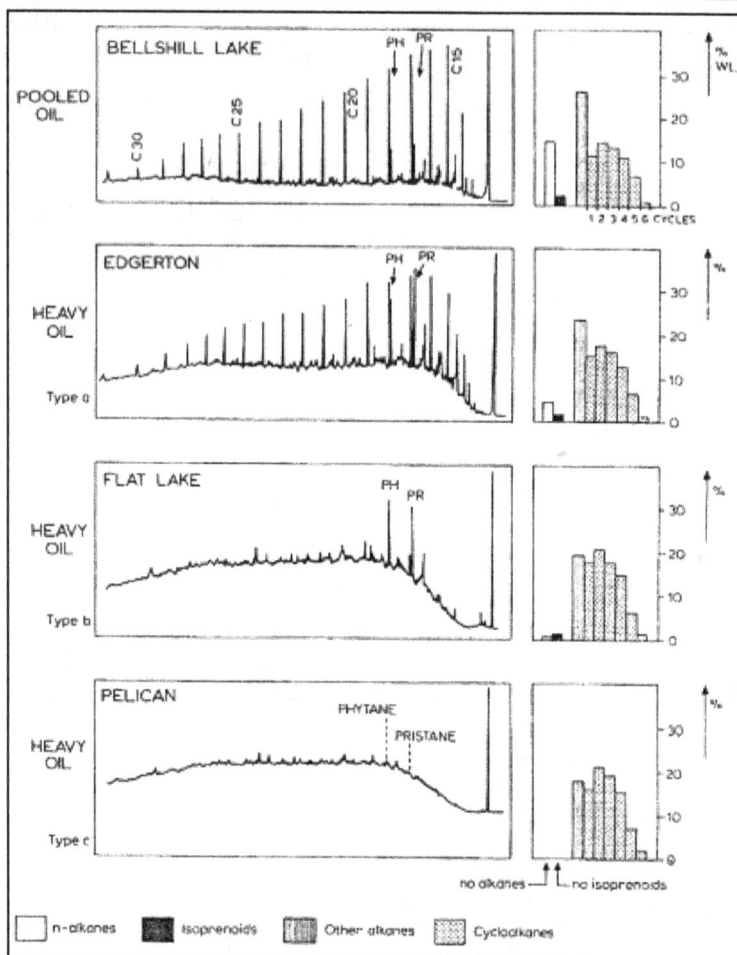

**Figure 9.6: Gas Chromatograms and Compositional Data
for Canadian Oils Showing the Progressive Changes
Produced by Bacterial Degradation. Unaltered oils are at the top,
and the most severely degraded ones at the bottom.
(*Source*: Barker, 1985)**

formation waters, they are saturated with them. The H_2S thus
produced is oxidized to sulphur by the reaction

$$SO_4^{2-} + 3\ H_2S + 2H^+ \rightarrow 4S^\circ + 4H_2O \qquad (9.3)$$

The free sulphur can in turn, dehydrogenate petroleum, producing high molecular weight cyclic hydrocarbons and additional H_2S. Gulf coast sulphur deposits were formed by such reactions. Anaerobic bacteria also consume preferentially paraffin hydrocarbons thus causing decrease in their content of degraded oil. In addition, while the bacteria destroy saturated and aromatic hydrocarbons, they and/or associated reactions add non-hydrocarbon (NSO) compounds and asphaltenes to the oil thus degrading it further.

A case study of Williston Basin (Table 9.2) indicated that water washing and biodegradation reduced crude oil API gravity more than 20 degrees, and increased sulphur content more than 2 per cent (Bailey *et al.*, 1973). There are distinct changes from SE to NW as the crude oil becomes heavier and acquired higher NSO content. This is evident in (*a*) decrease in the percentage of paraffins in C_{15}+ saturate fraction and that of saturates in the C_{15}+ oil fraction, and (*b*) increase of aromatics and asphaltenes in the C_{15}+ oil fraction.

Table 9.2: Quantitative Changes in Williston Basin Crude Oils

	Least Altered Oil (SE)	Most Altered Oil (NW)
API gravity	37.6	15.2
Percentage of sulfur (whole crude)	1.13	2.99
Percentage of paraffins in C_{15}–saturate fraction	46.5	6.9
Percentage of saturates in C_{15}+oil fraction	47.1	19.1
Percentage of aromatics in C_{15}–oil fraction	37.5	43.3
Percentage of asphaltenes in C_{15}+oil fraction	5.4	16.2

Source: North, 1990.

The end result of degradation in petroleum reservoirs is the formation of tar mats or tar belts at the edges of basins where the reservoir rocks reach or approach the surface. Notable examples are the Hit tar mat on the west side of the Persian Gulf Basin in Iraq in a Tertiary limestone ; the Bermudez or Guanoco pitch body and the Orinoco tar belt on the northern and southern margins of the eastern

Venezuelan Basin; and the Santa Maria tar mat in California (North, 1990).

Gases can be degraded like oils. They lose their gasoline and ethane contents while the nitrogen content increases probably due to its dissolution in invading meteoric waters. Many oil field waters produced in the reservoirs are reinjected into the formation to maintain pressure, either for secondary oil recovery or for enhanced (tertiary) oil recovery. The main bacterial problems found in oil field injection waters are caused by sulphate reducers, iron bacteria and slime formers, all of which cause severe corrosion and plugging problems. It is therefore essential to test for bacteria and to control them in injection water systems.

Chapter 10
Organic Geochemistry and Analytical Techniques

Organic geochemistry offers an important approach to petroleum exploration by defining various targets, which may occur in different geological sedimentary sequences and to locate the most favourable areas for exploring these targets. This is achieved by analysing cores, cuttings and crude oil or gas samples from exploratory wells, and rocks from outcrops. As described in the earlier chapters, hydrocarbon generation and migration processes in source rocks (Ch. 5), quantitative evaluation of their petroleum potential and oil/source rock correlation (Ch. 7) compliment the definition of plays and targets present in a sedimentary basin. Further, timing of petroleum generation can now be evaluated with the aid of mathematical models and compared with the age of the formation traps (Ch. 6). In addition, developments in the field of organic geochemistry such as application of biomarkers and introduction of fast and powerful analytical techniques that offer a systematic appraisal of the petroleum potential in a sedimentary basin are described in this chapter.

10.1 Biomarkers

A biological or biomarker is any organic compound detected in the geosphere whose basic skeleton suggests an unambiguous link with a known contemporary natural product. This link can easily be seen by comparing the structures of the vanadyl- porphyrin complex and its proposed natural product precursor, chlorophyll-a (Figure 10.1). Biomarkers exist because their carbon skeletons survive, in a recognisable form through the process of diagenesis and catagenesis which result mainly from temperature rise associated with burial of host sedimentary rocks. Use of biomarkers in petroleum geochemistry involves tracing of chemical and physical pathways takenup by organic compounds in the biosphere, subsequent to their death. Of predominence is the controls on the distribution of compounds incorporated into bottom sediments at the time of deposition. Such compounds are only relevant to petroleum geochemist, if their basic carbon skeleton remains intact, at least until a level of maturity corresponding to the early stages of petroleum generation by the thermal breakdown of kerogen. Many such compounds can now be recognized and their geochemical pathways followed with varying degrees of certainty. For example, triterpenoids, steroids and acyclic isoprenoids biosynthesised by organisms are steadily modified during early diagenesis by chemical and microbial removal of oxygen containing functional groups and carbon-carbon double bonds to yield saturated and aromatic hydrocarbons which retain the original carbon skeleton, and are observed in ancient sedimentary rocks and petroleum. Many of the large number of reactions and intermediates to these processes are now known.

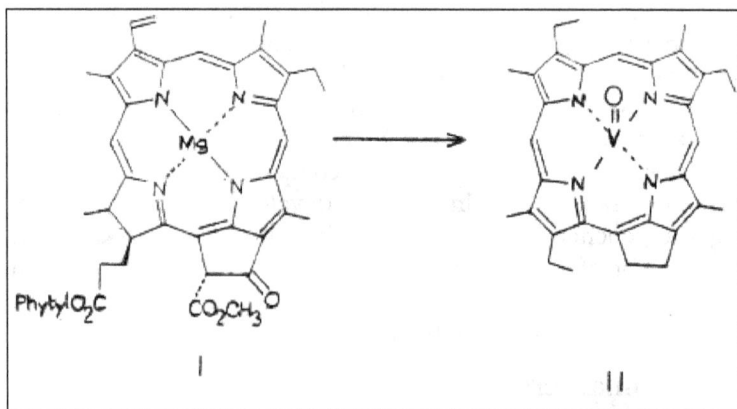

Figure 10.1: Structure of Chlorophyll-a (I) and Vanadyl Porphyrin Complex (II)

10.1.1 Distribution of Biomarkers

The distribution of biomarker compounds in sediments vary considerably with different depositional environments and the nature of their biological origin. They therefore respond in a sensitive way to changes in redox potential, salinity, water temperature and depth, nutrient supply and topography. The degree to which a diagenetic pathway is followed may strongly reflect the nature of mineral matrix present. All these factors superimposed on maturation and diagenetic effects, combine together to impart a specific set of fingerprints to sedimentary rock extracts (EOM) which are later inherited by crude oils when expulsion occurs. Thus the fingerprints of biomarker distribution revealed by GC-MS (10.7.1) are now routinely used for the classification of oils into groups and the correlation of these groups with potential source areas in a particular sedimentary basin.

Biomarkers are sometimes called geochemical fossils to designate molecules synthesised by living organism (plants, animals, bacteria etc) and preserved in the sediments without any major modification in their carbon skeleton. They can be used for characterization, and/or reconstruction of depositional environment in the same manner as macro or micro-fossils commonly used by geologists (Tissot and Welte, 1984).

10.1.2 Application of Biomarkers

Biomarkers have a variety of applications in petroleum exploration (Seifert *et al.*, 1980; Peters and Moldowan, 1993). Some of them include:

(i) Oil-oil and oil-source rock correlations (7.4.0).

(ii) Environments of deposition (marine, lacustrine, fluvio deltaic, hyper saline etc).

(iii) Elucidation of chemical transformations during maturation (diagenesis and catagenesis).

(iv) Characterization of source rock kerogen (oil or gas prone, and age of source rocks.

(v) Lithology of source rocks (carbonates, shales etc.).

10.2 Specific Biomarkers

Among the biomarkers isoprenoids, hopanoids, steroids,

porphyrins and diamondoids are most commonly studied. Specific members of each of these classes and their applications are comprehensively reviewed by Mackenzie (1984).

10.2.1 Isoprenoids

A wide range of saturated linear or cyclic compounds formally builtup from several isoprene units

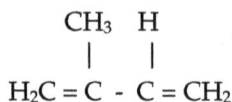

$$CH_3 \quad H$$
$$| \quad |$$
$$H_2C = C - C = CH_2$$

Isoprene basic unit

are known in living plants, bacteria and to a small extent, in animals. They occur mostly as long chain saturated hydrocarbons in sediments and crude oils, and constitute useful indicators of origin of oils. The isoprenoids are characterized by the presence of methyl (CH_3) group on every fourth carbon atom, irrespective of their total number. The commonest members among them are pristane (2,6,10,14 tetramethyl pentadecane) and phytane (2,6,10,14 tetramethyl hexadecane) as described earlier (2.1.2; Table 2.3). The ratio of pristane/phytane either in crude oil or in rock extract has several applications (Hughes *et al.*, 1995):

(*i*) It indicates the origin of the source rock. For example, very high (>3.0) value suggests terrestrial organic matter (coal sourced oil) in the source rock (Table 10.1).

(*ii*) It is extensively used as indicator of redox conditions of depositional environment (sedimentary rock). For example, very low value (<0.5) indicates hyper saline or anoxic depositional environment. On the other hand, relatively high value indicates oxic (oxidized) depositional environment (Table 10.1)

(*iii*) Very high pristane/n-C_{17} and phytane/n-C_{18} isoprenoid ratios suggests biodegradation of oils. One limitation however, of these ratios is that they are also affected by many other parameters.

10.2.2 Hopanoids

Hopanoids are pentacyclic triterpenoids found in sedimentary

Table 10.1: Biomarker Indicators of Depositional Environments and Age of Source Rocks

Source Information	Biomarker Parameter	Comments
Marine Source Rock	24-*n*-propylcholestane	Ubiquitous in oils derived from marine source rocks.
Lacustrine Source rock	Sterane/Hopanes	Low in oils derived from lacustrine source rock
	C_{26}/C_{25} tricyclic terpanes	> 1 in many lacustrine-shale-sourced oils
Higher plant input to Source Rock	Oleananes	Indicate flowering plant input to source
	C_{29} steranes	High relative to total C_{27}–C_{29} steranes
Coal Source Rock	Pristane/phytane	Very high in coal-sourced oils; *e.g.* > 3.0
	C_{31} homohopanes	High relative to total C_{31}–C_{35} in some coal-sourced oils
Hypersaline Depositional Environment	Gammacerane	High relative to C_{31} hopanes in oils derived from sources deposited under hypersaline conditions. High values indicate stratified water column during source deposition.
	Pristane/phytane	Very low (*e.g.* <0.5) in oils derived from source rocks deposited under hypersaline conditions.
Anoxic Depositional Environment of Source Rock	C_{35} homohopanes	High relative to total hopanes in oils derived from source rocks deposited under anoxic conditions.
		Further, higher C_{34} homohopanes relative to C_{31}–C_{34} is correlated with source rock Hydrogen Index.
	Pristane/phytane	< 1.0 can indicate anoxic conditions, but the ratio are affected by many other factors.
	V/V+Ni Porphyrins	High in reducing conditions

Contd...

Table 10.1–Contd...

Source Information	Biomarker Parameter	Comments
Carbonate Source Rock	30-norhopanes	High in carbonate-sourced oils; *e.g.*, C_{29}/C_{30} ratio.
	Diasteranes/steranes	Low in carbonate-sourced oils.
Age of Source Rock Deposition	Oleanane	Present in oils derived from Late Cretaceous or younger sources
	(24-norcholestane)/ (26-norcholestane)	High in many Tertiary sources. Low values are not age-diagnostic.
	Dinosteranes, triaromatic dinosteroids	Absence always means Pre-Mesozoic, while presence usually means Mesozoic or younger source rocks.
	C_{29} Monoaromatic Steroids	High in oils derived from sources older than 350 MY.BP.
	(24-isopropylcholestane)/ (24-*n*-propylcholestane)	High in oils from pre Ordovician sources

Source: Oil Tracers LLC, 2006.

rocks including petroleums. They are sufficiently immature to have allowed preservation over long (geological) time. Application of hopanes mainly include in the range C_{27} to C_{30} members, extendable upto C_{35} (excluding C_{28}). They originate from tetral (polyhydroxy bacterio hopane) present in several bacteria (Figure 10.2). In addition to basic tetral skeleton, a number of slightly modified skeletons, called norhopanes are found in petroleum source rocks (Figure 10.3). They exhibit steroisomerism and the isomers of 21α (H)-hopanoids are often called moretanoids. Some important pentacyclic triterpenoids other than hopanoids such as gammacerane and oleanane are found in shales and crude oils. In addition, a series of C_{20} to C_{26}, extended upto C_{45} tricyclic and tetracyclic alkanes are found in crude oils and organic rich ancient sediments. Following are some of the applications of this group of biomarkers as indicators of depositional environments and maturity of source rocks (Tables 10.1 and 10.2)

(*i*) Higher ratio of C_{29} norhopane/C_{30} norhopane (about 1.0) in oils indicates carbonate source rock (Subroto *et al.*, 1991)

(*ii*) Higher ratio of C_{35} homohopanes/total hopanes in oils suggests the deposition of source rocks under anoxic environment. Further, high C_{35}/C_{31} to C_{34} homohopanes in oils can be correlated to source rock hydrogen index, HI (Peters and Moldowan, 1991).

(*iii*) Lower ratio of Moretane/hopanes in oils increases thermal maturity of source rock. This is useful in identifying early oil window.

(*iv*) Higher abundance of gammacerane relative to C_{31} hopanes in oils indicates deposition of source rocks under hypersaline and stratified water conditions (Sinninghe Damste *et al.*, 1995).

(*v*) Presence of Oleanane in oils indicates terrestrial flowering plant input to source rocks. It also indicates Late Cretaceous or Younger source rocks (Moldowan *et al.*, 1994)

(*vi*) Higher ratio of tricyclic terpanes/hopanes indicates increase in thermal maturation of source rocks. It is useful in identifying the late oil window. However, the ratio is affected by biodegradation.

(*vii*) Ratio of C_{26}/C_{27} tricyclic terpanes (>1) in oils indicates their origin from lacustrine shale source rock (Zumberge, 1987).

Table 10.2: Biomarker Indicators of Source Rock Maturity

Petroleum Fraction (Compound Class)	Biomarker Parameter Measured in Petroleum Fraction	Effect of Increasing Maturity	Comments
Saturated Hydrocarbons	Moretane/Hopane	Decrease	Useful in early oil window
	Tricyclic Terpanes/Hopanes	Increase	Useful in late oil window; also increases at high levels of biodegradation.
	Diasteranes/Steranes	Increase	Useful in late oil window; also affected by source lithology (low in carbonates, high in shales); also increases at high levels of biodegradation.
Aromatic Hydrocarbons	Monoaromatic Steroids: $(C_{21}+C_{22})/[C_{21}+C_{22}+C_{27}+C_{28}+C_{29}]$	Increase	Useful in early to late oil window; resistant to effects of biodegradation
	Triaromatic Steroids $(C_{20}+C_{21})/[C_{20}+C_{21}+C_{26}+C_{27}+C_{28}]$	Increase	Useful in early to late oil window: resistant to effects of biodegradation.
	Monoaromatic/(Monoaromatic + Triaromatic Steroids)	Increase	Useful in early to late oil window; resistant to effects of biodegradation

Source: Oil Tracers LLC, 2006.

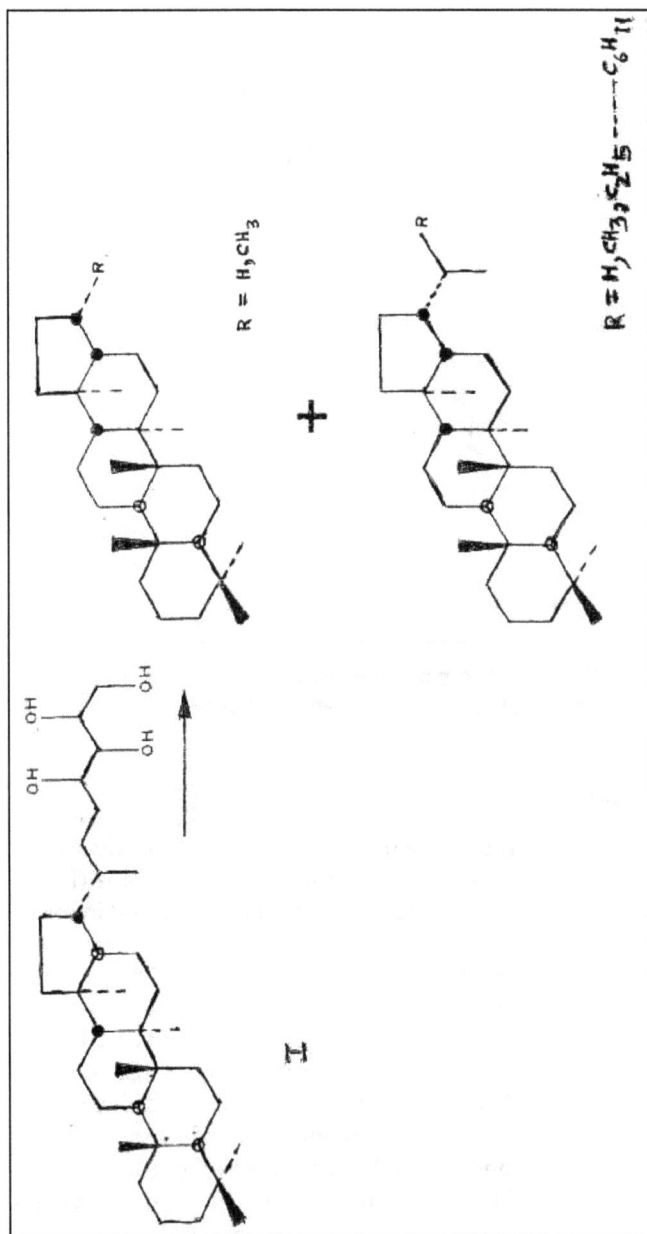

Figure 10.2: Basic Skeleton of Tetral I (Polyhydroxy bacterio hopane)

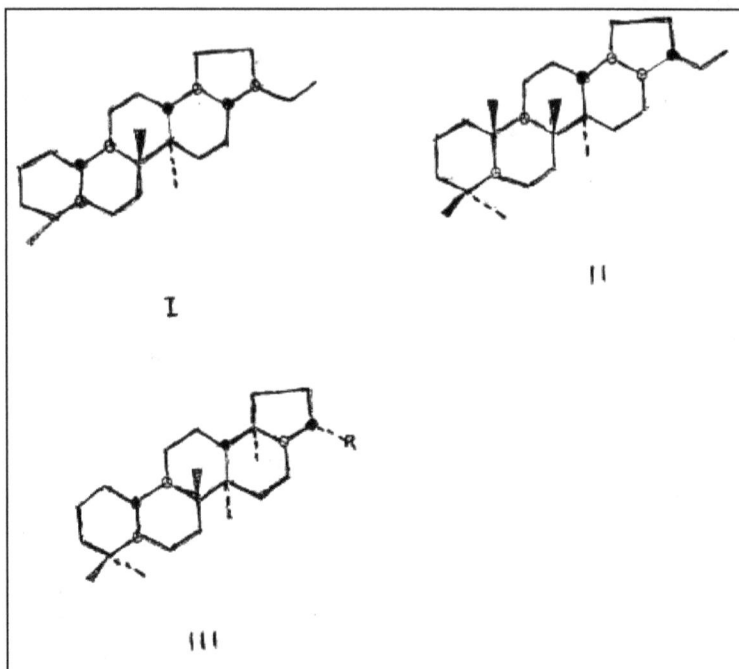

Figure 10.3: The Structures of Some Hopanes I: 25, 28, 30-trisnor moretane; II: 28, 30-bisnorhopane; III: 25-norhopane

10.2.3 Steroids

Tetracyclic triterpenoids when modified biochemically by the loss of methyl groups results in the formation of steroids (Hunt, 1996). They are found in terrestrial and marine plants, and in micro and macro organisms (blue green algae). Important biomarker compounds identified among steroids are steranes, diasteranes, mono-aromatic and tri-aromatic steroids. Diagenesis of natural product sterols (*e.g.*, cholesterol) yields cholestane and diasteranes (Figure 10.4). Alternatively on rearrangement and isomerisation, they get converted into diasteranes found in ancient sediments and petroleum. Similarly aromatisation (Figure 10.5) of sterols results in the formation of mono-aromatic steroids during diagenesis (Mackenzie, 1984) and further aromatization to di-aromatic steroids

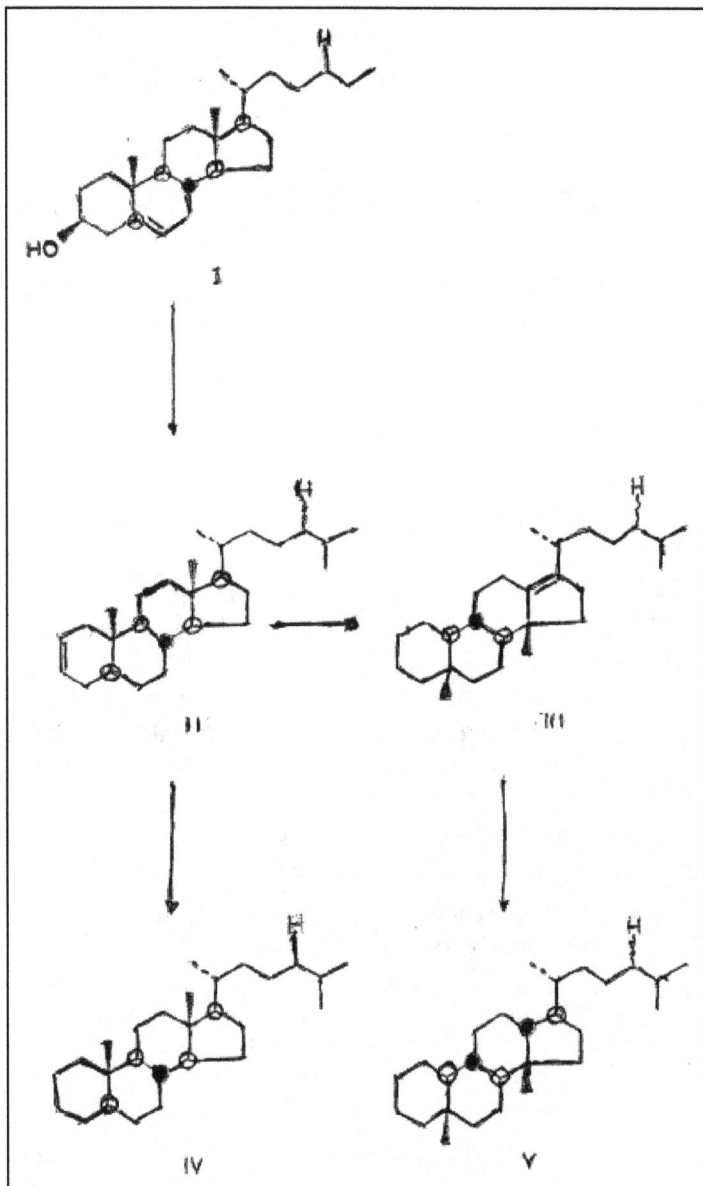

Figure 10.4: Simplified Reaction Scheme of Steroid Diagenesis; I: Cholesterol, II and III: Intermediate compounds, IV: Choestane, and V: Diasterane

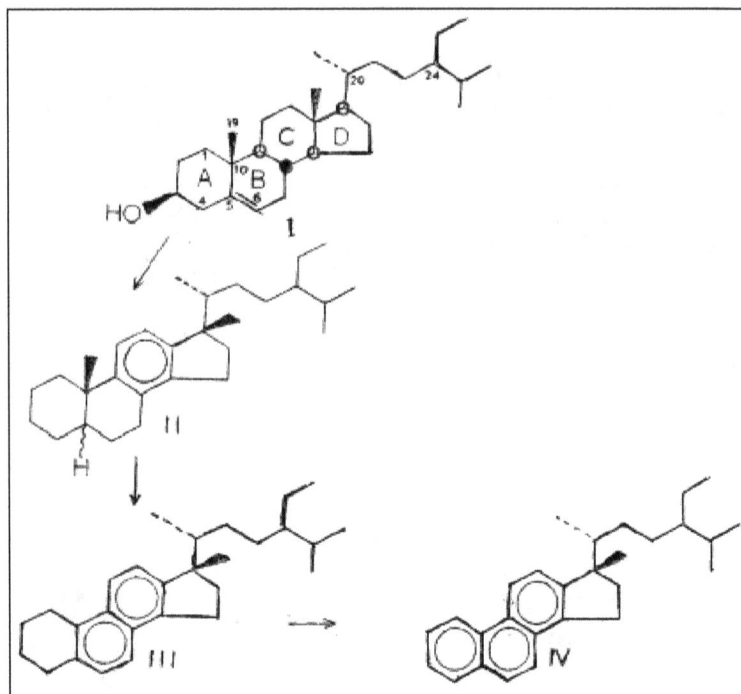

Figure 10.5: Simplified Reaction Scheme of Conversion of Sterol (I) to Monoaromatic (II), Diaromatic (III) and Triaromatic (IV) Steroids

and tri- aromatic steroids. Following are some of the uses of steroids as biomarkers in petroleum exploration (Tables 10.1 and 10.2)

(*i*) 24-n-propyl cholestane is ubiquitous in oils derived from marine source rock (Moldowan *et al.*, 1990)

(*ii*) Higher ratio of (24-iso propyl cholestane)/(24-n-propyl cholestane) in oils suggests their origin from Pre-Ordovician source rocks. Higher ratio of (24-norcholestane)/(26-norcholestane) suggests Tertiary source rocks (McCaffrey *et al.*, 1994 b).

(*iii*) Higher ratio (>1.0) of steranes/hopanes in oils suggests contribution of terrestrial organic matter and bacterial lipids to the source rocks. However, lower ratio (<1.0) indicates lacustrine source rocks.

(*iv*) The absence of dinosteranes, tri-aromatic dinosteroids always suggest Pre-Mesozoic or Younger source rocks (Moldowan *et al.*, 1996).

(*v*) Higher abundance of C_{29} mono-aromatic steroids suggests oils derived from sources older than 350 M.Y.B.P (Moldowan *et al.*, 1985)

(*vi*) The ratio of diasteranes/steranes in oils increases with increase of source rock maturity. This is helpful in identifying the late oil window. However, this ratio is affected by source lithology (low in carbonates and high in shales) and level of biodegradation.

(*vii*) Increase in the ratio of mono-aromatic steroids $(C_{21}+C_{22})/$ $(C_{21}+C_{22}+C_{27}+C_{28} + C_{29})$ in oils increase the thermal maturation of source rocks which is useful in identifying early to late oil window. This ratio is independent of the effects of oil degradation.

(*viii*) Increase in the ratio of tri-aromatic steroids $(C_{20}+C_{21})/$ $(C_{20}+C_{21}+C_{26}+C_{27}+C_{28})$ and that of (mono-aromatic)/ (mono-aromatic+ tri-aromatic steroids) in oils increase the thermal maturity of source rocks. Both these ratios are independent of the effects of biodegradation.

10.2.4 Porphyrins

Porphyrins are structurally related to chlorophylls. They mostly occur in sedimentary rocks and oils as vanadyl (Figure 10.1) and nickel (II) complexes. Copper (II) complexes are also reported in deep sea sediments. Complexes of vanadium and nickel with porphyrin are useful in identifying depositional environments and source rock maturity (Table 10.1). Typical examples are:

(*i*) Higher ratio of (V/V+ Ni) porphyrin complex in oils indicates an anoxic (reducing) depositional environment of source rock.

(*ii*) Higher ratio of V/Ni porphyrin complex (>1.0) in oils indicates Paleozoic while lower (<1.0) ratio indicates Mesozoic and Cenozoic source rocks (Lewin, 1984)

10.2.5 Diamandoids

Diamandoids are regid three dimensionally fused cyclo hexane ring alkanes that have a diamond–like (cage) structure (2.3.6). Typical examples of this group identified in crude oils and natural gas condensates include: (*i*) adamantane ($C_{10}H_{16}$), (ii) diamantane ($C_{14}H_{20}$), (*iii*) triamantane ($C_{18}H_{24}$), and (*iv*) tetramantane ($C_{22}H_{28}$) with the following structures:

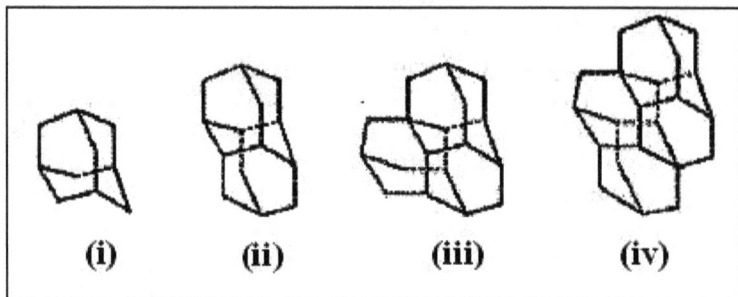

(i)   (ii)   (iii)   (iv)

The most important among them are adamantane, diamantane and their substituted alkyl homologs such as mono, di, tri- and tetramethyl adamantanes. The diamond structure endows these molecules with a high thermal stability and resistance to biodegradation. As such they find extensive applications in the study of high temperature thermal cracking of oils. For example, increase in methyl diamantane concentration is directly proportional to the extent of cracking of an oil (Wang *et al.*, 2006). Methyl adamantane index (MAI) and methyl diamantane index (MDI) exhibit excellent correlation with vitrinite reflectance (Ro) and their cross plots at different Ro values are frequently used for evaluation of maturity windows and depositional environments of high mature oil and condensates.

Adamantanes and diamantanes in oil and natural gas condensates can be identified by finger printing of their GC-MS Chromatograms (10.7.1) of saturated hydrocarbons in the range n-C_{10} to nC_{13} and n- C_{15} to n-C_{17} respectively and assigning the individual (m/e) peaks (Stout and Douglas, 2004). Measurement of diamondoid concentration levels using high resolution geochemical technologies (HRGT) and compound specific isotope analysis (CSIA) provides valuable information (10.7.2) on thermal cracking of crude oils and their correlation with source rocks. Such studies revealed the occurrence of multiple petroleum systems in some Brazilian and other South American Basins showing contribution of black oil from shallow source and highly cracked (light) oil from deep sources (Mello and Moldowan, 2006). This inturn, has opened new frontiers of exploration in areas considered to be mature petroleum systems.

10.3 Use of Biomarkers

Biomarkers in recent times, find several applications in petroleum explorations as mentioned earlier (10.1.3). In addition, they are extensively used in basin modeling and maturation of source rocks.

10.3.1 Basin Modeling

Source rock description and source rock maturity information derived from oil biomarkers are often key input data for basin modeling of a prospect or a block. For this purpose, specific biomarker parameters can be calibrated against specific kerogen quality parameters such as vitrinite reflectance (Ro) in a given basin. Then the biomarker ratios are measured in an oil sample from the basin and the values are calibrated to predict source rock characteristics. This approach allows explorationists to assess whether an oil was generated primarily from oil-prone or gas prone organic facies (Dahl *et al.*, 1994; McCaffrey *et al.*, 1994 a). The information gained from oil biomarkers (source, type, age, maturity, kerogen quality) when integrated into basin model has substantial economic impacts because it provides early estimates of oil quality and gas-oil ratio (GOR) for exploration targets in the area of interest.

10.3.2 Thermal Maturity of Source Rocks

The relative abundance of certain biomarkers in petroleum change as a function of source maturity. As a result, a variety of biomarker parameters have been identified that are useful for characterising the source rock maturity simply from analysis of the migrated oil (Peters and Moldowan, 1993). Biomarker maturity parameters (Table 10.2) make use of several processes that occur during source rock maturation. They include:

(*i*) Cracking–breaking of large molecules into smaller ones,

(*ii*) Isomerisation–changes in the 3-dimensional arrangements of atoms in molecules, and

(*iii*) Aromatisation–formation of aromatic rings (loss of hydrogen from naphthenes).

Several considerations should be kept in mind when using petroleum biomarkers to assess source rock thermal maturity. They include:

(*i*)   The exact relationship between a biomarker parameter and the source rock maturity is a function of heating rate, source lithofacies and source organic facies (kerogen type). As a result, the exact maturity (*e.g.*, vitrinite reflectance equivalent) associated with a given value for a biomarker parameter can change from basin to basin. Furthermore, the relationship between a biomarker maturity indicator and source rock maturity is generally non-linear.

(*ii*) With increasing maturity, many biomarker maturity indicators reach terminal values. Hence a given biomarker parameter is applicable only over a specific maturity range.

(*iii*) The concentrations of biomarkers decrease with thermal maturity.

Despite these limitations, biomarker indicators of source rock maturity can be extremely useful. For example, they can be used to determine the API gravity of a biodegraded oil prior to its degradation. This is accomplished by collecting a suite of non-degraded oils from the same petroleum system as the degraded oil. A correlation can be developed between a biomarker maturity parameter of a non-degraded oil and API gravity. The same biomarker parameter is then measured in a degraded oil, and the original API gravity is determined using the correlation developed from the non-degraded oil suite. Moldowan *et al.*, (1992) provided an excellent example of this approach in which they determined the original API gravity of degraded Adriatic oils. For this application, the most effective biomarker parameters are based on compounds that are highly resistant to biodegradation, such as [(tri-aromatic)/(mono-aromatic + tri-aromatic sterols)].

Biomarkers in petroleum are analyzed by gas chromatography-mass spectrometry (GC-MS) technique (10.7.1). Analysis are typically performed on saturated hydrocarbon fraction or the aromatic hydrocarbon fractions. These oil fractions are separated by liquid (column) chromatography (10.5.0) prior to GC-MS analysis. Biomarker organic geochemistry thus made substantial contributions to the understanding of thermal history and evolution of sedimentary basins. The present knowledge of structural aspects of biomarker hydrocarbons and the GC-MS technology currently available have sufficiently advanced so that biomarker analysis can form an integral part of the geochemical analysis of oil and sediment samples from exploration wells.

10.4 Analytical Techniques in Petroleum Exploration

Many geoscientists involved in oil exploration do not directly participate in laboratory geochemical analysis. It is nevertheless important that they be familiar with the techniques and methods of geochemical analysis. Any analytical method has its own merits and limitations and an understanding of possible limitations can often help geoscientists to deal with apparently anomalous data. A variety of analytical errors can occur, and unless data quality is carefully monitored the errors will not be noticed. The most important errors encountered in geochemical analysis are associated with sampling, storage and analytical methods.

10.4.1 Sampling, Storage and Analysis

It is generally difficult to bring to the surface a representative sample of petroleum from the sub-surface reservoir. This is because of differences in temperature and pressure at subsurface of the reservoir from where the samples normally present in the form of fluids are pumped to the surface. The volatile lighter fraction of petroleum can be lost in variable and largely unpredictable way. At the other extreme, parts of the heavier fraction may be left behind in the reservoir. Although devices are available for obtaining samples under reservoir conditions, they are expensive, difficult to use, and are not employed regularly. Additional sampling problem arises in oil fields that produce from multiple pay zones because of mixing of oils from different reservoirs which may or may not have same or similar composition. If the average composition of mixed oil widely varies, it is of little value from geological or geochemical point of view. It is, however, still important to the refiner (Barker, 1985).

Contamination during sampling is always a potential source of uncertainty in determining petroleum composition. A wide variety of chemicals such as mud additives, gases, lubricants, elastomer seals, corrosion inhibitors etc. used in drilling, testing and production operations are all possible contaminants to different extents. Oils should not be stored or transported in plastic bags or bottles (other than Teflon) because of likely dissolution of lighter ends in the samplers resulting in their loss, and partial dissolution of plasticisers (used while drilling) that will contaminate the aromatic fraction in particular, in the subsequent analysis. Canned, sealed samples are best, these precautions are seldom taken in routine operations of drilling.

Samples obtained from drilling *e.g.*, conventional cores, sidewall cores and cuttings are all susceptible to contamination with materials used in drilling fluids. The potential for contamination is much severe for cuttings which consists of small fragments of rocks broken off during drilling than others. One of the most serious contamination problems arises from mixing of different lithologies in the course of transport of cuttings to the surface. Careful microscopic examination is required to identify problem samples (Waples, 1981). Microbial activity which occurs during storage often can affect organic material (bitumen, oil or hydrocarbon gases) particularly in unlithefied sediments or those which are very rich in organic matter. However, its impact is not significant for rock samples because of inaccessibility of the organic matter.

In addition to errors that may likely to occur during sampling, preservation and storage as mentioned above, errors may also associate with analytical measurements. These include impurities in chemicals and solvents used as reagents, faulty functioning of instruments and systematic errors associated with analytical methods. They can however, be largely eliminated or reduced by stringent tests. These include: use of reference materials as standards for calibration of instruments and quality checks of analytical data by statistical methods.

Once the samples have been obtained and their authenticity and suitability have been checked, the particular analytical scheme to be followed in the laboratory depends upon the type of information required. Detailed analysis of petroleum generally involves an initial separation into fractions. For some studies, distillation and collection of narrow boiling point cuts is useful. For others, different fractions of related compounds such as saturates, aromatics, resins and asphaltenes may be required. For special purposes, further fractionation *e.g.*, saturates into n-alkanes and the rest (iso plus cycloalkanes); aromatics into mono, di, tri and polynuclear aromatics etc., may be needed. This can be achieved by employing separation techniques such as column chromatography and high pressure liquid chromatography (HPLC). Fractions obtained in these various separation schemes can be characterized in detail by gas chromatography (GC), mass spectrometry (MS) or a combination of these two (GC-MS) and several additions to this technique. Identification and quantitative estimation of individual constituents or compounds in bitumen can be achieved by instrumental techniques such as spectroscopy (uv-visible, infrared, electron spin resonance, nuclear magnetic resonance), thermogravimetry, differential thermal analysis etc. Principles and application of these analytical techniques in organic geochemistry are briefly described in the following pages.

10.4.2 Kerogen and Bitumen Analysis

Many methods involved in kerogen and bitumen are based on their preliminary separation from source rock (6.1.0). The separation of kerogen and its subsequent analysis by petrographic and geochemical methods for maturation and source rock evaluation are discussed earlier (Chs. 6 and 7).

Analysis of bitumen requires its prior separation from source rock by extraction. There are three kinds namely (i) solvent, (ii) ultrasonic, and (iii) thermal extraction procedures available for separation of bitumen from source-rock. The traditional one is solvent extraction which is usually carried out with a soxhlet condenser. This allows continuous extraction with solvent for a time period of 8-24 hours. A variety of solvents such as chloroform or a mixture of benzene and methanol are widely used. Ultrasonic extraction was popular for some time but it has not been widely used now. One definite advantage with this method is its speed (extraction can be completed within 20 minutes), but the results are not reproducible compared to those obtained by soxhlet extraction. The third type is thermal extraction, a relatively recent development which shows great promise. In this method, a crushed rock sample is exposed to high temperature which mobilises its bitumen and allows it to be swept out of the sample by a gas flow. It is then collected and analyzed. The temperature used for thermal extraction should be lower than that required to decompose kerogen (6.5.0).

The next step involved in solvent and ultrasonic extraction techniques is removal of the solvent. This is usually accomplished by low temperature evaporation, which often is accomplished by passing a stream of nitrogen gas over the solution. During this process, the more volatile bitumen molecules evaporate with the solvent, and the remaining bitumen is severely depleted in compounds that have fewer than 15 carbon atoms. It is therefore commonly referred to as C_{15}+ bitumen. If one would like to compare extraction data obtained from different groups of samples, it is essential that standard conditions of solvent type, extraction time, particle size of source rock etc. should be maintained. Even then, results are not always as reproducible as one might desire.

After removal of solvent from the extracted bitumen, the next step is to separate different classes (fractions) of compounds in the bitumen. The asphaltenes are first separated by the addition of an excess of pentane and filtered out. The remaining fractions in the bitumen (saturates, aromatics and resins) are then separated by column chromatography.

10.5 Column Chromatography

The column consists of a long vertical glass tube with a stop cock at the bottom just like a burette. It is filled with a slurry containing silica gel or alumina (stationary phase), both of which have considerable adsorptive powers. The sample solution is introduced at the top of the column and it is eluted successively with solvents (mobile phase) of increasing polarity (heptane, benzene and methan). The most polar organic compounds are adsorbed more strongly on the column packing. A non-polar solvent like heptane will elute only non-polar compounds, which are weekly adsorbed; thus the heptane fraction will contain the saturated hydrocarbons. Subsequent elution of the column with benzene will remove the aromatics and some sulphur compounds. Methanol will elute most of the remainder (resins) except asphaltenes which are already separated by precipitation with pentane. It is thus easy to separate the four fractions–asphaltenes, saturated hydrocarbons, aromatics and resins, and to carry out analysis on each of them individually.

Saturated hydrocarbon fraction is often treated with crystalline urea or synthetic zeolite molecular sieves. This treatment yields two fractions. One containing alkanes which from adducts with urea (Schiessler and Flitter, 1952) or adsorbed on the molecular sieves (O'Connor *et al.*, 1962), and the second fraction comprising branched and cyclic hydrocarbons remaining in the solution. Most bitumen analysis involves either the whole C_{15}+ bitumen or the saturated hydrocarbon fraction only.

High pressure liquid chromatography (HPLC) is similar to column chromatography except that the mobile phase (solvent) is pumped under high pressure through a column containing fine grained reactive solid (stationary phase). An ultraviolet (UV) detector is used to detect a compound as it emerges from the column after separation. It can be qualitatively identified from its retention time and quantitatively estimated from the area of its peak (10.6.0) in the chromatogram. HPLC is used as a nondestructive preparative technique to separate individual compounds from a complex mixture of petroleum.

10.6 Gas Chromatography (GC)

Gas Chromatography (GC) technique is similar to column chromatography except that the mobile phase is an inert gas (helium) rather than a liquid. The column is very long thin tube of metal or glass which is packed with a thin film of liquid coated on a inert porus support material like diatomaceous earth (Kieselguhr) acting as a stationary phase. It is housed in a coil form in an oven which can be heated to the desired temperature upto 300-400°C. The rate of movement of the sample (gas) through the column depends upon the nature of the column, sample polarity and column temperature. Often the column temperature is changed during the run according to a fixed programme. The most mobile (volatile) component passes through easily at low temperature, while the least mobile requires higher temperature in order to move at reasonable rate. Changing column temperature in gas chromatography accomplishes the same thing as changing solvent polarity in column chromatography. As the individual compounds emerge from the gas chromatograph, they pass through a detector (preferably flame ionization), and their presence is recorded on a chart in the form of peaks. The area under each peak is proportional to the number of molecules of that particular compound present in the sample. Identification of individual compounds is accomplished by coinjection of a standard (reference) compound such as n-pentane or methyl palmitate. The individual compounds are qualitatively identified by their retention time which is defined as the time elapsed between the injection of sample in to the chromatographic column and emergence of its peak in the detector. It is characteristic of a compound separated in the column.

Gas chromatography is generally employed in the analysis of light hydrocarbons (C_1- C_7) collected during mud logging which is used to detect hydrocarbon pay zones with rotary drilling. The principle involved in this technique is that during breaking of the rock into small fragments or cuttings by the drill bit, the gas present is released into the mud-stream. It is then fed to a gas chromatograph wherein individual hydrocarbons are analyzed. Light hydrocarbon compounds (including aromatics and alicyclics) produced by thermal processes are usually analyzed as two sub-groups: (i) head space or cutting gas (C_1- C_4) and (ii) gasoline range (C_4- C_7) of cuttings or cores. They are of particular interest in organic maturation studies since their abundance and composition have been found to be maturity indicators of source rocks (Bailey *et al.*, 1974; Hunt and Whelan, 1978).

Gas chromatography technique is most commonly employed for the analysis of hydrocarbons present in kerogen or bitumen after extraction from source rock. If bitumen analysis with either urea adduction or molecular sieve adsorption has been carried (10.5.0), the two saturated fractions (n-alkanes, and branched plus cyclic alkanes) can be analyzed separately. Seifert and Moldowan (1978) have developed a method for the separation of aromatic fraction. Figure 10.6 shows a typical n-alkane chromatogram. The n-alkanes emerge from the chromatograph at more or less regular intervals. A series like this is called a homologous series and the individual members of the series are called homologs. Figure 10.7 shows a gas chromatogram for the branched-cyclic fraction of the same oil. No homologous series are apparent in this figure. The urea adduction separation is very useful in producing more easily interpretable chromatograms because the dominant n-alkanes no longer obscure some of the smaller branched-cyclic hydrocarbon peaks. Modern chromatographs are designed with interchangeable columns and detectors to analyze saturates and aromatics in the C_{10}–C_{30} range. The most prominent development in GC is the improved high resolution glass capillary columns which are capable of direct separation of highly complex mixtures of hydrocarbons (Douglas. 1969). The branched- cyclic fraction is difficult to analyze inspite of urea adduction because it contains many kinds of molecules.

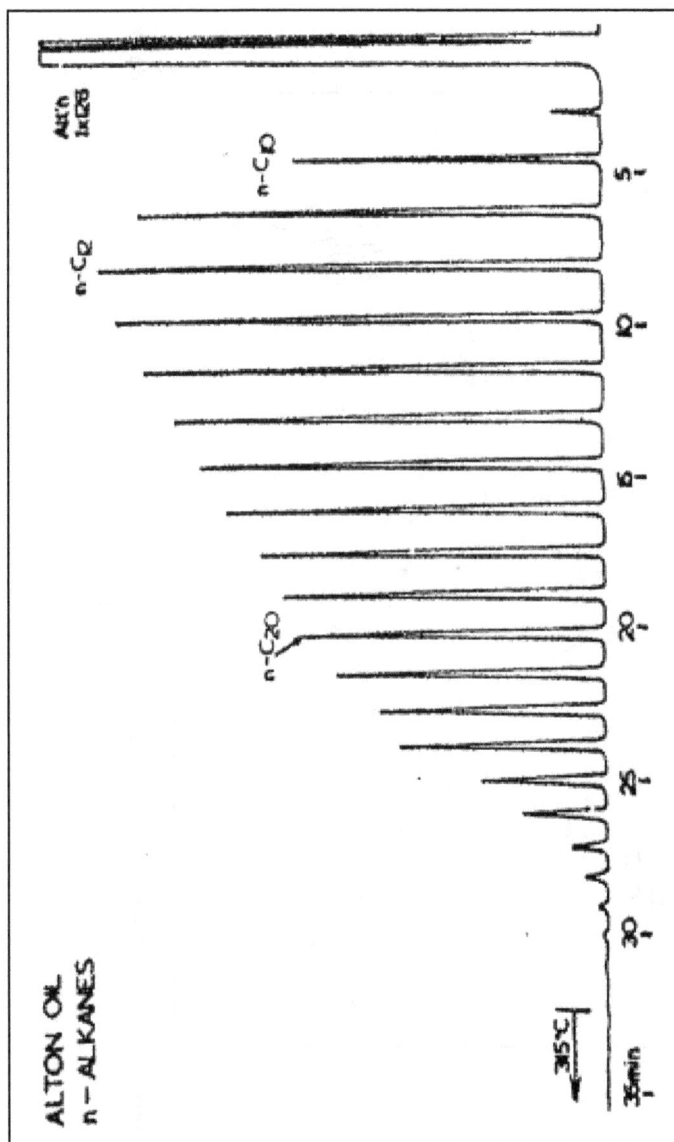

Figure 10.6: Gas Chromatogram of n-alkanes of an Australian Crude oil
(Source: Waples, 1981)

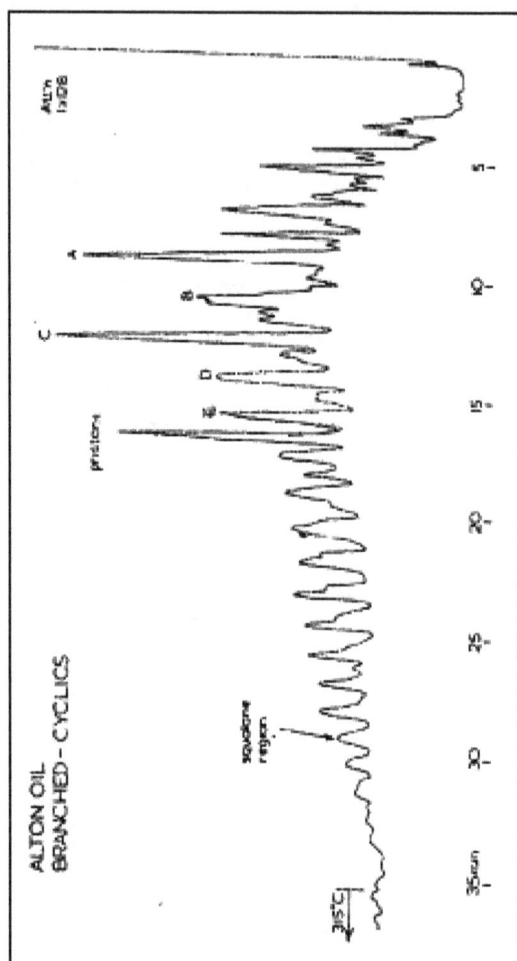

Figure 10.7: Gas Chromatogram of the Branched and Cyclic Hydrocarbons of an Australian Crude Oil (*Source:* Waples, 1981)

However, combined gas chromatography and mass spectrometry (GC-MS) is very helpful in this case (10.7.1).

10.7 Mass Spectrometry

In the mass spectrometry method a small aliquot of the sample (to be analyzed) is initially heated to produce gaseous molecules and bombarded with a beam of high energy electrons to break them into charged ions. Some of the resultant ions will decompose to give rise to a series of charged fragments of various masses. The fragmentation pattern of each compound, which is closely related to its chemical structure, provides a unique fingerprint for that compound. The charged fragments are then passed through a magnetic field, where they are separated according to their mass to charge (m/e) ratios. The number of particles of each mass is then recorded. Samples smaller than nanogramme (10^{-9} g) can be analyzed for masses upto 2,000 in few seconds. Figure 10.8 shows a typical mass spectrum of 3-methyl pentane. It shows the molecular ion M^+ of the parent compound with m/e of 86 followed by three peaks with m/e of 71, 57 and 43 corresponding to the breaking of the parent ion into smaller charged fragments. The height of each peak is proportional to the number of ions with that particular m/e value. Interpretation of mass spectra of petroleum hydrocarbons is

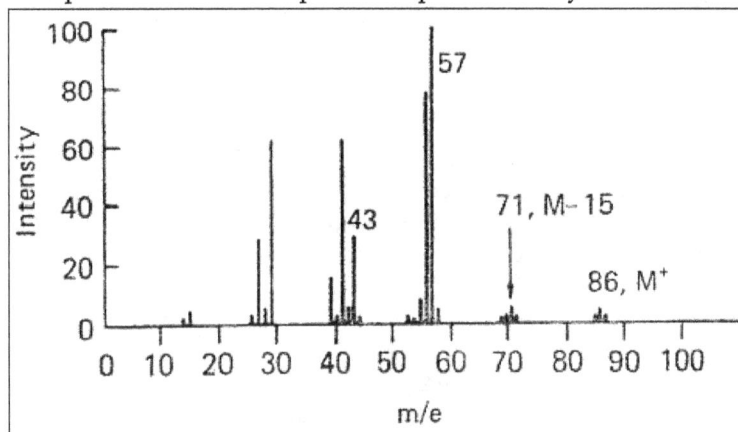

Figure 10.8: Mass Spectrum of 3-methylpentane
(*Source*: Waples, 1981)

somewhat complex and can be done by a specialist with the aid of a catalogue of mass spectra of different hydrocarbon compounds.

10.7.1 Combined GC-MS

In a combined GC-MS analysis, the sample containing the mixture of hydrocarbons is first run through a gas chromatograph which then automatically enters into the mass spectrometer where they are analyzed individually. Interfacing the mass spectrometer with the gas chromatograph eliminates human handling of extremely small amounts of material emerging from the latter. The mass spectrometer is usually connected to a computer to handle the huge amount of data generated. Combined GC-MS is very useful in analyzing complex compounds such as steranes and triterpenes as well as elucidation of their structures (Gallegoes *et al.*, 1967; Burlingame and Schnoes, 1969). The fact that compounds with similar chemical structures have similar mass spectra proved very useful to petroleum geochemists in exploration. For example steranes (cholestane, ergostane and sitostane) all have a characteristic large m/e peak at 217 derived by the fragmentation of their parent molecular ions (Figure 10.9). Similarly most triterpanes have a large m/e peak at 191; mono-aromatic steranes at 239 and 253; and triaromatic steranes have a large m/e peak at 231. These characteristic peaks are most useful in their identification. Detailed analysis of GC-MS chromatograms can be used as fingerprints in correlating samples. Alternatively detailed analysis of the individual compounds is of considerable significance in the study of their source, diagenesis, maturity and biodegradation.

10.7.2 High Resolution Geochemistry Technology (HRGT)

Recent developments in high resolution geochemistry technology (HRGT) involve the use of sophisticated instrumental techniques such as continuous flow (CF)–IRMS and metastable reaction monitoring (MRM)–GC–MS which are recent developments of conventional GC-MS. Isotope ratio mass spectrometer (IRMS) works on the same principal of mass spectrometer (10.7.0) but it is specifically designed to measure the different proportions of a particular isotope. When it is coupled with GC, a continuous flow of gases from a mixture can be fed to IRMS whereby they can be effectively separated and possible configurations of each isotope (isotopologues) can be measured rapidly at very low (nanogram) levels. Some of the petroleum hydrocarbon gases measured as their isotopologues (given in brackes) include: H_2 (2,3), CO_2 (44,45,46), N_2

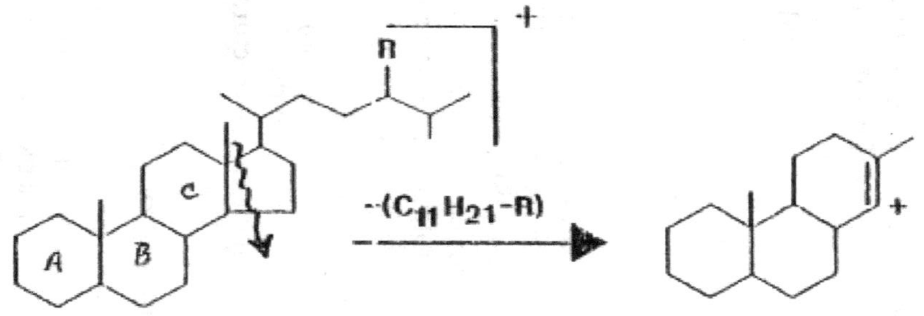

m/e 371 + R

m/e 217

R	TYPE OF COMPOUND	MOLECULAR ION	FRAGMENT ION
H	CHOLESTANE (C_{27})	372	217
CH_3	ERGOSTANE (C_{28})	386	217
C_2H_5	SITOSTANE (C_{29})	400	217

Figure 10.9: Principal Fragmentation Pattern for Steranes in a Mass Spectrometer (*Source*: Waples, 1985)

(28,29), O_2 (32,34) and SO_2 (64,66). This technique is widely used for assessment of maturity, origin and genetic classification of petroleum and coal gases. Compound specific isotope analysis of oil and gas helps in understanding petroleum systems even at micro level variation of isotope ratios. This technique is useful in carrying out finer correlations of oil-oil and oil-source rock, and understanding paleo depositional environments.

Metastable reaction monitoring (MRM)–GC–MS is useful for accurate measurement of high temperature thermal maturity of compounds such as diamondoid molecules (10.2.5) in oils and condensates for understanding their cracking pattern. It can also be employed for measurement of specific biomarker compounds in oils such as steranes and hopanes (Table 10.1) indicative of the age of their source rocks.

10.8.0 Spectroscopic Methods

Among Spectroscopic methods for the analysis of kerogen and bitumen hydrocarbons, the most important are: Ultraviolet (UV), Visible, Infra red (IR), Electron spin resonance (ESR) and Nuclear magnetic resonance (NMR). Application of spectroscopic methods to composition and structural analysis of petroleum hydrocarbons has received considerable attention in recent years. As composition studies in petroleum advanced to higher molecular weight ranges, physical nature of the material often requires that structural analysis be performed by a more convenient and suitable technique. It is perhaps for this reason that spectroscopic methods of structural analysis have come into fairly common usage.

10.8.1 Ultraviolet (UV) Spectrophotometry

Ultraviolet (UV) spectrophotometry is one of the powerful techniques to determine the types of aromatic systems in petroleum asphaltenes (Rao, 1961). The technique can add valuable information about the degree of condensation of poly-cyclic aromatic ring systems when used in conjunction with high performance liquid chromatography (HPLC). This approach not only confirms the complex nature of asphaltene fraction, but also allows further detailed identification of the individual functional constituents of asphaltenes (Lee *et al.*, 1981).

One of the consequences of absorption of UV radiation is the phenomena of fluorescence which is widely employed in organic maturation studies of kerogen. In the fluorescence method, an aliquot of sample is irradiated with UV light which excites the electrons into unstable higher energy levels. After a short time lag, the excited electrons return back to the ground state by re-emitting the excess energy as fluorescent light. The exact wavelength and intensity of the fluorescent light is related to the nature of the compound and its concentration respectively. Fluorescence is used to characterize the evolution of kerogen as it is initially high in shallow immature samples and decrease to a minimum in the zone of oil generation (Alpern *et al.*, 1972). It drops to zero at the end of wet gas zone. Furthermore, the fluorescent light moves progressively to longer wavelengths with increasing depth of burial. Fluorescence is a promising technique for the characterization of kerogen evolution. Its range of variation is mostly located over the zone where vitrinite reflectance increases little with increase of depth *e.g.*, beginning of oil generation zone. It is valuable in the analysis of vitrinite poor samples (such as Type I and II) of algal origin (Ting, 1975). The method is routinely used in the field for source rock analysis.

10.8.2 Visible Spectrophotometry

Visible absorption spectrophotometry is used as a fast and inexpensive method for monitoring the purity of samples separated by liquid chromatography. It is also used extensively for the estimation of porphyrins and their metal complexes (Brooks, 1977). Basic porphyrins show intense absorption in the visible region (400 nm) which is used for their qualitative identification and quantitative estimation. Unlike basic porphyrins, their metal complexes, particularly vanadium and nickel show different absorption maxima. For example, vanadyl-porphyrin complex shows two absorption maxima at 569 and 535 nm. They shift to lower wavelengths (545 and 512 nm) in case of nickel-porphyrin complex. The metal (V and/or Ni)–porphyrin complexes are extensively used as maturation and correlation indicators (Eglinton *et al.*, 1980).

10.8.3 Infrared (IR) Spectroscopy

Conventional infrared (IR) spectroscopy yields information about the functional groups of various petroleum constituents, and aids in the identification of N-H and O-H groups, the nature of polyethylene chains, C-H out of bending frequencies and the nature of any polynuclear system. The main zones of interest in IR spectra of petroleum fractions (Tissot and Welte, 1984) include:

☆ A wide systematic band centered at 3430 cm^{-1} related to hydroxy (OH) groups (phenolic, alcoholic, carboxylic).

☆ A strong absorption maxima at 2920 and 2855 cm^{-1} related to CH_2 and CH_3 aliphatic groups.

☆ A wide band with a maximum around 1710 cm^{-1} attributed to various C=O groups (ketones, acids, esters).

☆ A wide band with a maximum around 1630 cm^{-1} mostly related to aromatic C=C with partial contribution of olefinic C=C.

☆ An absorption band at 1455 cm^{-1}, either due to CH_3 or due to linear and cyclic CH_2 groups, another band at 1375 cm^{-1} related to exclusively CH_3 group.

☆ A very wide band from 1400-1040 cm^{-1} that includes C=O stretching and O-H bending.

☆ A succession of weak bands from 930 to 700 cm^{-1} related to various aromatic CH (bending out of plane) and dependent on the number of adjacent protons.

☆ A band at 720 cm^{-1} due to aliphatic chains of four or more carbon atoms.

With the development of Fourier transform infrared (FTIR) spectroscopy, quantitative estimates of various functional groups are also made possible. This is particularly important for higher molecular weight solid constituents of petroleum (*e.g.*, asphaltenes). It is possible from IR data to deduce structural parameters such as (*i*) saturated hydrogen to saturated carbon ratio, (*ii*) paraffinic and naphthenic character and (*iii*) paraffin chain length.

Advantages of IR technique are : use of very small samples, avoidance of complicated and time consuming treatments. Organic maturation processes produce kerogen residues with increasing carbon contents whose chemical composition appears to be independent of the nature of its precursors. IR spectra obtained from samples of increasing organic maturation show main chemical alterations such as removal of carbonyl and carboxyl groups, removal of saturated hydrocarbon (C-H) groups, formation and removal of aromatic C-H groups, and evolution of hydroxy (O-H) groups. IR spectra combined with other geochemical analysis of kerogen from a sedimentary basin can give information on (*i*) identification of precursors, (*ii*) evaluation of level of organic maturation, and (*iii*) estimation of oil potential of source rocks and identification of oil generation basement (Brooks, 1981).

10.8.4 Electron Spin Resonance (ESR) Spectroscopy

Electron spin resonance (ESR) has been proposed as a routine tool for measuring the degree of evolution of source rocks or even the paleotemperatures (Pusey, 1973). The parameter which is responsible for paramagnetic susceptibility and thus ESR signal is the abundance of free radicals from kerogen and bitumen molecules. Free radicals appear in them due to changes in their chemical structure as a result of splitting of bonds (Figures 10.10 and 10.11). In general, the abundance of free radicals increases with the depth

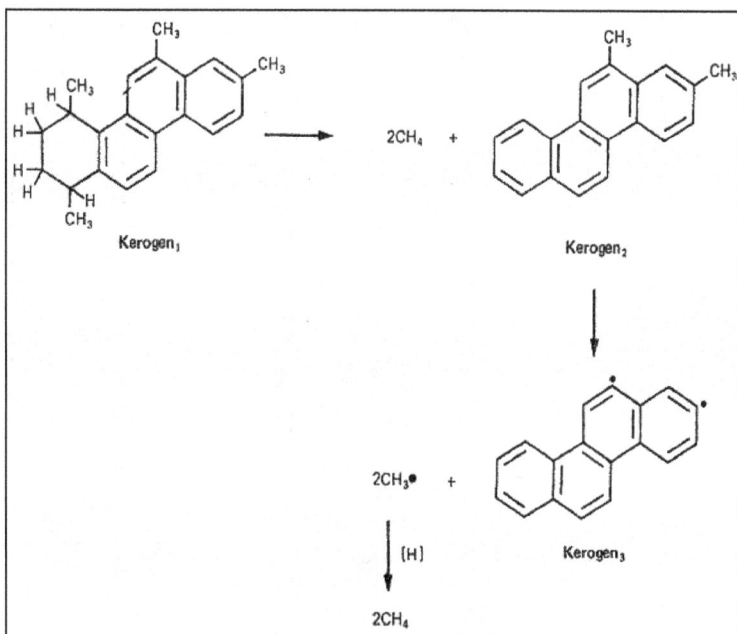

Figure: 10.10: Formation of CH$_4$ from Kerogens.
• Indicates free radicals
(***Source***: Waples, 1981)

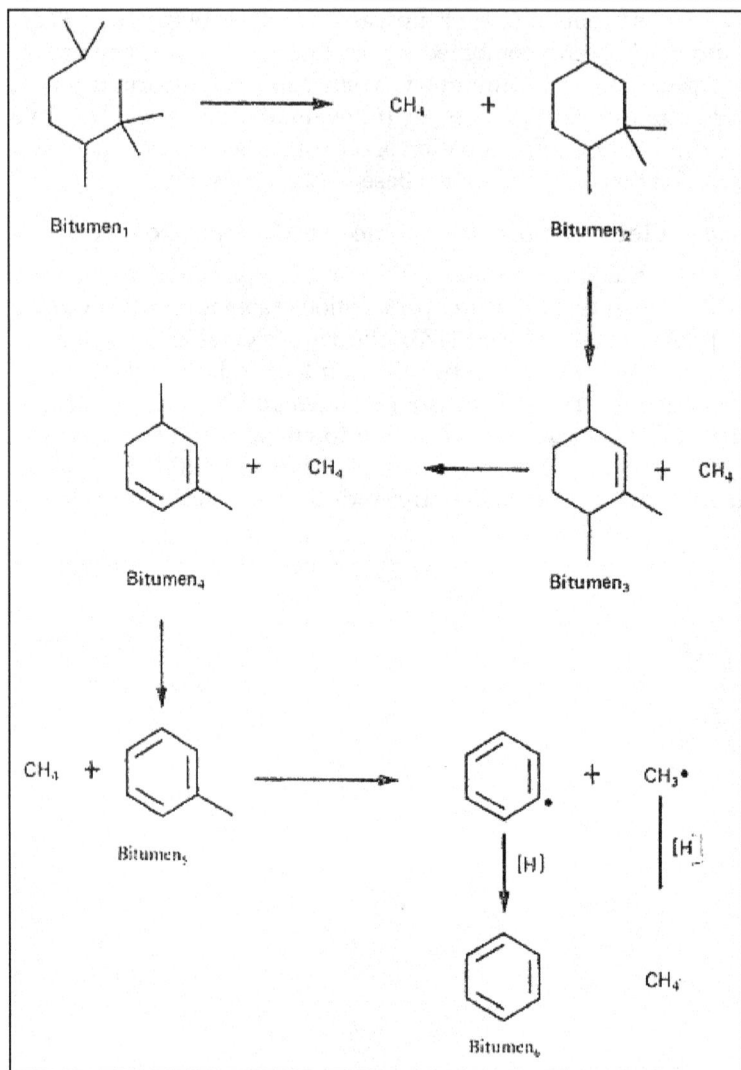

Figure 10.11: Formation of CH₄ from Bitumen.
• Indicates free radicals
(*Source*: Waples, 1981)

of burial and temperatures to a certain stage of evolution, which approximately corresponds to a 2 per cent vitrinite reflectance level. At that stage the paramagnetic susceptibility is maximum. With

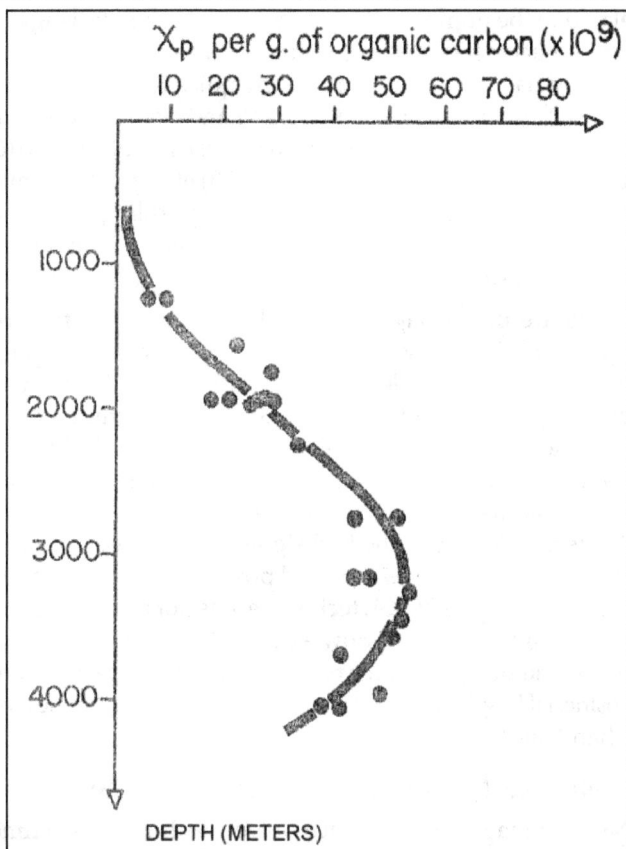

Figure 10.12: Variation of Paramagnetic Susceptibility λ_p in Type III Kerogen (Douala Basin) as a Function of Depth (*Source*: Tissot, 1977)

further burial and temperature increase, the signal decreases markedly (Figure 10.12). Beyond the maximum, which broadly corresponds to the end of wet gas and condensate generation, the decrease of signal may be due to coalescence of the polyatomic nuclei to form larger aggregates, resulting in the disappearance of free radicals (Tissot, 1977).

ESR technique can be used for measurement of paleo-temperatures. It is based on the fact that the atomic fraction of kerogen contains free electrons (radicals) whose number and distribution are related to the number and distribution of aromatic rings. These electron characteristics change progressively as the kerogen matures. The method has the advantage of being quick (once a laboratory is established), and it can be used on small and even impure samples of kerogen. As with most paleothermometers *e.g.*, carbon ratio (6.3.1), pyrolysis (6.5.0), vitrinite reflectance (6.2.3) etc., great care must be taken that the shale samples collected during drilling are not caved. Ideally, conventional core or side wall core materials should be used (Durand, 1980) for this purpose.

There are three major drawbacks in using ESR for organic maturation studies. These include: (*i*) the influence of kerogen type. For example, at the same level of maturation, vitrinite has a higher concentration of free radicals than exinite and liptinite due to its inherent chemical structure which favours free radical formation and stabilization, (*ii*) measurement of free radicals in the thermally altered organic matter is affected by the presence of trace amounts of metal ions thus lowering the ESR signal, and (*iii*) reduction of ESR signal over long geological time and post-mature rocks. Because of these reasons, the use of ESR technique was not much in favour for organic maturation studies now a days. However, limited use can be made in the study of source rocks containing kerogen of constant composition if large number of sample are available to create a profile (Marchand and Conard, 1980).

10.8.5 Nuclear Magnetic Resonance (NMR) Spectroscopy

Nuclear magnetic resonance (NMR) technique is claimed to have great precision in the determination of composition (aromatic, paraffinic and olefinic) and H/C ratios in petroleum products. Information on the hydrocarbon type, composition of gasoline and related compounds can be obtained by NMR (Myers *et al.*, 1975). Proton magnetic resonance (PMR) spectroscopy has been used to measure the maturity of petroleum (Alexander *et al.*, 1980). Aromatic hydrocarbons from a series of petroleum bearing sediments were determined by PMR. The percentage aromatic protons (PAP) in aromatic fraction increases uniformly with increase in organic maturation (burial depth). Further, the PAP values are proportional to the level of maturity of the organic matter.

The ability of high resolution PMR to distinguish hydrogen atoms in different chemical environments and to estimate their amounts directly from the intensities of signals makes it a convenient technique to study high molecular fractions such as asphaltenes present in petroleum. Three hydrogen types can be estimated in petroleum extracts and related materials: (i) aromatic and phenolic hydrogen, (ii) aliphatic hydrogen adjacent to an aromatic ring, and (iii) aliphatic hydrogen remote from an aromatic ring (Speight, 2007). It is also possible to estimate the hydrogen present in methylene bridges between aromatic rings. High resolution PMR clearly affords an alternative to GC-MS and other spectroscopic methods for compounds either present in small amounts or relatively high molecular weights. Construction and fabrication of NMR devices particularly suitable to petroleum hydrocarbons would greatly reduce the time (2 minutes per sample) in routine analysis.

Developments of carbon–13 magnetic resonance (CMR) technique has extended its application to structural types in petroleum and allowed more precise identification of carbon atoms in various locations (Tissot, 1984). It is used to indicate differences in aromatic carbon content of asphaltenes from different crude oils. With the introduction of solid state high resolution CMR technique, quaternary and tertiary aromatic carbons, and ratios of primary to quaternary and secondary to quaternary aliphatic carbons can be estimated.

10.9 Thermogravimetry (TG) and Differential Thermal Analysis (DTA)

Thermogravimetry (TG) and Differential thermal analysis (DTA) have become increasingly useful for characterizing kerogen (Brooks, 1981). TG is a technique for measuring weight changes in a sample as a function of temperature. DTA technique measures the enthalpy (heat of reaction) changes of a sample as a function of temperature. These techniques may give some indication of the structure of kerogen but their main contribution is to follow the evolution of kerogen. When combined with other techniques such as mass spectrometry (10.7.0) for characterizing the products from pyrolysis, TGA provides valuable information on the kerogen structural types and their changes during catagenetic evolution.

TG analysis has been used to follow the organic maturation of different types of kerogen during burial. Based on TG analysis, it is possible to deduce that (*i*) heavy loss of weight in the early stages of heating corresponds to the release of oxygenated products (CO_2, H_2O) and to a kerogen rich in oxygen (vitrinite), and (*ii*) heavy loss of weight in later stages corresponds to the release of hydrocarbons and to a kerogen rich in hydrogen (liptinite and exinite). Such observations are used to classify kerogen according to its nature and degree of organic maturation. This classification closely resembles that obtained from elemental analysis (6.3.3).

TG analysis of the Paris basin type II kerogen in nitrogen atmosphere (Figure 10.13) shows a weight loss amounting to 70 per cent (sample A) corresponding to the stage of diagenesis. During catagenesis the weight loss decreases regularly to 55 per cent (sample B), then 40 per cent (sample C) and finally to 20 per cent (sample D). In very deep samples, where metagenesis is reached, the weight loss decreases to about 10 per cent. The TGA temperature corresponding

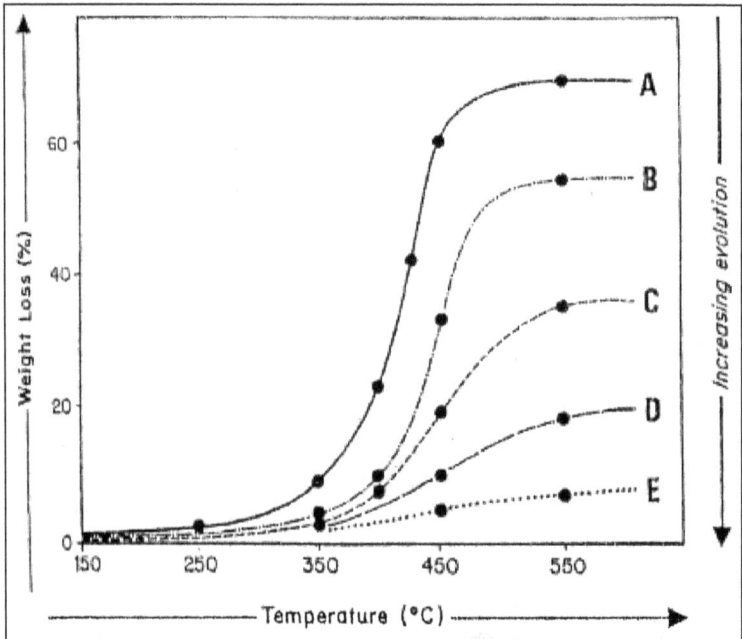

Figure 10.13: Thermogravimetric Analysis of Kerogen
Type II (Temperature increase of 4°C min⁻¹). Kerogens
from the Paris Basin and from the Sahara area
arranged in order of increasing evolution.
(*Source*: Tissot and Welte, 1984)

to the maximum degradation rate also increases with depth. This evolution is in agreement with the hypothesis of progressive elimination, through bond breaking, of the non-condensed constituents of kerogen (*e.g.*, aliphatic chains and saturated or aromatic cycles, single or associated in small number), the residue becoming progressively more condensed and less sensitive to degradation (Tissot *et al.*, 1974, a).

Differential thermal analysis (DTA) provides additional information on the evolution of kerogen. Combustion in an oxygen atmosphere shows a strong endothermic peak in the temperature region 350-450°C followed by exothermic peak at higher temperature in the range 720-1000°C. The exothermic peak shifts towards higher temperatures with increasing depth (maturity) of the sample (Von Gaertner and Schmitz, 1963; Tissot and Welte, 1984).

Chapter 11

Geochemical Basin Modeling: Application to Petroleum Exploration

---

Application of geochemistry for petroleum exploration is realised to the maximum extent when it is integrated with geology and utilize a comprehensive model for entire hydrocarbon system of a basin.

11.1 Integrated Models of Hydrocarbon System

There are primarily two types of models namely qualitative and quantitative. Qualitative models are mostly descriptive and use standard geological techniques of mapping and construction of cross-sections. They provide information on probable locations of hydrocarbon generation, their pathways and likely areas of accumulation and preservation. The quantitative models are broadly of two types: (a) deterministic and (b) probabilistic.

Deterministic models (Yukler and Welte, 1980; Welte and Yukler, 1981; Nakayama and Vansiclen, 1981; Ungerer *et al.*, 1984) seek to identify and quantify all variables of the system and thus predict its behaviour by establishing values or limits for each system. They require large amount of input data and therefore worthwhile only when extensive exploration has already been carried out. If data are abundant they may work well and even have advantages because properly formulated models suits well to describe the unique aspects of a basin.

Probabilistic models, in contrast, do not presume to have identified all possible variables. They overcome this limitation by fitting the model to a selected, well studied and understood set, before application to a new system. In this approach, it is believed that the effects of many unidentified or poorly understood variables can be taken into account indirectly. In probabilistic models both input and output data are presented as probability distributions. This approach allows one to take the natural heterogeneity of the system into account in the input data and select one's own confidence level in interpreting the output data. Several workers (Bishop *et al.*, 1983; Sluijk and Nederlof, 1984; and Barker *et al.*, 1984) have employed these models in hydrocarbon systems. Retrospective analysis of drilling results indicates significant improvement in exploration efficiency by using the probabilistic models (Murris, 1984). Further details on these models are beyond the scope of this chapter and the reader is advised to refer to reviews (Yukler and Kokesh, 1984; Waples, 1984) and books (Ungerer *et al.*, 1984; Murris 1984; Dore *et al.*, 1990; Welte *et al.*, 1997).

11.2 Qualitative Models of Hydrocarbon System

The methodology to be followed in developing a qualitative geochemical model for hydrocarbon system of a basin involves the following steps:

(*i*) Tectonic style and evolution of the basin (Bois *et al.*, 1982; Kingston *et al.*, 1983, b).

(*ii*) Distribution and maturity of source rocks (North, 1979 and 1980, a, b).

(*iii*) Deposition and diagenesis of carrier beds and development of tectonically controlled permeability zones (faults, fractures etc.)

(*iv*) Effective migration pathways on the basis of (*a*) distribution of source rocks, (*b*) their juxtaposition with adequate carrier beds or other migration pathways, and structural contours controlling direction and limiting the extent of migration.

(*v*) Preservation of oil and gas based on thermal considerations, biodegradation and presence of possible sulphate rich reservoirs.

Steps (*ii*) through (*iv*) involving time and space, require an understanding of the tectonic factors that control subsidence rates, lithology, type of organisms, organic and inorganic diagenesis, paleoclimate, structural development and heat flow. The more complete the understanding of a basin or the closer comparison we can draw between it and another well studied basin, the more is the confidence we will have in our model. Several researchers (Bois *et al.*, 1982; Cohen, 1982; Harding, 1984) have suggested ways in which general principles of basin analysis can be applied for petroleum exploration.

11.3 Quantitative Geochemical Models

A quantitative approach allowing computation of the amount of oil and gas generated and migrated from any place of a basin requires the following information.

(*i*) Amount of oil and gas generated in each source rock and in every part of the basin.

(*ii*) Time, temperature conditions of hydrocarbon formation (6.6.0).

(*iii*) Amount of oil and gas expelled out of the source rock into the porous reservoir rock and location of their accumulation in traps.

(*iv*) Evaluation of the ultimate oil and gas reserves of a sedimentary basin (7.3.0).

As petroleum and natural gas result from kerogen degradation, various indices have been proposed (7.2.0) to characterize the quantity, quality and evolution stage (maturity) of kerogen. However, these indices do not allow a truly quantitative evaluation of hydrocarbon generation as a function of time. This is because of the fact that most of the indices do not account for the respective kinetics of degradation of different types of organic matter and also for complex burial histories. However, this can be achieved by mathematical models based on the kinetics of kerogen degradation (Tissot and Welte, 1984).

11.4 Mathematical Model of Kerogen Degradation and Hydrocarbon Generation

A mathematical model of petroleum generation, accounting explicitly for geological time, was first introduced by Tissot (1969) and later developed by Tissot and Espitalie (1975). Several researchers (Braun and Burnham, 1987, 1990; Tissot *et al.*, 1987; Ungerer, 1990; Schenk *et al.*, 1997) carried out extensive work on kinetic models of petroleum generation, expulsion and cracking. The model proposed by Tissot (1969) based on kinetics of kerogen degradation is described in detail.

Kerogen is a macromolecule composed of polycondensed nuclei bearing alkyl chains and functional groups, the links between nuclei being heteroatomic bonds or carbon chains. As the burial depth and temperature increase, the bonds are successively broken, roughly in the order of increasing rupture energy. The products generated are first heavy heteroatomic compounds, carbon dioxide, and water; then progressively smaller molecules; and finally hydrocarbons. At the same time, the remaining kerogen becomes progressively more aromatic and evolves towards a carbon residue (Figure 11.1). The purpose of the mathematical model is to represent the kinetics of the parallel and successive reactions shown in the figure which are not reversible. In fact, when source rocks have been buried to a certain depth, then brought again to surface by subsequent folding and erosion, the organic content keeps the composition and physico-chemical properties corresponding to the maximum burial depth. Furthermore, some of the by-products of kerogen evolution, such as water and carbon dioxide, are highly mobile in sub-surface conditions and are not available for recombination, should it be the case.

For simulating the system of reactions (Figure 11.1), it is more preferable to consider only a limited number of steps which are given below:

(*i*) A, B_1, B_{n-1}, which was the first version proposed by Tissot (1969).

(*ii*) A, B_{n-1}, B_n, which accounts for the successive oil and gas formation (Tissot and Espitalie, 1975).

The second formulation corresponds to Eq.11.1

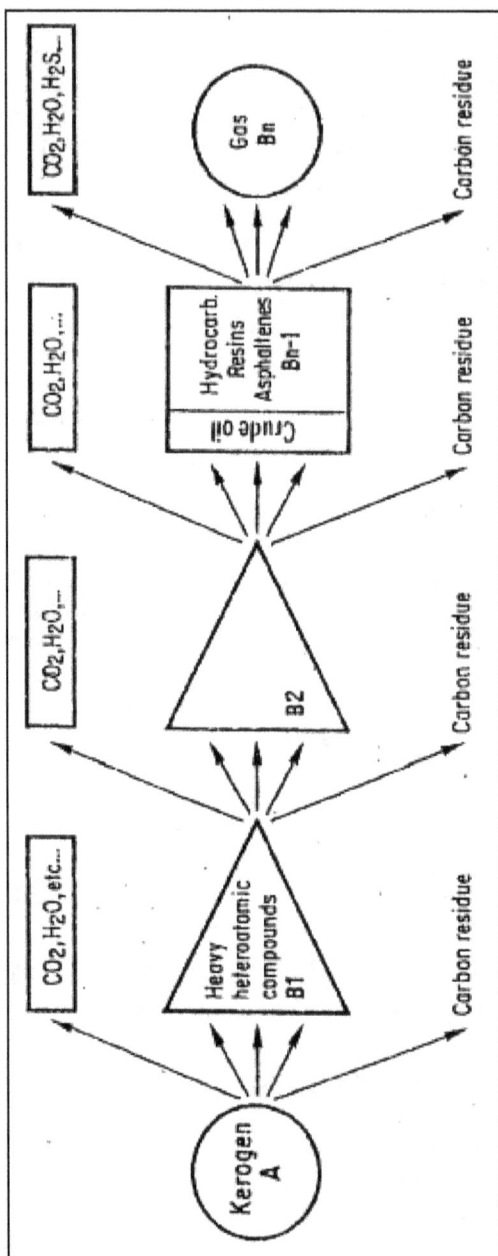

**Figure 11.1: General Scheme of Kerogen Degradation
(*Source:* Tissot and Espitalie, 1975)**

$$
\begin{array}{l}
\text{Kerogen} \\
\text{(A)}
\end{array}
\Longleftarrow
\left\{
\begin{array}{l}
CO_2, H_2O, \text{etc.} \\
\text{Crude} \\
\text{Oil} \\
\text{(Bn-1)} \\
\text{Carbon residue}
\end{array}
\right.
\Longrightarrow
\left\{
\begin{array}{l}
CO_2, H_2O, H_2S \text{ etc.} \\
\text{Gas (Bn)} \\
\text{Carbon residue}
\end{array}
\right.
$$

$$(11.1)$$

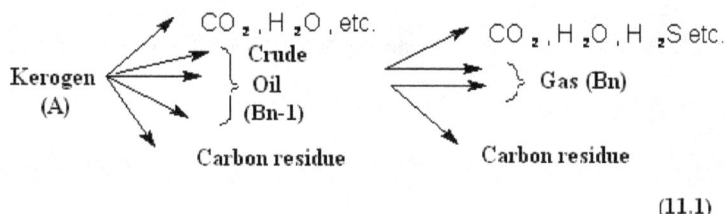

where, A is the kerogen, comprising n_i bonds of a given type i, at the time t; x_i is the amount of organic matter reacting in the ith reaction (breaking of i-type bond). The B_{li} to B_{lm} are the products of the first step of reactions (formation of oil), their respective amount at time t being y_1 to y_m. B_{21} to B_{2n} are the products of the second step of reactions (cracking), their respective amounts at time t being u_1 to u_n (Eq. 11.2).

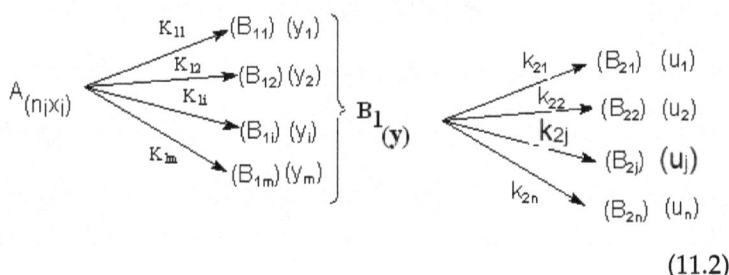

$$
A_{(n_j x_j)}
\Longleftarrow
\begin{array}{l}
K_{11} \rightarrow (B_{11})\ (y_1) \\
K_{12} \rightarrow (B_{12})\ (y_2) \\
K_{li} \\
\rightarrow (B_{1i})\ (y_i) \\
K_{1m} \\
(B_{1m})\ (y_m)
\end{array}
\Bigg\} \ B_{1(y)}
\Longleftarrow
\begin{array}{l}
k_{21} \rightarrow (B_{21})\ (u_1) \\
k_{22} \rightarrow (B_{22})\ (u_2) \\
k_{2j} \\
\rightarrow (B_{2j})\ (u_j) \\
k_{2n} \\
(B_{2n})\ (u_n)
\end{array}
$$

$$(11.2)$$

The first set of transformations represents a certain number of parallel and/or successive reactions. It is assumed that the probability of breaking a bond of type i is independent of the abundance of bonds of the other types and also independent of the time elapsed. Then breaking of i-type bond obeys the Poisson law:

$$-\frac{dn_i}{n_i} = k_{li}dt \qquad (11.3)$$

k_{li} being a constant at a given temperature. If the structure is homogeneous *i.e.*, the density of bonds is statistically same in the kerogen volume, we have:

$$x_i = \mu n_i \qquad (11.4)$$

Where μ is a constant, and, with regard to parallel reactions:

$$x_{i0} = x_0 P_i \qquad (11.5)$$

P_i being the frequency of i-type bond at the origin $t = 0$, and $x_0 =$ Σx_{i0} being the total amount of labile-or pyrolysable-organic matter. Then Eq. 11.3 becomes:

$$-\frac{dx_i}{dt} = k_{li} x_i \tag{11.6}$$

Therefore, we obtain for each of the i reactions a kinetic equation similar to that of the first-order reactions. It has been shown that such formulation is able to account for kerogen degradation in geological conditions (Tissot, 1969).

If we make the additional hypothesis that the B_{li} constituents of oil have the same general behaviour in respect to cracking, we can consider for the second step of reactions $B_1 = \Sigma B_{li}$. The respective amounts x_i, y_i and u_j can be obtained from the following system of differential equations.

$$\frac{du_j}{dt} = k_{2j} y \tag{11.7}$$

$$y = \sum_i y_i \tag{11.8}$$

$$\sum_i x_{io} + \sum_i y_{io} + \sum_j u_{jo} = \sum_i x_i + \sum_i y_i + \sum_j u_j \tag{11.9}$$

Equations 11.6 and 11.7 express the kinetics of the system, whereas Eqs. 11.8 and 11.9 express the mass balance. The variation of k_{li} and k_{2j} with temperature may be described by using the Arrhenius equation:

$$k_{1i} = A_{1i} e^{-\frac{E_{1i}}{RT}} \tag{11.10}$$

The Eq. 11.10 which is basically valid for fast laboratory or industrial reactions may be extended to slow reactions occurring under geological situations (Tissot, 1969; Connan, 1974). E_{li} is the activation energy of the ith reaction, A_{li} is a constant and T is the absolute temperature (Kelvin). In geological situations, temperature is a function of time through subsidence and related burial. As burial is reconstructed by geological history on the basis of successive time intervals (*e.g.*, Lower Cretaceous, Upper Cretaceous, Paleocene, etc.), depth is considered to be a linear function of time during each interval. Thus temperature also is approximated as a succession of linear function of time during the successive time intervals. Under these conditions the system [Eqs. 11.6-11.9] cannot be easily integrated as k_{1i} becomes a complex function of the time t. The easiest way of solving the system is by numerical integration, with successive Δt increments. This is done by using a computer, and the program includes the adjustment of the various parameters x_{i0}, E_{1i}, A_{1i} and y_0, which is in fact the calibration of the model to the particular type of organic matter.

For this calibration, only the first set of reactions (formation of oil Eq. 11.1) is considered. At the beginning, values measured on comparable recent sediments (*e.g.*, Black Sea sediments for Type-II kerogen) are used for y_0, and approximate values for A_{1i} (Tissot, 1969). In respect to the E_{1i}, it is necessary to consider all activation energies from a few kcal mol^{-1}, corresponding to the rupture of weak bonds, as in adsorption, to about 80 kcal mol^{-1}, corresponding to breaking of carbon-carbon bonds. The objective of adjustment should be to determine the frequency, P_i, of each i-type of bonds, or the amount x_{i0} of labile organic material involved in the ith reaction. Calibration is based on values of extractable organic matter (hydrocarbons + resins + asphaltenes) naturally present in cores taken at different depths in the sedimentary basin and/or on comparable figures resulting from laboratory experiments at higher temperature. The adjustment is based on the method of least squares. Table 11.1 and Figure 11.2 shows the results concerning the three main types of kerogen I, II and III.

Figure 11.2 is based on the data obtained from the source rocks of the Uinta, Paris and Douala basins which were originally used to define the three main types of kerogen. They offer a first approach to calculate the amount of oil and gas generated from other source rocks of comparable type. However, due to the variability of kerogen composition, a specific adjustment based on laboratory assays, which provide a direct calibration of the model for each particular source rock, is generally preferable.

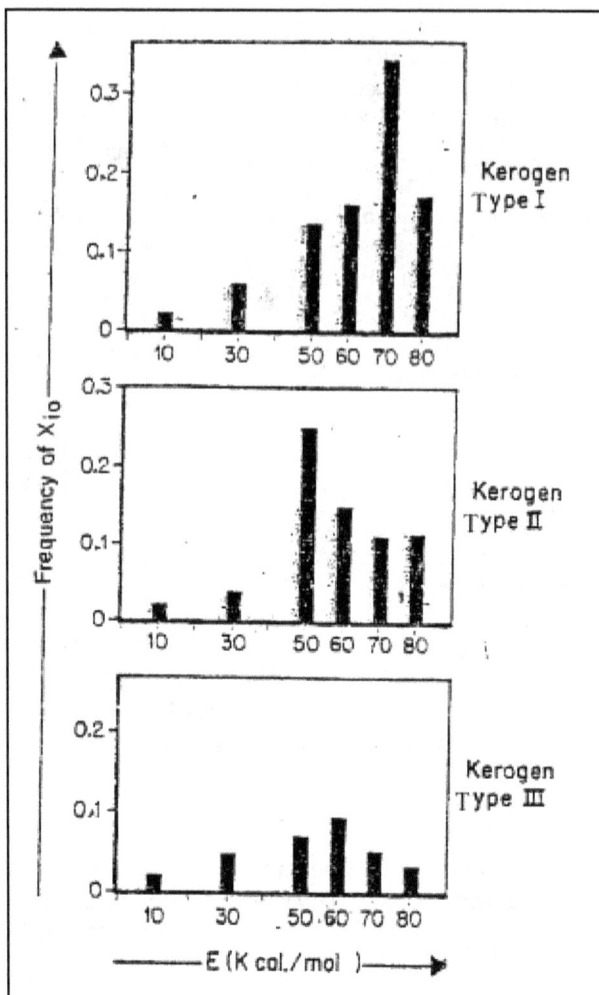

Figure 11.2: Distribution of the Activation Energies Involved in the Degradation of the Three Main Types of Kerogen (*Source*: Tissot and Espitalie, 1975)

Calibration of the second set of reactions (Eq. 11.1) namely, formation of gas has been made from laboratory experiments on cracking, including the results of McNab *et al.*, (1952), and Johns and Shimoyama (1972). Consideration of a single reaction with an

Table 11.1: Distribution of Activation Energies and Genetic Potentials of the Principal Kerogen Types

Activation Energies		Kerogen Types					
Class	Average Value (kcal mol⁻¹)	Type I		Type II		Type III	
		x_{io}	A	x_{io}	A	x_{io}	A
E_{11}	10	0.024	4.75×10^5	0.022	1.27×10^5	0.023	5.20×10^3
E_{12}	30	0.064	3.04×10^{16}	0.034	7.47×10^{16}	0.053	4.20×10^{16}
E_{13}	50	0.136	2.28×10^{25}	0.251	1.48×10^{27}	0.072	4.33×10^{25}
E_{14}	60	0.152	3.98×10^{30}	0.152	5.52×10^{29}	0.091	1.97×10^{32}
E_{15}	70	0.347	4.47×10^{32}	0.116	2.04×10^{35}	0.049	1.20×10^{33}
E_{16}	80	0.172	1.10×10^{34}	0.120	3.80×10^{35}	0.027	7.56×10^{31}

Genetic potential of kerogen

$$x_0 = \sum_i x_{i0}$$

	Type I	Type II	Type III
x_0	0.895	0.695	0.313

A is expressed as 10^6 yr⁻¹

Value of y_0 Type I: 0.051
Type II: 0.035
Type III: 0.018

Source: Tissot and Welte, 1984.

activation energy of 50 kcal mol^{-1} and Arrhenius constant (A) of 5 ×
10^{13} sec^{-1} seems convenient to account for gas generation in deep
parts of the sedimentary basins.

11.4.1 Genetic Potential of Source Rocks and Transformation Ratio

The total amount of hydrocarbons X_0 which can be produced
by a certain kerogen, provided it is heated to a sufficient temperature
over a sufficient time, can be given by Eq. 11.11 (shown in Table
11.1).

$$x_0 = \sum_i x_{io} \tag{11.11}$$

This quantity is equivalent to the Genetic Potential of the kerogen
defined in 6.5.0. The value depends on the type of kerogen, *i.e.*, on its
original chemical composition. A source rock could be defined as a
rock whose genetic potential is above a threshold value, *e.g.*, 0.25 or
0.30 related to the unit weight of kerogen, or (if we consider an average
value of 1 per cent organic carbon in rock) 0.25 to 0.30 per cent
related to unit weight of rock.

At any time, the stage of evolution is measured by using the
Transformation Ratio r, (or Transformation Index 6.5.0) which is the
ratio of the kerogen already transformed to the genetic potential:

$$r = \frac{\sum x_{io} - \sum x_i}{\sum x_{io}} = \frac{x_0 - \sum x_i}{x_0} \tag{11.12}$$

Transformation Ratio (r) is zero at shallow depths and
progressively increases to 1, which is reached when all labile organic
material has been expelled leaving a carbonaceous residue.

11.4.2 Validity of the Model

In various sedimentary basins, comparison between the figures
computed by using the model and the corresponding amounts of
petroleum generated through burial shows excellent agreement, with
a quadratic deviation lower than 10^{-2}, and a correlation coefficient
better than 0.9. The model has been used to simulate experimental

heating-either isothermally or with a regular rate of temperature increase-during various times from one hour to one year, with a satisfactory agreement. Furthermore, the same set of constants A_i and E_i (Table 11.1) is sufficient to account for all conditions of kerogen degradation (Tissot and Espitalie, 1975) including (*a*) natural evolution in sedimentary basins at relatively low temperature (50-150°C) over a period of 10 to 400 million years; (*b*) artificial evolution through laboratory experiments (180-250°C), and (*c*) high temperature (400-500°C) retorting of oil shales. The fact that a single model with the same set of constants is able to simulate such different situations is a confirmation of the validity of the hypothesis made on the basis of kinetic parameters.

11.4.3 Significance of Activation Energies in Relation to the Type of Organic Matter

The parameters E_{1i} used in the model are called activation energies for the sake of simplification. They have a role similar to that of activation energy, but they are not strictly activation energies, as the latter are normally defined in respect to a particular and single reaction. This is not the case in the model, where we consider the "formation of petroleum" from numerous parallel and/or successive reactions. Many types of bonds are originally present in kerogen, with distinct rupture energies, particularly:

☆ Weak bonds corresponding to physical or chemical adsorption (hydrogen bonds, etc.).

☆ Carbonyl and carboxyl bonds.

☆ Ether and sulfur bonds.

☆ Carbon-carbon bonds.

Furthermore, the rupture energy of most types of bonding may vary according to neighboring functional groups or substituents, length of chains, etc. Consequently, consideration of a distribution of the activation energies E_{1i} from 0 to 80 kcal mol^{-1} is probably closer to the effective mechanisms than a hypothetical measurement of each individual type of bond. Therefore, the best representation of kerogen composition may be the histogram of activation energies, derived from Table 11.1 and presented in Figure 11.2 for Type-I, –II and –III kerogens. With increasing burial and temperature (and decreasing 1/T) the various bonds corresponding to the successive activation energies E_i are progressively broken, roughly in order of increasing E_i. This is suggested (Figure 11.3) by the temperature dependence of the reaction parameters k_i.

The Genetic Potential (Eq. 11.11) and the distribution of the activation energies change according to the type of kerogen (Tissot and Espitalie, 1975) as detailed below:

   (*i*) Type-I kerogen contains a large proportion of labile organic material x_0, and thus its Genetic Potential is high. The

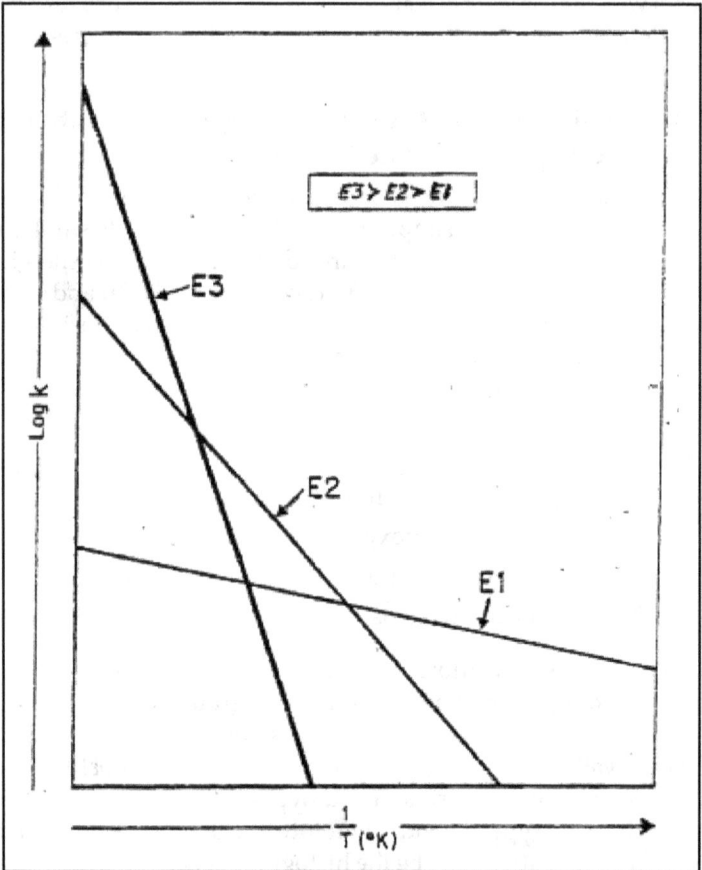

Figure 11.3: Temperature Dependence of the Reaction Parameters ki Involved in Kerogen Degradation (*Source*: Tissot and Espitalie, 1975)

distribution of activation energies include few with low values corresponding to weak bonds. Most values are grouped around 70 kcal mol^{-1}, and may correspond to carbon –carbon bonds. Therefore, this particular kerogen requires higher temperatures for generating oil than the other two types (II and III).

(ii) Type-II kerogen contains slightly less of the labile organic material resulting in a Genetic Potential slightly lower than in Type I. However, the distribution of activation energies is wide and includes lower values than Type I with a mode at 50 kcal mol^{-1}. Thus, the main formation of petroleum starts at somewhat lower temperature than in Type I.

(iii) Type-III kerogen contains still less of the labile organic material than Type I and II and its Genetic Potential is low. As a result, the total amount of oil generated is comparatively small. The distribution of activation energies is smooth with a maximum frequency at 60 kcal mol^{-1}.

A high concentration of organic matter related to Type I or II results in a rich oil shale with a high oil yield. The Green River shales belong to Type I and their values of x_0 range from 0.8 to 0.9. This means that 80 to 90 per cent of the organic matter is able to be converted to oil. The Toarcian shales of Western Europe belong to Type II, and the corresponding value of x_0 is 0.6, *i.e.*, 60 per cent of the organic matter can be converted to oil. The Type–III organic matter, on the other hand, may be concentrated in certain coals or carbonaceous shales, but with a low oil yield ($x_0 = 0.25$) and no commercial oil. The relationship between the three main types of kerogen, as defined by their chemical properties, and the respective distribution of activation energies, may be used for a quick calibration of the model. If one knows the specific kerogen type (I, II or III) from petrographic methods (6.2.0) or elemental analysis (6.3.0), the appropriate values of A_i and E_i (Table 11.1) can be used in the model.

A review of the apparent activation energies proposed in the literature for petroleum generation, carbonisation of coal, or oil-shale retorting shows that they range from 8 to 65 kcal mol^{-1}. These apparent activation energies were computed on the basis of a single reaction, although it deals in fact with a set of successive and/or parallel reactions. As the true activation energies E_i, corresponding to the breaking of various bonds, are affected roughly in order of ascending values with increasing temperature, the apparent activation energy is close to the lowest E_i at low temperature and close to the highest E_i at high temperature. This behaviour is confirmed by experimental data obtained from laboratory pyrolysis of oil shales by Weitkamp and Gutberlet (1968). They observed a progressive increase of the apparent activation energy from 20 to 60 kcal mol^{-1}, when the conversion rate of kerogen increased from 0 to 80 per cent. This consideration is sufficient to explain that apparent activation energies related to the beginning of oil formation are on the average 10-15 kcal mol^{-1} (Tissot, 1969; Connan, 1974) whereas apparent activation energies related to oil shale pyrolysis or carbonisation of coal are about 50 to 65 kcal mol^{-1} (Abelson, 1963).

11.5 Application of the Model to Petroleum Exploration

The mathematical model of kerogen degradation provides quantitative estimation of oil and gas generated as a function of time. Therefore, it is directly applicable for evaluation of the oil and gas potential of a sedimentary basin, and for determination of the timing of petroleum formation for comparison with the age of structural or stratigraphic traps.

Figure 11.4 presents the data required for use of the model: (*a*) a direct calibration of the E_i distribution on laboratory assays, which in fact represents the chemical composition of organic matter, or alternatively identification by optical or chemical methods of the kerogen type and subsequent use of the corresponding distribution (shown in Table 11.1); (*b*) the burial curve, *i.e.*, the depth versus time relationship, from the time of source rock deposition to the present time; and (*c*) the geothermal gradient, for computation of the temperature versus time relationship. The gradient may vary with geological periods according to geotectonic conditions. Provided such data are available, the amount of oil and gas per tonne of rock generated in any place of the basin can be computed as a function of time (Figure 11.4). In order to cover the whole sedimentary basin, it is convenient to divide volume of the basin into a three-dimensional grid. The most effective way to use this geochemical modeling is obviously to introduce this simulation into the basin geological modeling.

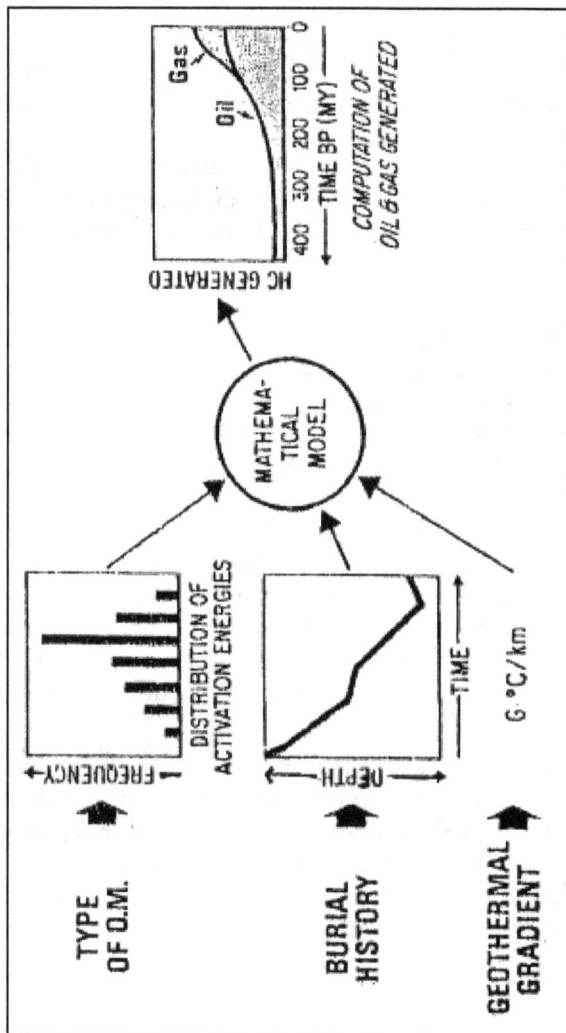

Figure 11.4: Principle Use of the Kinetic Model of Oil and Gas Generation.
The model calculates the quantities of oil and gas generated, as a function of time
(Source: Tissot et al., 1980)

Figure 11.5 shows the application of mathematical model to oil generation in Jurassic source rocks of the Paris Basin. It can be expressed by the fraction of kerogen, which has been converted to petroleum, in grams per kilogram of organic matter. Alternatively, if the distribution of organic carbon across the basin is also known, it can be related to the whole rock and expressed as grams petroleum per tonne of rock. From the figure, it is clear that all known oil fields fall inside the area, where more than 110 g petroleum has been generated per kilogram of organic carbon, and more than 1500 g petroleum has been generated per tonne of rock. Regarding the timing of oil generation most petroleum has been formed in this basin during Cretaceous and Tertiary period.

Figure 11.6 provides an example of oil and gas generation in Paleozoic source rocks of the Illizi basin, Algeria. Geological history of the basin includes two periods of sedimentation and subsidence separated by Hercynian folding and erosion. Except the southwestern part, kerogen degradation continued generating abundant oil and also gas in the northern and eastern parts of the basin where Cretaceous sedimentation was very thick. The amount of petroleum formed has been calculated, with consideration of the respective timing of generation. Thus, oil and gas potential may be described as: (*a*) low for both oil and gas in the south west, due to erosion after generation; (*b*) good for oil in the central and eastern parts of the basin with some gas associated in the east; (*c*) good for gas in the northeast, with some oil associated in the less buried or colder areas.

The example of the Hassi Messaoud area in northern Sahara, Algeria, has been earlier presented in 6.6.5 (Figure 6.18 and 6.19) where purely geological considerations allowed an independent confirmation of the timing of petroleum generation. The present mathematical model reaffirms that the main phase of oil generation from Silurian source rocks is not reached until Cretaceous time, and extends mostly over Cretaceous and Tertiary period (Tissot and Welte, 1984).

There are several applications of geochemical modeling coupled with geological basin modeling of hydrocarbons. Some of them include:

(*i*) Determination of prospective areas for oil and gas in sedimentary basins.

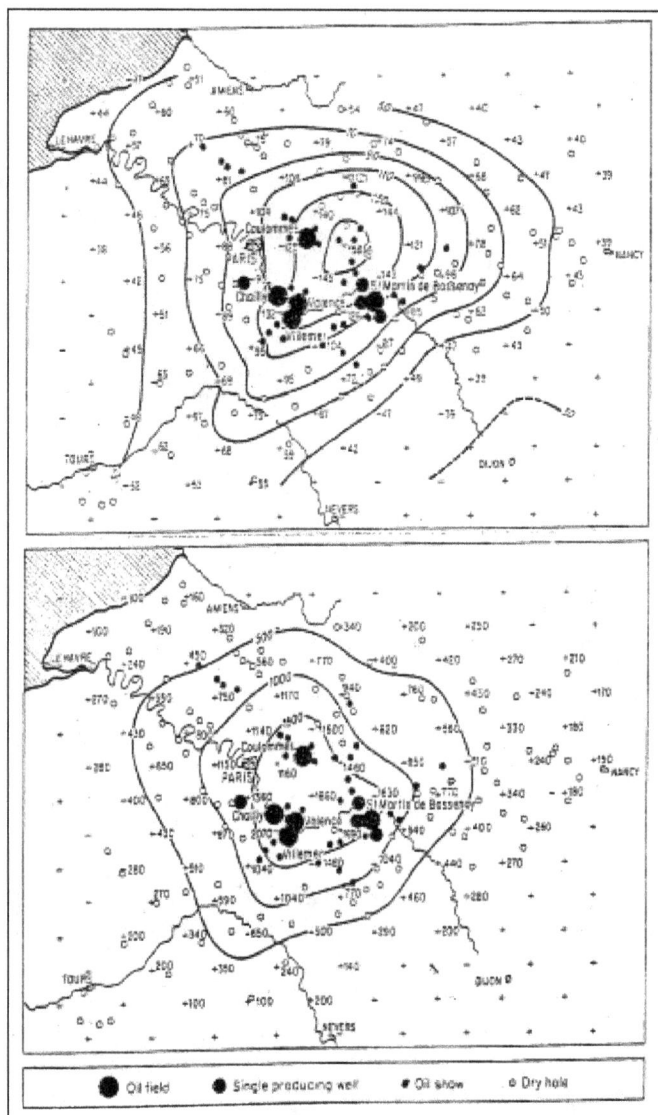

**Figure 11.5: Application of the Mathematical Model
to the Generation of Oil from Lower Jurassic Source Rocks
of the Paris Basin. Top: The oil generated is expressed in g
petroleum per kg organic carbon. Bottom: The oil generated is
expressed in g petroleum per tonne of rock
(*Source*: Tissot and Weite, 1984)**

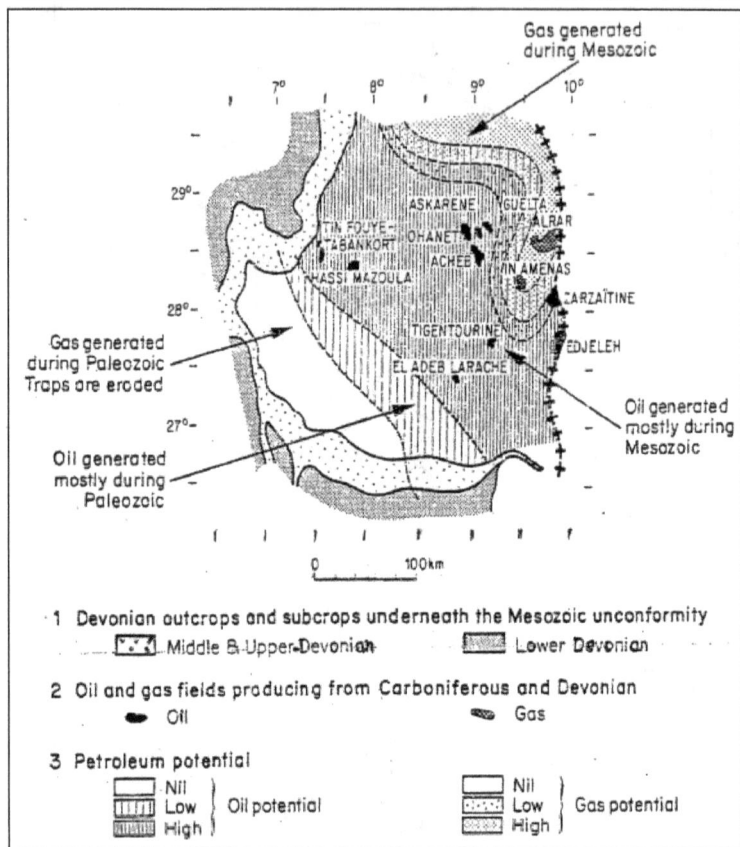

Figure 11.6: Hydrocarbon Potential of the Lower Devonian Beds in the Illizi Basin (Algeria), as Determined by Use of the Mathematical Model (*Source*: Tissot and Espitalie, 1975)

(*ii*) Timing of oil and gas generation, and migration, for comparison with the age of formation of traps and impermeable seals.

(*iii*) Evaluation of the ultimate reserves of a sedimentary basin.

(*iv*) Distribution of abnormal pressures; history of water flow in the basin.

(*v*) Reconstruction of ancient geothermal gradients and thermal history of Sediments.

(*vi*) Simulation of retorting conditions of oil shales.

Basin simulation programmes or systems are constantly improving and changing at a rapid pace to incorporate many more geological and geochemical processes. New developments in basin modeling appear to proceed in the following directions:

(*i*) Increasing integration with seismic interpretation for model development,

(*ii*) Inclusion in the simulation of processes that are not due to the normal mechanical compaction (*e.g.*, diagenesis, igneous intrusions, gas saturation, etc); and

(*iii*) Development of real three-dimensional model systems.

However, one should keep in mind three important aspects when applying a modern basin simulation: (*a*) default input data especially for boundary conditions and physical parameters, (*b*) unforseen effect of interaction of many processes being modeled, and (*c*) improvements and changes to be incorporated in every new version (Poekhau *et al.*, 1997).

Chapter 12
Unconventional Petroleum Resources

Unconventional oil and gas systems are those typically independent of conventional systems comprising source, migration fairway, reservoir and trap. The term unconventional also refers to naturally occurring oil or gas deposits of potential commercial significance not recoverable by conventional drilling methods. Law and Curtis (2002) referred unconventional gas resources as those of regionally pervasive natural systems with high gas saturations, typically independent of structural and stratigraphic traps, and commonly characterized by low permeability host rock formations.

12.1 Typical Unconventional Petroleum Systems

These includes:

    (*i*) Basinal (basin centred) gas and tight gas sands,

    (*ii*) Shale gas and oil shale,

    (*iii*) Bituminous sands,

    (*iv*) Coal bed methane (CBM), and

    (*v*) Gas hydrates.

12.1.1 Need for Exploration

As per the world oil statistics, the oil and gas reserves discovered so far are 1.13 trillion barrels (Tb) and 178 trillion cubic metres (Tm^3) respectively. With the present world average consumption of oil (82 Mb/d) and gas (2.7 Tm^3/d), the reserves would hardly last 40 and 65 years, respectively. But this is only a gross assumption not taking into account the increase in oil and gas demand by 3-4 per cent per annum. However, one reassuring feature is the assessment of the world's remaining oil potential. According to US Geological Survey (US GS), the world endowment of recoverable oil stands at 3 Tb out of which about 24 per cent has been produced and 29 per cent has been discovered and booked as reserves. This is in addition to one trillion barrels oil equivalent of gas (OEG) yet to be discovered. Coming to India, the consumption of crude oil was 148 million metric tones (MMt) per annum (2006-2007) while the production from indigenous sources was about 32 MMt. Similarly, the demand for natural gas was 231 MMm^3/day against the domestic production of 65 MMm^3/day. As per the India's Hydrocarbon Vision Statement (2001), the requirement for oil and gas by 2025 will be 368 MMt/y and 391 Mm^3/d. respectively. Keeping this supply–demand position of oil and natural gas, there is an imperative need for exploring and exploiting unconventional reservoirs to supplement the domestic energy requirements.

12.2 Basinal (Deep) Gas (Tight Gas) Sands

The deep basinal gas sands comprise an all-encompassing category that include many different types of reservoirs of different depositional systems, including sands and limestones, and at varying burial depths. Some of the most prolific basin–centred gas accumulations are produced from depths of less than 1 km (Law, 2002). The characteristic features are regionally pervasive, gas saturated, low permeability (less than 1 milli darcy) sandstones or occasionally limestones, commonly (but not exclusively) over pressured with an updip stratigraphic or regional seal (Price, 1997). The cause of low permeability can reflect original depositional composition, compaction or extreme diagenetic cementation. Seals are a combination of relative permeability in the tight formation (capillary effects), often accompanied by stratigraphic (Pinchouts), structural (closure and faulting) and diagenetic seals (Law, 2002). During burial, the rate at which gas generated exceeds that at which it leaks. As gas accumulates in the pore system, the capillary pressure of the water-wet pores is overcome and the mobile water is expelled from it, resulting in an over pressured, gas saturated reservoir with little or no free water.

Currently, most basinal gas resources are recognized in the USA contributing about 15 per cent of total gas production. Other basins under appraisal include: Gippland, Cooper and Barrow basins in Australia; Sichum basin in China; and Ahnet basin in Algeria (Hirst *et al.*, 2001). The best known resource of tight gas sands occurs in several Rocky Mountain region foreland basins in South West USA, Greater Green River and Wind River basins of Wyoming, Utah and Colorado (Law and Spencer, 1993). Tight gas sands occur in India at Mandapeta (Gondwana system) in Andhra Pradesh and Navagam (Eocene) field in Gujarat. The potential basin-centered gas resource around the world is enormous (about 500-700 x 10^{12} Scft) although accurate data is not available.

For economic productivity all basin centered gas accumulations requires the presence of natural fractured network which usually have to be supplemented by artificial (hydraulic) fracturing to create commercial production. Extensive high resolution seismic data is required to define fracture networks and locate high porosity zones for well placements. Large number of wells in close proximity on a regular grid have to be drilled for production at low but economic rate. Enhanced gas recovery from tight gas sands can be achieved through the use of specific proppant fluids and CO_2 injection. Many of the production problems associated with tight gas sands are comparable to those encountered with shale gas (12.3.0) and coal bed methane (12.6.0) accumulations.

12.3 Shale Gas

Shale gas systems are similar to coal bed methane (CBM) systems (10.6.0) in that they constitute widely dispersed gas accumulations in low permeability (typically less than 0.1 md), fine grained, organic carbon–rich shales characterized by regionally high gas saturations (Curtis, 2002). As with CBM, shale gas occurs as adsorbed on kerogen and clay particle surfaces in the matrix of the host shale, or as free gas in intergranular porosity and natural fractures in the shale. Shale gas systems constitute viable accumulations because they contain either natural fractures or artificially fractured to stimulate production. The production is often of low volume with low recovery factor. Typically about 20-500 x 10^6 Scft/d of gas can be obtained by drilling many closely spaced low cost wells with hydraulic fracturing. High resolution imaging of gas within shale formations, better fracture prediction techniques and enhanced stimulation are likely to improve the shale gas production (Eagler and Perry, 2002).

Shale gas is only produced commercially in USA (Hill and Nelson, 2000) accounting about 4 per cent of its total gas production (400x10^9 Scft per annum from 37,000 wells). Western Canadian sedimentary basins of Alberta and British Columbia are identified as other major shale gas fields. Global resources of shale gas (estimated as potential reserves) could exceed 1600x10^{12}Scft.

12.4 Oil Shale

Most oil shales are actually bituminous, non-marine limestones or marlstones containing kerogen. Only few marine examples can be properly described as shales. The oil shale contains about 10-50 per cent of algal organic matter (OM). In commercially significant deposits, the OM content is at least 30 per cent, much higher than in most oil-source sediments, but because of low level maturation, oil cannot be expelled from the shale. The OM in oil shale may be either sapropelic or humic or a mixture of the two. In all true shales, however, sapropelic OM and the constituent alginite and exinite macerals are dominant. Other than coals, oil shales constitute the most abundant fossil fuel resource. However, they are far more variable in their quality than coals. In rich sapropelic shales like the Green River shales, upto 65-70 per cent of the kerogen content is converted to oil on distillation. However, in humic shales less than 30 per cent is only convertible. Ultra-rich shales, like torbonites have oil content less of 40 per cent.

The total world potential of shale oil resources (Yen and Chilingar, 1976) was estimated at 30x10^{12} barrels. More cautious estimates by US Geological Survey and the United Nations, average about 3.2x10^{12} barrels of recoverable reserves. Except the oldest and largest deposits in the Lower Paleozoic of Eurasia and North Africa which are of marine origin, nearly all important Post Devonian shales are of lacustrine origin. Economically most important deposits occur in Green River oil shales of Western USA, Irati oil shales of

Brazil, Mid-Land valley of Scotland, Fusion deposits in China and Kukkersite oil shale of Estonia. Fractured oil shale (Lower Eocene) occurrence was reported in India in Indroda field (Gujarat) and (Middle Eocene) in Wadu field of Cambay basin.

There are many technological problems involved in the recovery of oil from shales. Among them the most important is the necessity for enormous input of heat to distil oil out of the shale because even a rich oil shale yields on an average, less than one barrel of oil per tonne of rock. It has not been so far demonstrated whether this can be achieved with a net gain of energy without depriving large territories of potable water. Even if it is achieved, the second problem is the disposal of spent shale which is an environmental hazard. To avoid this problem, techniques of underground (*in situ*) retorting are under development wherein mining of shale and the waste disposal can be largely eliminated.

12.5 Bituminous (Tar) Sands

Bituminous sands are those impregnated with oil too heavy and viscous to be extracted by conventional drilling techniques even under artificial stimulation. They occur commonly at shallow depths (less than 600 m) and not far above major unconformities. Bituminous sands are popularly named as tar sands. By definition, a tar sand is any consolidated or unconsolidated rock that contains a hydrocarbon material with a gas free viscosity (measured at reservoir temperature) greater than 10,000 m Pa or that contains a hydrocarbon material that is extractable from the mined or quarried rock. By far the largest and most prospective tar sand deposits, however, are exposed at the Earth's surface or buried beneath little or unconsolidated surfacial sediments.

Compilation of about 20 large tar sand deposits (Phizackerley and Scott, 1967; Demaison, 1977) indicates more than 3×10^{11} m^3 of oil in place. About 98 per cent of this volume is contained in tar sands in Venezuela, the Alberta province (Canada), and the former Soviet Union. The tar sand deposits accessible to surface exploration are dominated by three gigantic accumulations–those of Orinoco oil belt of Eastern Venezuela, the Athabasca tar sands of Alberta in Western Canada, and the Olenek tar sands in central Siberia. It is estimated that at least 66 per cent of the world's petroleum reserves are preserved in tar sands with 32 per cent (1.7×10^{12} barrels) in

Athabasca and 34 per cent (1.8 x 10^{12} barrels) in Orinoco fields. Heavy oil and tar sands occur in India in three areas (*i*) the northern part of the intra cratonic Cambay basin around Mehsana horst, (*ii*) the frontal folded regime of Himalayas, and (*iii*) the geosynclinal regime of Surma Valley near the India–Bangladesh border. Nearly all tar sand deposits share a common feature (Demaison, 1977) namely, the reservoir sands themselves are in almost all cases, aerially wide spread fluvial–deltaic–coastal complexes deposited between emergent land masses and large offshore sedimentary basins. They are therefore able to draw oil by migration from large source areas.

Though tar sands and bituminous (oil) sands represent a vast hydrocarbon resource, the potential of their conversion to reserves is a function of technology, economics and environmental impacts. The commonest method of recovery is a cyclic (*in situ*) high temperature steam stimulation at about 325° C. The steam takes a long time–several years to sweep through the reservoir rock at a typical depth of 500-600 m. The oil is recovered from the producing wells by conventional pumping. Alternatively the tar sands are mined and oil is extracted by

☆ Hot water floatation to remove thin coating of oil from sand grains,

☆ Addition of naphtha to the resulting tar like material for thining and subsequent pumping.

The efficiency of recovery in the extraction is one barrel of oil for every two tonnes of mined tar sand. There are several limitations for large scale adoption of the above process namely: (i) capital costs, (ii) high demand for water, (iii) net energy gain (the difference between the energy return and the energy invested should be positive), (iv) environmental impacts, and (v) need for drilling large number of closely spaced wells. The high demand of water cannot be met from areas of low rain or snow fall. Liquid waste generated has to be pumped into huge ponds for storage which create serious environmental problems. Further, as the oil in the reservoirs is too viscous to move towards the depleted wells, they have shorter lives than do in conventional oil wells, necessicitating drilling of large number of closely spaced wells.

12.6 Coal Bed Methane (CBM)

Natural gas mostly comprising methane occurring within the coal deposits is referred as 'coal bed methane' (CBM). CBM which till recently was looked upon as an evil leading to safety problems in the mining of coal, has been recognized as a viable energy option. Since organic matter associated with porous coal has large surface area (10^9 Scft/tonne), coal seams can hold much more methane gas than an equivalent thickness of sandstone. As such they are self-sourcing reservoirs containing biogenic, thermogenic or mixed gas (Ayers, 2002). CBM is fundamentally different from natural gas found in conventional petroleum reservoirs. In CBM reservoirs, the gas molecules are physically attached (adsorbed/absorbed) within the coal's molecular structure but not compressed into conventional pore spaces as in the natural reservoirs.

12.6.1 Generation of CBM

Methane is generated during the formation of coal through "coalification process" of vegetable matter (Figure 12.1). The process can be broadly divided into two stages namely biochemical and physico-chemical, involving the following five steps in succession (Stach *et al.*, 1982).

1. Peatification (anaerobic degradation of organic matter in the peat swamp).

2. Humification (formation of dark coloured humic substances).

3. Bituminisation (generation of hydrocarbons with increase in temperature and pressure).

4. Debituminisation (thermal degradation of organic matter and the generated hydrocarbons), and

5. Graphitisation (formation of graphite).

Levine (1993) gave an excellent summary of the coalification process. Many physical and chemical changes, governed by biological and geological factors, occur during the above processes. Whereas darkening in colour and increase in hardness and compactness are the main physical changes, loss in moisture and volatile contents, and increase in carbon content are the main chemical changes. Many acids (humic, fatty, tannic, gallic etc.) and,

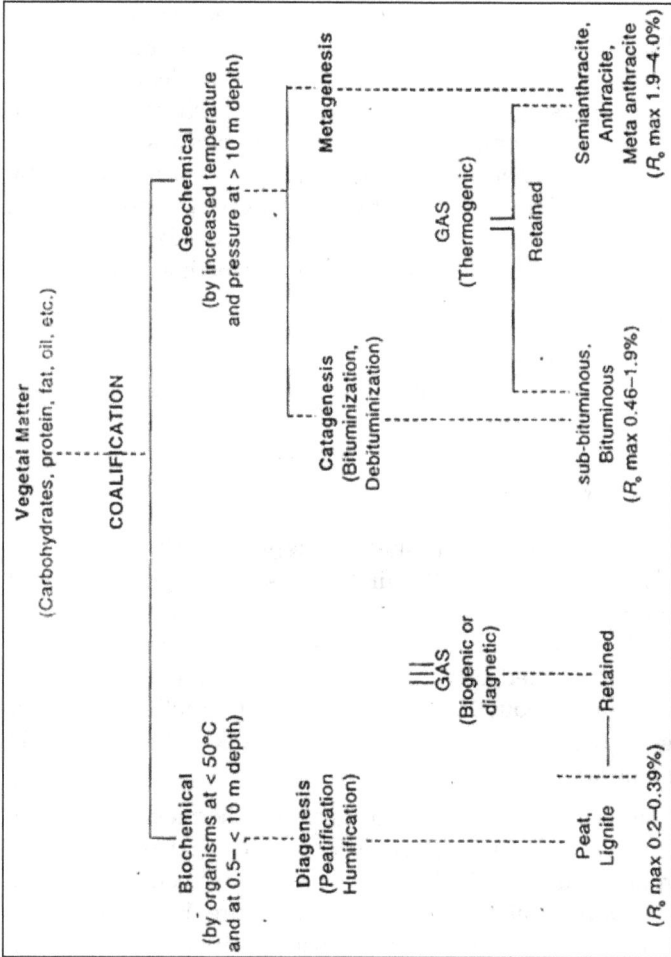

Figure 12.1: Schematic Representation of Methane in Terrestrial and Marine Environments (*Source*: Singh and Singh, 1999)

dry and wet gases (CH_4, CO_2, N_2, N_2O, H_2S, ethane, propane, butane etc.) are formed during the decomposition of organic matter.

Biochemical stage (Figure 12.1) of coalification begins with the accumulation of vegetable (organic) matter and its partial degradation by microorganisms (fungi, aerobic bacteria, insects etc.) leading to wide range of products (organo-petrographic entities termed as macerals) in water saturated wetlands (basins/grabens). Further decomposition by anaerobic bacteria results in release of methane or marsh gas. During subsequent geochemical stage of coalification, rise in temperature and pressure due to subsidence of basins/grabens, either by growing thickness of overburden or by tectonic activities, generate hydrocarbons (hydrogen rich constituents). Thermal cracking of free lipid hydrocarbons and/or cracking of kerogen fraction of coal generates methane gas.

Thus, the generation of CBM during coal formation occurs in two principal ways:

1. By metabolic activities of biological agencies (biological process), and

2. By thermal cracking of hydrogen rich substances (thermogenic process).

Methane generated at shallow depths (< 10 m) and low temperature (< 50° C) by the first process in low rank stage of coalification (sub bituminous) is termed as biogenic or diagenetic methane. Methane generated in the second step (catagenesis and metagenesis) is called as thermogenic methane. Biogenic methane constitutes only about 10 per cent of the total methane whereas themogenic methane constitutes the rest of bulk CBM generated in subsequent steps of coalification.

In marine environment both biogenic and thermogenic methane are produced with a distinct difference to that of terrestrial environment (Figure 12.2). Diagenesis begins in recently deposited sediments under ambient conditions of pressure (Water depth of < 2,000 m) and temperature of sediment below seafloor (25-50° C), and under anaerobic conditions leading to biogenic methane. Subsequent increase of hydrostatic pressure and sediment temperature (> 150° C) leads to formation of methane by H_2: CO_2 (carbonate reduction) and decarboxylation of acetate (fermentation) in the complete absence of sulphate. Methane generated remains as solution in interstitial

	TERRESTRIAL ENVIRONMENT			MARINE ENVIRONMENT	
	with increase in depth, time and pressure →			with increase in depth and hydrostatic pressure →	
Conditions	temperature < 50°C depth < 10 m R_o max 0.2–0.39%	> 50–170°C > 10 m depth 0.4–1.9%	> 170°C increased > 1.9%	Open water and upper part of sediment column	Ocean bottom sediments
				Plenty of oxygen	lesser oxygen, high concentra- tion of sulphates	lesser or no sulphate but high concentration of carbonates
Process	diagenesis	catagenesis	metagenesis	aerobic respira- tion using O_2 of water and organic matter	anaerobic respiration by sulphate reduction	anaerobic respiration by carbonate reduction
Products	peat, lignite, gas (a mixture of CO_2, N_2, wet gases and CH_4)	sub-bituminous, bituminous coal oil, condensates, gas containing	anthracites, semi-anthracite, gas containing	carbon dioxide	hydrogen sulphide carbon dioxide	carbon dioxide
	Biogenic methane	Thermogenic methane	Thermogenic methane			Biogenic and thermo- genic methane
	Part					
	Released as ← marsh gas	Entrapped as → CBM				

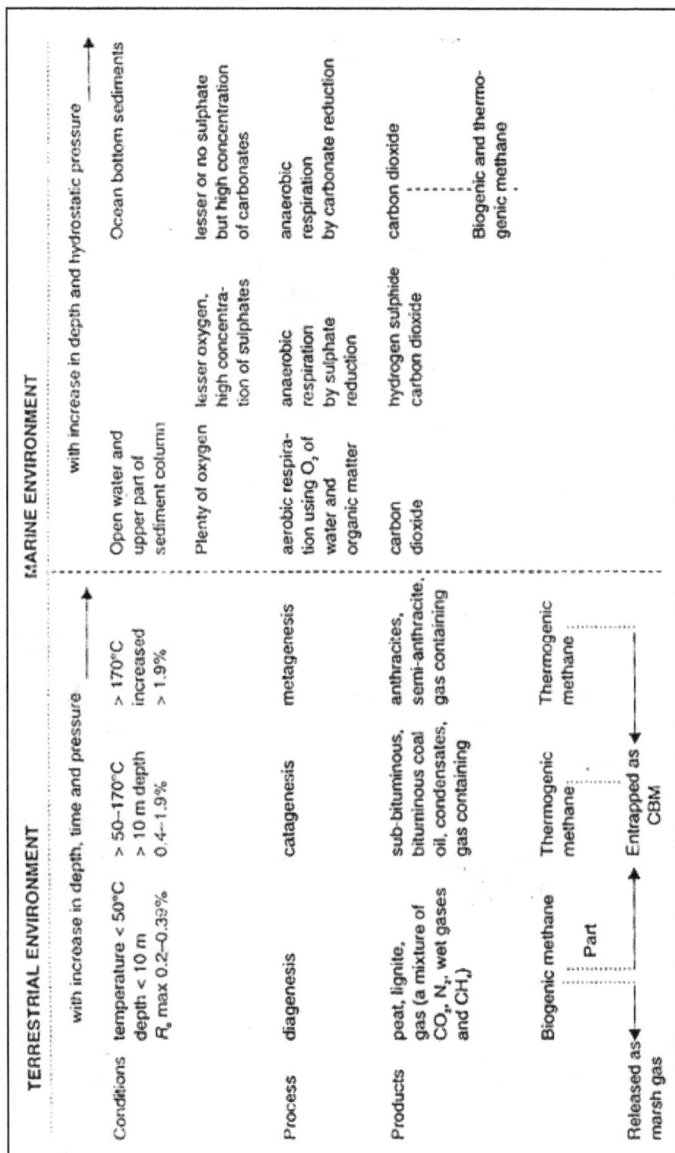

Figure 12.2: Schematic Representation of Methane in Terrestrial and Marine Environments
(*Source*: Singh and Singh, 1999)

porewaters because of its high solubility at higher hydrostatic pressures (depths) caused by the overlying water column. However, it exudes as free gas from interstitial porewaters either on supersaturation or by the release of hydrostatic pressure.

12.6.2 Biogenic Methane

As already mentioned, diagenesis is a low temperature transformation of organic sedimentary matter by micro organisms (aerobic/anaerobic) bacteria leading to chemical rearrangements at shallow depths resulting in (kerogen type) peat/lignite, and generation of gas termed as biogenic methane. Production of biogenic methane by metabolisation of organic matter using methanogenic (a special form) bacteria is known as methanogenesis. Two important metabolic pathways are inferred for the generation of biogenic methane (Figure 12.3). They are (*i*) H_2/CO_2 (carbonate reduction), and (*ii*) decarboxylaion (fermentation) of acetate, represented by the following reactions (Whiticar *et al.*, 1986; Ferry, 1992).

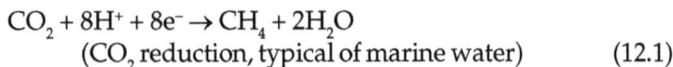

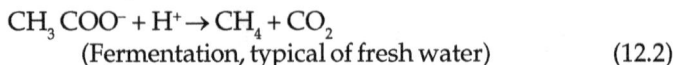

$$CO_2 + 8H^+ + 8e^- \rightarrow CH_4 + 2H_2O$$
(CO$_2$ reduction, typical of marine water)   (12.1)

$$CH_3COO^- + H^+ \rightarrow CH_4 + CO_2$$
(Fermentation, typical of fresh water)   (12.2)

Generation of biogenic methane is controlled by temperature at which the methanogenic bacteria thrive, namely 4–45° C, the optimum growth temperature being 35°C. However, special type of methanogenic bacteria (Methenopyrus, M. Thermo aggregans, Methanohalophilus etc.) are capable of living in extremes of temperature (113°C), pressure (300 atmos.), salt content (4.4 M NaCl) and pH (9.7) in fresh, estuarine, coastal waters and deep marine sediments. Methanogenic bacteria have been identified from oil reservoirs and coal fields. Notable among the latter is the San Juan coalfield in New Mexico, the most prolific CBM basin in the world where subsurface groundwater flow transports bacteria and probably nutrients, through permeable rock strata to the coal bed. The bacteria metabolise hydrocarbons and other organic compounds in the coal to produce biogenic methane and CO_2.

12.6.3 Thermogenic Methane

Kerogen (peat/lignite) formed by the diagenesis of sedimentary organic matter undergoes thermal maturation (catagenesis) under

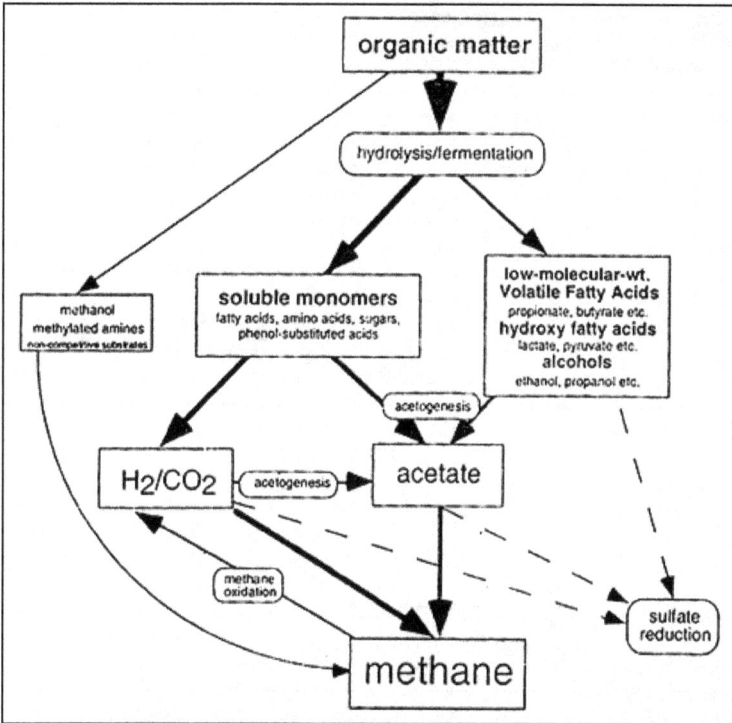

**Figure 12.3: Pathways Involved in Methanogenesis for
Bacterial Decay of Organic Matter in Marine Sediments
(Solid arrows show pathway to methane,
dashed arrows show substrates to sulphate reduction)
(*Source*: Wellsburry and Parkes, 2000)**

relatively low energy (pressure > 10 m depth, and temperature > 50-170° C) conditions to generate sub-bituminous/bituminous coal. Subsequent maturation of coal at high energy (temperature > 170°C) conditions (metagenesis) leads to the formation of anthracite. Both catagenesis and metagenesis processes inturn, generate methane termed as thermogenic methane (Figures 12.1 and 12.2).

Methane generation by thermogenic process, begins at vitrinite reflectance (Ro) range of 0.5–0.8 per cent, peaks near the boundary between medium volatile bituminous to low volatile bituminous coal stage (Ro 1.5–1.9 per cent, maximum at 1.7 per cent), and temperature 120–180° C (Figure 12.4). However, further rise in

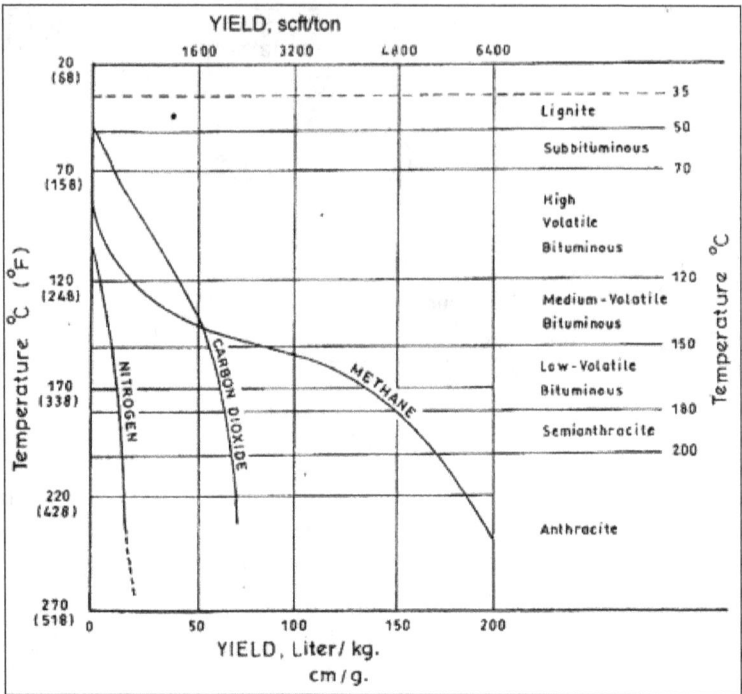

**Figure 12.4: Relationship Between Gas Generation and Coal Rank
(*Source*: Singh and Singh, 1999)**

reflectance values (Ro above 2.0 per cent) and temperature (above 180° C) results in slow increase of methane yield (Tang *et al.*, 1991; Killops *et al.*, 1994). Thus it could be presumed that the prospect of CBM generation is more in the regions of high geothermal gradient as well as in the vicinity of intrusive bodies.

Continuous gas production is essential in any commercial CBM field. It many be either biogenic or thermogenic or mixed. A distinction between biogenic and thermogenic methane (7.8.0) can be made based on (*i*) carbon numbers, (*ii*) stable carbon isotope and (*iii*) stable hydrogen isotope signatures of methane (Table 12.1) as follows (Waples, 1981):

☆ Since biogenic methane consists predominantly CH_4 (C_1) while thermogenic methane contain considerable amounts of ethane, propane, butane etc. along with methane, the ratio of carbon numbers (C1/C2 + C3 etc.) known as wetness is higher (»1.0) for biogenic and lower for thermogenic methane.

☆ Biogenic methane tends to be depleted (–50 to –100‰) in $\delta^{13}C$ relative to thermogenic methane (–20 to –50‰).

☆ Biogenic methane is more depleted (–150 to more negative than –250‰) in δ D values.

Further, within biogenic methane, one can distinguish its origin (pathway) via, H_2:CO_2, carbonate reduction (Eq. 12.1) or through decarboxylation of acetate (Eq. 12.2) based on the differences in the respective $\delta^{13}C$ and δ D values (Table 12.1).

Table 12.1: Distinction Between Biogenic and Thermogenic Methane

Sl.No.	Parameter	Biogenic CH_4	Thermogenic CH_4
1.	$C_1/C_2 + C_3$ etc.	Higher (Since it contain only CH_4)	Lower (Since it contain other higher hydrocarbons)
2.	$\delta^{13}C$	-50 to –100 ‰ (relatively depleted)	–20 to –50 ‰ (relatively enriched)
		H_2: CO_2 origin: –60 to –110 ‰ (more depleted)	–
		Acetate origin: –50 to –65 ‰ (relatively less depleted)	–
3.	δ D	H_2: CO_2 origin: –150 to –250 ‰	–
		Acetate origin: more negative than -250 ‰	–

Source: Waples, 1981.

12.6.4 Retention of Methane on Coal Beds

Although methane is the major component of coal gases, others such as CO_2, N_2, ethane, propane, butane (wet gas) and liquids such as water and other hydrocarbons are also released during coalification (Table 12.2). Total amount of methane generated during the coal formation (Ro max 0.5–1.8 per cent) approximately range between 2000–5000 Scft/ton. However, part of methane generated is retained in coal beds/seams which is termed as CBM, and the excess above retention capacity of the coal bed, tend to migrate to the surrounding reservoir rock (*e.g.*, sandstone). Retention of methane in the coal beds occurs by any of the following ways (Biswas, 1995):

☆ As adsorbed gas molecules on the internal surfaces or absorbed within the molecular structure of the coal.

☆ As gas molecules held within the matrix porosity (micro or macro).

☆ As free gas within cleat and fracture net work of coal seam, and

☆ As gas dissolved in ground water within the coal bed.

Table 12.2: Gas Generation During Coalification
(upto medium bituminous; temp. < 150° C, Ro 1.10 per cent)

Sl.No.	Gas	Generation Capacity (Scft/ton)
1.	Methane	2,000–3,000 +
2.	Carbon dioxide	2,000 +
3.	Wet gases*	100–1,000 +
4.	Nitrogen	250–500 +

* Ethane, propane, butane, and pentane.

Scft = Standard cubic feet.

Source: Scott, 1993.

Methane gas sorbed on coal particles can be liberated by desorption of coal seams by releasing the gas pressure either by dewatering or by drilling borewells which facilitate the flow of gas through fractures. As pressure is released in the coal seam, the low energy physical bonds are broken and the gas molecules diffuse through the coal matrix until they reach a natural fracture (cleat), and subsequently flow through towards well bore (Figure 12.5). The amount of methane produced depends on desorption capacity of coals which varies from coal to coal, depending upon its physical and chemical properties, especially the type of coal (maceral composition).

12.6.5 Methane Generating Capacity

Methane generating capacity of coals in general, depends on three factors:

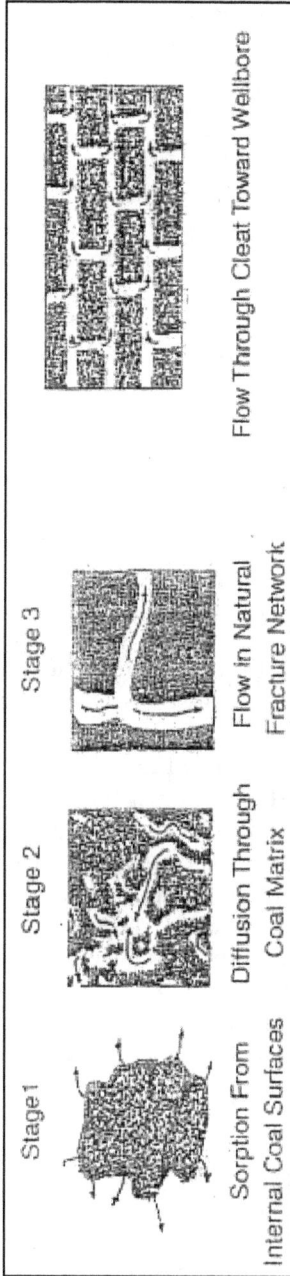

Figure 12.5: Occurrence and Flow Behaviour of Coal Bed Methane (*Source:* Mallick and Raju, 1999)

1. Organic matter and its type.
2. Nature of coal.
3. Geological conditions of deposition

(*i*) Organic Matter and its Type

Methane-generating capacity of coal depends on its origin and is related to kerogen macerals (particles). Whereas macerals of vitrinite group is the greatest contributor of methane, those of inertinite is of relatively little hydrocarbon generation. Therefore, vitrinite rich coals of high rank are the most important sources of CBM because they have more micro-porosity (higher absorbing capacity) than the other two (alginite and exinite) maceral groups (Kuldeep Chandra, 1997).

(*ii*) Nature of Coal

Since methane is generated during coal formation process, all coals invariably contain methane. However, the gas content of a coal normally increases with (*i*) rank of coal, (*ii*) depth of burial of coal seam, provided the roof and overburden are impervious to methane, and (*iii*) thickness of coal seam. Since economically important quantities of methane (> 300 Scft/ton or 8.5 cc/gm) are generated by thermogenic process, this requirement is met in the medium to low volatile bituminous rank (Figure 12.4) at Ro max between 1.5 and 1.8 per cent (Table 12.3). Ash content of coal has an influence on the CBM, the lower the ash, higher is the gas content of a coal seam. The coal should also have minimum gas saturation of 80 per cent. Optimum production of CBM from a coal seam is obtained when permeability is greater than 3.0 md (milli darcy). The permeability in turn is governed by the cleats, their frequency, spacing and tectonically induced, relaxed fractures. Investigations, worldover, have shown that high rank coals buried at greater depths (> 300-1200 m) are suitable for CBM exploration, provided certain other geological and inherited coal seam characteristics are favourable as well.

(*iii*) Geological Conditions

In general, gas is more concentrated in geologically active areas, such as folded and faulted regions, as well as surrounding areas of the faults. Well developed cracks and faults in coal seams owing to tectonic disturbances, provide them permeability. Sealing capability and thickness of seam roof, and floor rocks play a significant role in methane accumulation. The fluvial basins, having higher rate of subsidence accompanied by thermal events and moderate tectonics, are the prospective sites for exploration of CBM.

Table 12.3: Rank of the Coal in Terms of Per cent Total Carbon (TC), Vitrinite Reflectance (R_o) and Volatile Matter (VM)

Rank of Coal	Per cent TC (Daf)*	Per cent R_o	Per cent VM
Lignite	63-72	0.2-0.4	>40
High volatile bituminous coal	76-84	0.5-1.1	32-40
Medium volatile bituminous coal	85-88	1.1-1.5	23-31
Low volatile bituminous coal	89-91	1.5-2.0	15-22
Anthracite	92-98	> 2.0	8-14

*Daf: Dry ash free basis.

Source: Kuldeep Chandra, 1997; Pandey, 1999.

Ground water movement is considered as a critical parameter for evaluating CBM potential of an area. It is responsible for transporting methanogenic bacteria which generate biogenic methane and maintain high reservoir pressure. The best conditions would be coal seams having artesian flow conditions which indirectly indicate permeable nature of the coals.

12.6.6 Estimation of Methane in Coal Beds

Methane gas content in coal seams is measured by two methods: (*i*) qualitative and (*ii*) quantitative (Kuldeep Chandra *et al.*, 1999)

(*i*) Qualitative Method

It is based on the relationship between methane generated and percent (dry ash free, daf) volatile matter (VM) at 20° C and 1 atmosphere pressure, given by the equation.

$$\text{Volume of CH}_4 \, (\text{cc/gm}) = -325 \log \text{VM}(\%)/37.8 \qquad (12.3)$$

This equation is based on the fact that thermogenic methane begins to form when coal attains 37.8 per cent VM or high volatile bituminous boundary. As the VM decreases, the volume of methane generation increases. Eq. 12.3 is however, not applicable to inertinite rich coals.

(ii) Quantitative Methods

There are two quantitative methods of estimation of methane: (a) direct and (b) indirect method.

(a) Direct Method

It involves the following steps:

(i) Measurement of gas lost (Q_1) from the coal sample between the time elapsed from its removal by drill bit in the well and its arrival on to the surface in a sealed container,

(ii) Measurement of desorbed gas (Q_2) from the sample in the laboratory when it attains equilibrium, and

(iii) Measurement of residual gas (Q_3) that remains in the sample after cessation of desorption.

Lost (Q_1) and desorbed (Q_2) gases are measured from a desorption isotherm (Figure 12.6). Residual gas (Q_3) is measured by

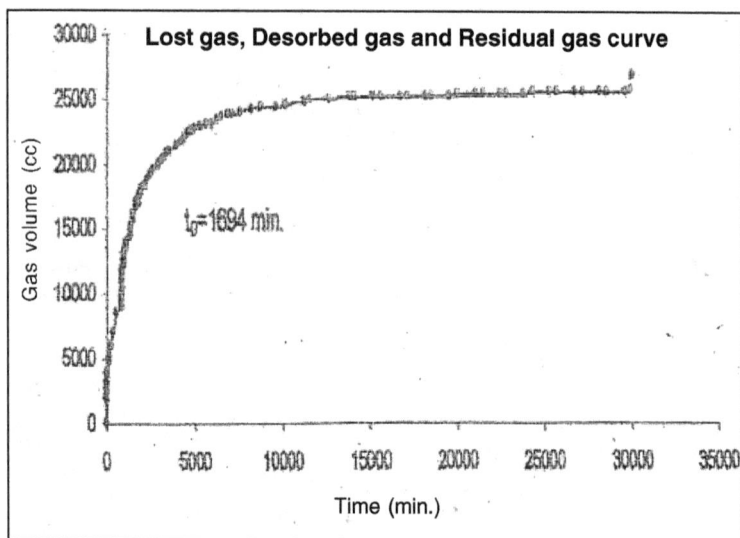

Figure 12.6: Desorption Time Constant Determination from Desorbed Gas Data
(*Source*: Kuldeep Chandra *et al.*, 1999)

crushing the coal sample within the sealed canister and again measuring the desorbed gas. The gas content in the coal sample is obtained by Eq. 12.4.

$$\text{Gas content Q (c.c/gm)} = \frac{Q_1 + Q_2 + Q_3}{W} \qquad (12.4)$$

where, W is weight of the coal sample taken for analysis.

(b) Indirect Method

In this method, a gas free coal sample at a fixed temperature is subjected to varying (increasing) pressures, and the quantity of methane adsorbed is measured (Scft/ton or c.c/gm). Once the sample reaches equilibrium (at initial reservoir pressure), the pressure is decreased at regular intervals and the volume of gas that desorbs is again measured. The adsorption and desorption data are fitted to a Langmuir's adsorption isotherm (Figure 12.7) and the total gas content can be estimated.

12.6.7 Production of CBM

CBM production in a coal basin depends on the following factors: (*i*) coal resource, (*ii*) thermal maturation, (*iii*) burial history of the basin and depth of coal seam, and (*iv*) permeability, hydrology and reservoir pressure (Kuldeep Chandra, 1997). These factors help in defining fairways or sweet spots which quantify the CBM resource and productivity. The traditional approach to CBM production is to de-water the formation so that the gas desorbes and migrates to the fracture system (12.6.4) and flows into the well bore (Mallic and Raju, 1999). As a result, the CBM reservoirs initially produce more water and little commercial gas (Figure 12.8) unlike natural gas reservoirs where the gas production is more than water. Large quatities of water during initial production in CBM reservoirs can be a problem especially if the produced water is saline with high pH. Sophisticated pumps were developed to handle large volumes of water at high rates (Leach, 2002). Over the time, water production decreases while gas production increases. The most challenging aspect of the initial production is minimising the environmental impact and ensuring that local aquifers are not contaminated. Deviated and horizontal wells are more efficient producers in tight formations, especially after hydraulic fracturing. However, multi

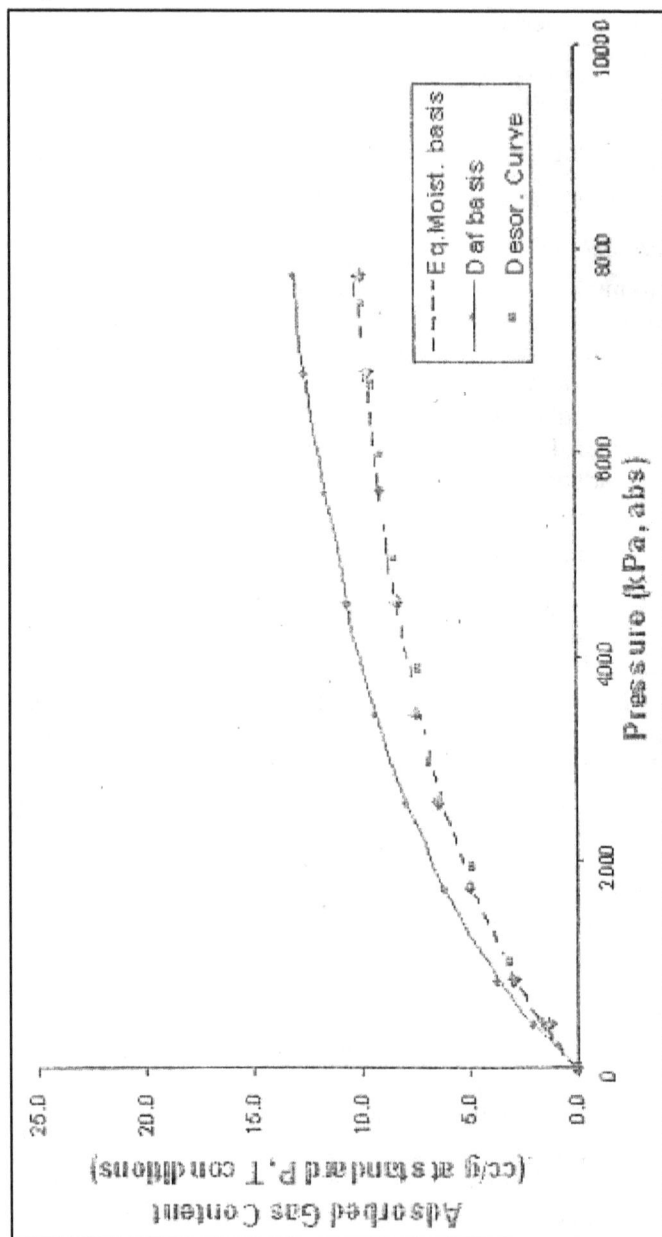

Figure 12.7: Langmuir Isotherm Curves of a Coal Sample
(*Source*: **Kuldeep Chandra *et al.*, 1999**)

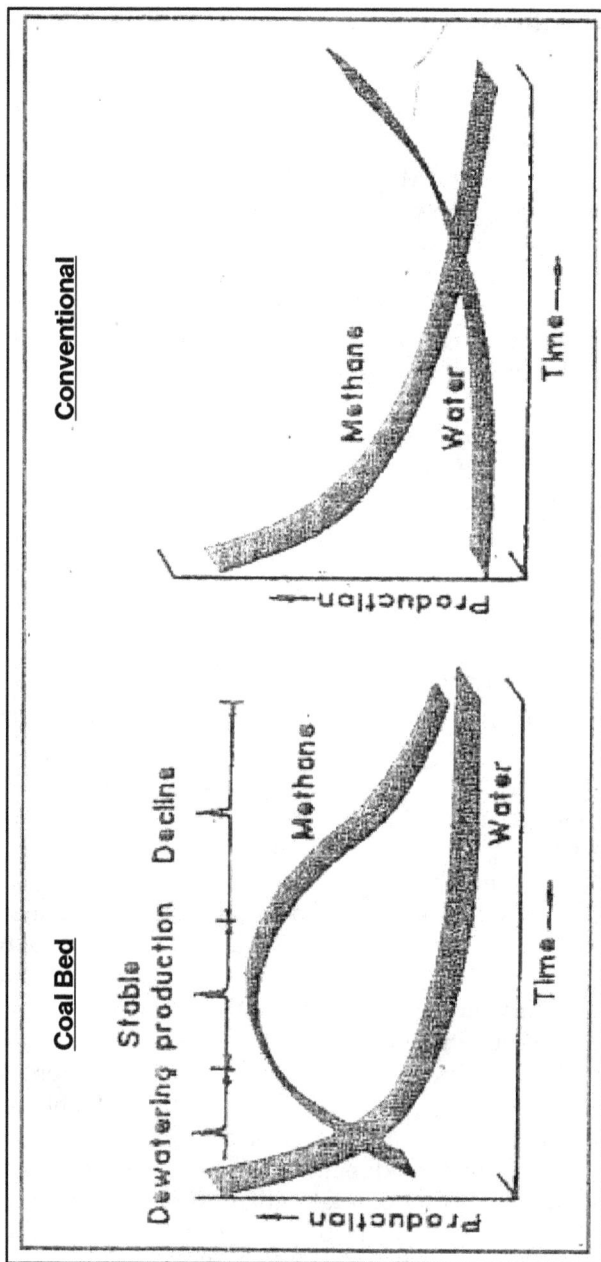

Figure 12.8: Characteristic Production Behaviour of Coal Bed Methane Gas and Conventional Gas Reservoir (*Source:* Mallick and Raju, 1999)

seam completions give better recovery factors (upto 85 per cent) than normally attainable (about 30 per cent) from traditional techniques.

12.6.8 CBM Resources: Global

Major CBM resources are recognized in 70 countries (Landis and Weaver, 1993) with CBM wells drilled in 17 countries (Ayers, 2002). Canada has huge potential with estimates of $3,000 \times 10^{12}$ Scft of CBM, followed by USA with around 700×10^{12} Scft, both of them accounting to about 50 per cent of the global potential (Schwochow, 2002). However, virtually all CBM Production (1.6×10^{12} Scft per annum) comes from eleven coal basins (particularly San Juan) in the USA. Australia produces around 20×10^9 Scft and Canada around 10×10^9 Scft gas per annum. Plans for CBM production are going on in Europe, India, China, Poland and Russia. The estimated global resources are large in the range $5,000–10,000 \times 10^{12}$ Scft (Kuuskraa *et al.*, 1992). Obviously CBM is poised to be a major contributor to global gas production. The challenge is to convert these resources to reserves by increased understanding of coal bed gas systems, better reservoir characterization, and improved production technology (Odedra *et al.*, 2005).

12.6.9 CBM Resources: Indian Scenario

India is the 4th largest country in proven coal reserves (467 billion tones) and 3rd largest producer (300 million tonnes per annum) in the world (Sharma, 2003). Lower Gondwana coals (mainly Permian) are more important for CBM exploration since they account for bulk (99.9 per cent) of coal resources in our country. They mostly occur in Peninsular India along prominent river valleys of eastern and central India, namely Damodar, Sone, Koel, Mahanadi, Kanhan, Pench, Wardha, Pranahita-Godavari and Narmada in the states of Bihar, Jharkhand, West Bengal, Orissa, Madhya Pradesh, Andhra Pradesh and Maharashtra (Figure 12.9). Major part of them (upto 300 m depth) are sub-bitumninous in rank with high ash content (13-45 per cent), far below the threshold value (300 scft/ton) of methane. However, high rank (bituminous) coals amenable for economic generation of CBM mostly occur in deeper parts (300-1200 m) of the basins by younger sediments. Though tertiary coal fields account for only 0.5 per cent of total Indian resource, they are of better quality in terms of lower ash content (8-10 per cent), inspite of the fact that the seams are thinner with relatively high sulphur content than Gondwana coals. The tertiary coal seams are mainly

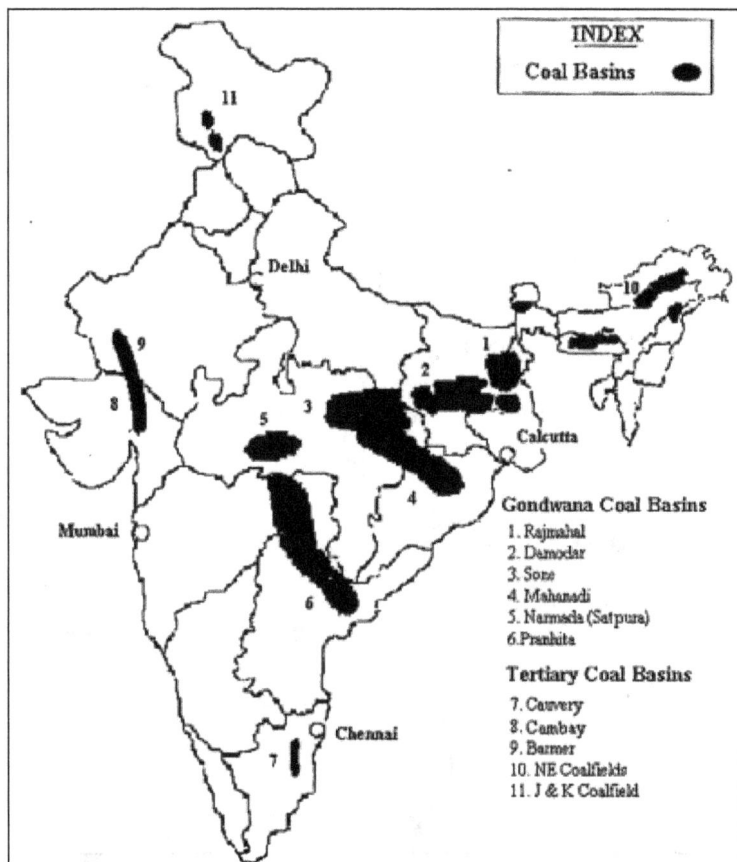

Figure 12.9: Gondwana and Tertiary Coal Basins in India
(*Source*: Sharma, 2003)

confined to Upper Assam, Jammu and Kahsmir, Tamil Nadu and Gujarat. They are mainly of Eocene and Oligocene age and of lower rank (lignite).

Qualitative estimates in the absence of actual core desorption data, indicated quite variable CBM resources *e.g.*, 1.5 Tm^3 (Biswas, 1995); 1.25 Tm^3 (Jamal and Peters, 1997); 850 Bm^3 (Bastia *et al.*, 1995) and 1347 Bm^3 (DGH, 2006, a). India so far has awarded 26 CBM blocks under New Exploration Licensing Policy (NELP) to several oil industries including multi-national companies (Figure 12.10)

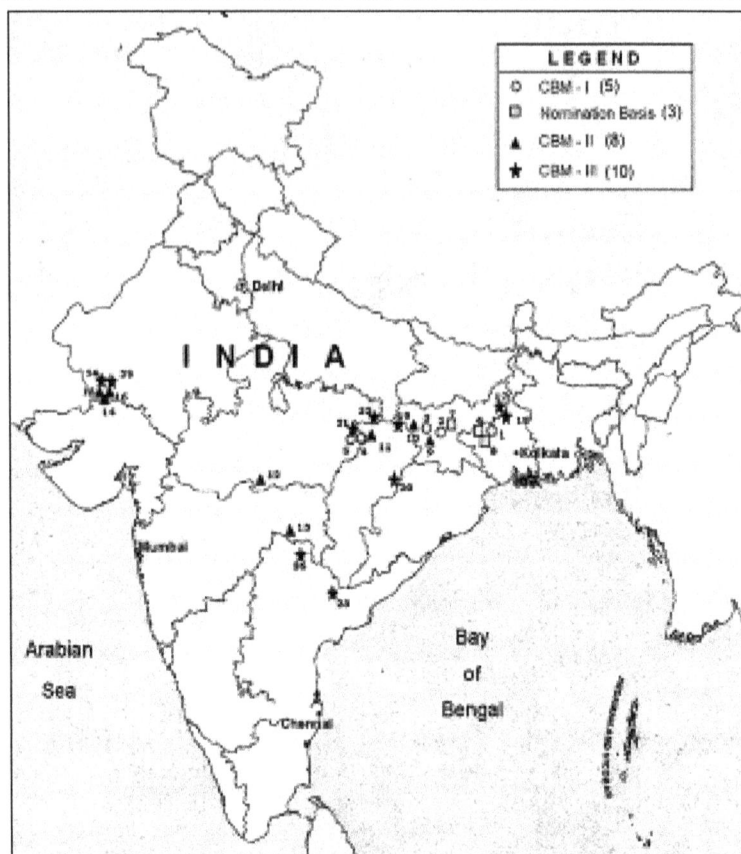

Figure 12.10: Map Showing CBM Blocks Awarded
(*Source*: DGH, 2006a)

with an estimated production potential of 38 MMm³ per day. Table 12.4 incorporates data on some typical CBM coal fields. The geological and geochemical settings of these coal fields (Special Report, 2001) are briefly described:

12.6.10 Typical Indian CBM Coal Fields

(*i*) Raniganj Coalfield

The Raniganj Coalfield is the eastern most member of the Damodar Valley (Figure 12.9) group of Gondwana coalfields. In this coalfield, regionally developed coal seams occur in the Barakar

(Lower Permian) and Raniganj (Upper Permian) formations. The Barakar Formation contains six major coal seams, which are considerably thick (upto 60 m) and of high rank. The Raniganj formation, which is a widespread coal measure contains 10 persistent coal seams, the total thickness of which ranges form 60 to 70 m. The coals of Raniganj seams are high volatile, bituminous A rank and they are quite gassy. There are evidences of uninterrupted emission of methane from fault zones and some of the old boreholes. The seams show marked increase in thickness towards the eastern part of the coalfield where they also show merging tendency. The gas content of seams of the Barakar formation is 3 to 8 m³/tonne, and the Raniganj formation is, 2 to 6 m³/tonne.

Table 12.4: Coal Resource and CBM Potential of Typical Coal Fields

Field	Coal Reserves (Billion Metric Tones)	CBM Potential (Billion Cubic Metres)
Eastern Raniganj	25.6	42.48
Bokaro	4.4	45.03
N. Karanpura	9.8	61.73
S. Karanpura	6.01	28
Eastern Sohagpur	8.20	49.0
Western Sohagpur	6.60	36.82
Satpura	3.68	18.44
North Barmer (lignite)	4.80	9.0

Source: Special Report on CBM, 2001.

(*ii*) Bokaro Coalfield

The Bokaro Coalfield, containing high rank coal reserves, is considered to possess significantly large quantity of methane gas. The Gondwana sequence in this coalfield consists of Barakar (Lower Permian), Barren Measures (Middle Permian), Raniganj (Upper Permian), Panchet and Supra-Panchet (Upper Triassic) formations in ascending order.

The Barakar Coal Measure is well developed (about 1000 m) in this field. In the East Bokaro field, the Coal Measure contains 26 coal seams of various thicknesses, the prominent ones being

3.8 to 36.6 m. The West Bokaro field consist of 13 persistent seams, the prominent ones vary in thickness from 3 to 23.8 m. The Bokaro seams are high volatile bituminous A to medium volatile bituminous rank, and show vitrinite reflectance of 0.8 to 1.2 per cent. The seams are very gassy, and there are several cases of methane explosions in the underground mines in this field. The gas content of the seams vary from 4 to 10 m^3/tonne. The central part of the basin, being a synclinal low, preserves younger sediments of Triassic age and holds high prospect for CBM. A total coal thickness of about 140 m is developed in this part of the field.

(*iii*) North Karanpura Coalfield

The North Karanpura Coalfield is the westernmost member of the Damodar Valley (Figure 12.9) group of Gondwana Coalfields. Barakar Formation of lower Gondwana sequence is the main coal bearing stratigraphic unit in this area. The Lower Gondwana sequence is unconformably overlain by the Mahadeva Formation of Upper Triassic age. The coalfield is situated in a half graben, the southern boundary of which is defined by a prominent E-W trending fault.

Barakar Formation contains 5 regionally persistent coal seams showing wide variation in thickness, quality and rank across the field. The lower most seam-I (generally 2.0-25 m thick) and the succeeding seam-II (3.0 to 24.5 m thick) are considered to be the main targets for CBM. The seams are high volatile bituminous A to medium volatile bituminous rank, with vitrinite reflectance of 0.9 to 1.44 per cent which are well cleated. Gas content of the seams is variable, but increases with depth to as high as 13.5 m^3/tonne. However, for the estimation of CBM resource, a value of 6 to 8 m^3/tonne is considered.

(*iv*) Sohagpur Coalfield

The Sohagpur Coalfield covering an area of 3000 sq. km. forms a part of the South Rewa Gondwana Basin. The Gondwana sedimentary succession has a maximum thickness of about 4 km. The Lower Gondwana sequence consists of the Barakar (Lower Permian), Barren Measures (Middle Permian), Raniganj (Upper Permian) and Pali (Permo-Triassic) formations, and the Upper Gondwana sequence consists of the Parsora and Taki formations. Gondwana sediments in this area are intruded by dolerite dykes.

Barakar Formation contains five regionally persistent coal seams, of which Seam-III (2.5-8.2 m) and Seam-V (3.8-11.31 m) are considered to be suitable for CBM exploration and production. A prominent E-W trending fault which shades towards north, runs along the axial part of the basin. In the northern down thrown block, the coal seams attain higher rank showing vitrinite reflectance of 0.78 to 0.99 per cent and considered suitable for CBM exploration. There are evidences of gas seepages though a number of fault zones in the northern part of the coalfield. Gas content of the seams is about 4 m^3/tonne upto a depth of 600-700 m. However, there is a tendency for the gas content to increase towards the northern part of the field due to added overburden of younger sediments.

(v) Satpura Coalfield

The Satpura Coalfield (Figure 12.9) is located along with Narmada valley, bounded in the north by the Son-Narmada lineament, an important geo-fracture on the Pre-Cambrian platform. Lower Gondwana sequence comprises the Talchir, Barakar, Motur (Lower Permian) and Bijouri formations (Upper Permian). This sequence is unconformably overlain by the Pachmari (Lower Triassic) and Denwa (Lower-Mid Triassic) formations. The Bagra formation (Triassic) is the upper most unit of the Lower Gondwana sequence. This succession is unconformably overlained by the Jabalpur formation of Lower Cretaceous age. The thickness of the total Gondwana sequence is estimated to be about 4 kms in the deepest part of the basin. Gondwana sediments are intruded by dolerite dykes and sills. Most of the working collieries are located in the southern part of the basin.

The Barakar Coal Measures which crops out in the south-eastern part of the basin is designated as Pench Valley Coalfield, whereas its extension to the west is known as Kanhan-Tawa Valley Coalfield. The Barakar sequence in Kanhan-Tawa Valley contains 3 or 4 seams which range in thickness from 2.5 to 5.6 m. Coal seams of the Satpura Basin exhibit significant regional variation in rank from Peach Valley in the east to Tawa valley in the west. Coals of the Kanhan-Tawa Valley are high volatile bituminous A to medium volatile bituminous rank with vitrinite reflectance varying from 0.70-1.02 per cent. As such, they hold high prospect for CBM. There are also evidences of gas emission through fault planes. The gas content of the seams is about 5 m^3/tonne.

(vi) Barmer Basin

The Barmer Basin (Figure 12.9), which lies in the Thar Desert of Rajasthan, is the northern extension of the Cambay Basin of Gujarat. Sedimentary sequence of the basin ranges in age from Cretaceous to Recent. Tertiary and Quaternary sequences with a total thickness of about 4 km are the important sediments containing coal seams.

In the Barmer basin, a 29-30 m thick lignite bed was encountered at a depth of 1312 to 1407 m in the two boreholes drilled in Guda area. But in the northern part, where the Tertiary section is much attenuated, lignite is being worked out by open mine at Giral. However, regionally the lignite bed has been encountered at a depth of 450 m below ground level in different boreholes. The Middle Eocene lignite has indicated emission of methane. Adsorption isotherms of lignite samples from the Guda boreholes show a maximum gas content of 6.5 to 7.2 m^3/tonne at a pressure of 41.5 kg/sq. cm. The deep-seated lignite seam in Barmer basin having good porosity holds significant promise for commercial exploration of CBM.

12.7 Gas Hydrates

Methane (CH_4), the most common form of natural gas associated with petroleum reservoirs, is in gaseous state at standard temperature and pressure conditions. But under high pressure and low temperature, it combines with water molecules to form an icy-white compound called methane hydrate (Carroll, 2001). Other gases such as ethane, propane, butane, CO_2, H_2S etc., also form compounds with water and in general, they are called as gas hydrates. In nature such high pressure and low temperature conditions are prevalent in areas of permafrost (both offshore and onshore), on the outer continental margins and shelves, and beneath sea floors. Gas hydrates comprise more than 99 per cent of methane and in an ideally saturated gas hydrate the molar ratio of methane to water is about 1:6. Thus it contains huge volumes of gas *e.g.*, 1 m^3 of solid hydrate (theoretically) contains upto 164 m^3 of methane gas at standard pressure and temperature. In nature, the more typical value is 40 m^3 of methane. Though the hydrates are least understood unconventional energy resource as of now, they hold the greatest volume of the global methane endowment (Carroll, 2001).

Methane produced from gas hydrates is truly unconventional because of its origin, trapping mechanism and production technologies. Unlike conventional traps, which are a prerequisite for the accumulation of oil and natural gas, gas hydrates can form their own trap within the pore space and expand themselves against the sediment load in the form of massive hydrates.

12.7.1 Discovery of Gas Hydrate

Davy (1811) was the first to discover gas hydrate when chlorine gas and water reacted to give rise to a solid compound at cold weather conditions in the laboratory. A variety of other molecules are later found to form hydrates called as clathrates in which they are considered as guests while the water being the host molecule. In general, a clathrate is a compound formed by the inclusion of molecules of one kind (guest) within the cavities of the crystal lattice of another (host). Clathrates, also known as "inclusion or container compounds," display no chemical bonding between the host and the guest molecules. They can form spontaneously under certain pressure, temperature conditions when a host molecule (*e.g.*, water) crystallises into an open lattice structure, and a guest molecule of suitable size (*e.g.*, methane) fits into the lattice voids. Table 12.5 incorporates typical examples of clathrates.

Table 12.5: Common Clathrates: Various Hosts and Guests

Host	Guest
Urea	Straight chain hydrocarbons
Thiourea	Branched chain and cyclic hydrocarbons
Dinitrodiphenyl	Derivatives of diphenyl
Phenol	Hydrogen chloride, sulfur dioxide, acetylene
Water (ice)	Halogens, noble gases, sulfur hexafluoride, low molecular weight hydrocarbons, CO_2, SO_3, N_2, etc.
Nickel dicyanobenzene	Benzene, Chloroform
Clay minerals (molecular sieves)	Hydrophilic substances
Zeolites	Wide range of adsorbed substances
Graphite	Oxygen, hydrocarbons, alkali metals (in sheet like cavities and buck/balls)
Cellulose	Water, hydrocarbons, dyes, iodine

Source: Max, 2000.

12.7.2 Occurrence

Gas hydrates occur ubiquitously in three distinct areas: Polar (permafrost) regions, shallow offshore sediments in Arctic and sub-arctic regions, and deep ocean sediments in tropical regions. Naturally occurring hydrates exist in four forms: disseminated or dispersed, nodular, layered and massive hydrates. Except layered hydrates, others occur in unconsolidated sediments (Kuldeep Chandra, 1997). The most favourable conditions for occurrence of gas hydrates are:

(*i*) Areas of hemipelagic sedimentation of Pleistocene age,

(*ii*) Rapid sedimentation rate (> 3 cm/1000 years),

(*iii*) P & T conditions, pressure (> 50 atms) and temperature (4-6°C) envelop,

(*iv*) Geothermal gradient (<6°C/100 m), and

(*v*) Sediment organic carbon content (>0.5 per cent) and methane concentration in porewater (>10 ml/l)

12.7.3 Crystal Structure

Methane (gas) hydrate is a crystalline ice-like solid compound (Figure 12.11). It occurs in three structures I, II and H, the most predominant being the structure I (Figure 12.12A and B). In this structure, the cages are arranged in a body centered cubic packing, and are large enough to include methane, ethane and other gas molecules of similar molecular diameter such as CO_2 and H_2S. In structure II, face centered (diamond) cubic packing, some cages are larger enough to include bigger molecules such as propane and isobutane. Structure H being least common in nature, contains cages still larger than structure II in a hexagonal crystallographic system (Sloan, 1998).

12.7.4 Physical Properties

Some typical physical properties of methane hydrate are given in Table 12.6. The density of methane hydrate (0.91 g/cm³) is less than water. It may vary slightly depending on the methane saturation of the hydrate lattice and incorporation of other molecules such as H_2S in the lattice. The heat of hydrate formation and heat of hydrate dissociation are equal in magnitude but of opposite sign. When hydrate forms heat is released from the system (exothermic) and

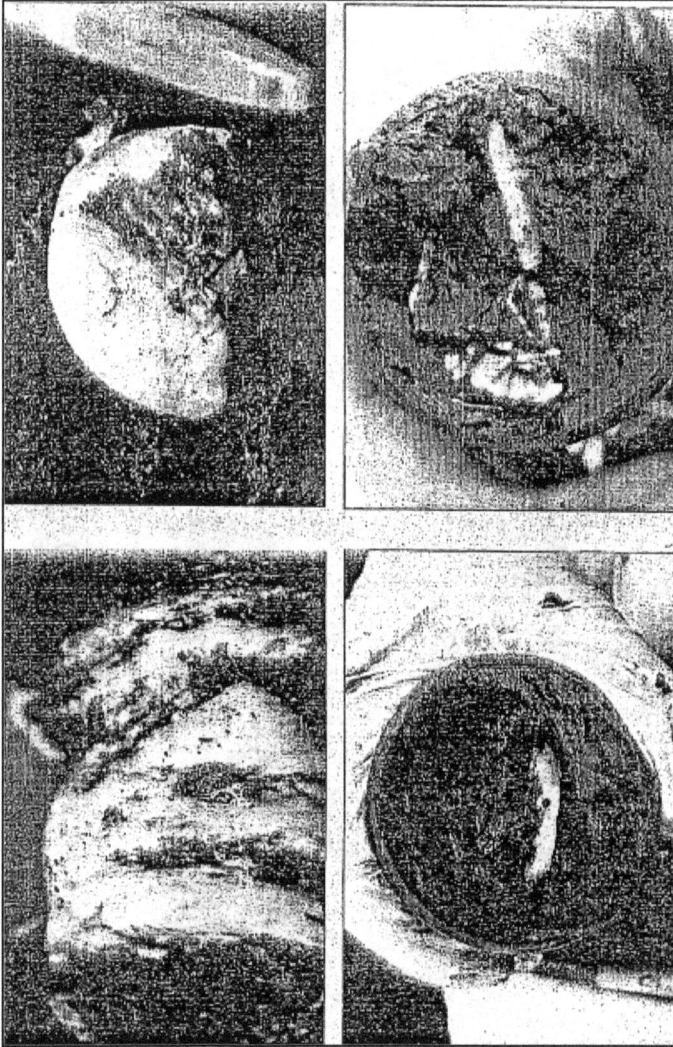

Figure 12.11: Gas Hydrates Recovered from Krishna-Godavari (KG) Basin (*Source:* DGH, 2006b)

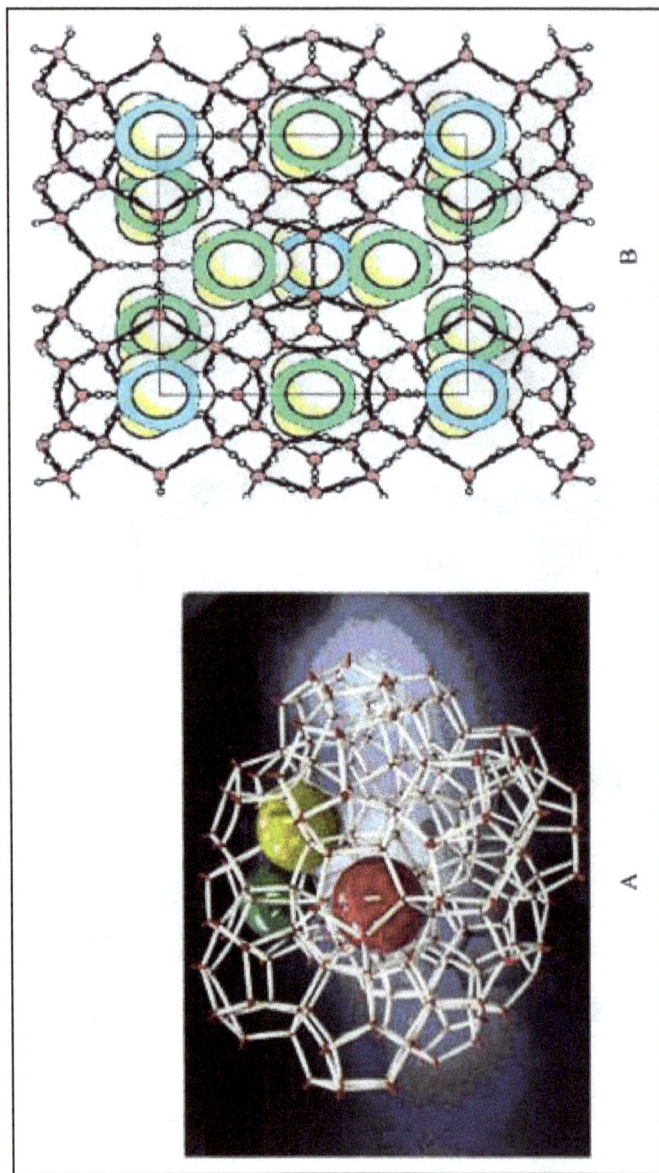

Figure 12.12: Structure I (Body centred cubic lattice) of gas hydrates. A. Cavities occupied by different guest molecules such as methane, CO, and H$_2$S; B. Cavities occupied by methane gas molecules (*Source:* Gupta, 2000)

when hydrate dissociates, heat is takenup into the system (endothermic). The measured value of methane hydrate formation enthalpy at 273°K is 54 KJ/mole (Sloan, 1998).

Table 12.6: Some Physical Properties of Methane Hydrate

S.No.	Property	Magnitude
1.	Bulk density (gms/cm³)	0.912
2.	Thermal conductivity at 263°K (w/m-k)	0.49 ± 0.02
3.	Adiabatic compressibility at 273°K, 10-11 pa	14
4.	Heat of fusion at 273°K (KJ/mole)	54
5.	Heat of dissociation at 273°K (KJ/mole)	–54
6.	Heat capacity at constant pressure (KJ/mole)	257
7.	Dielectric constant at 273°K	58
8.	Poisson's ratio	0.33

Source: Max, 2000.

12.7.5 Composition of Gas Hydrates

All over the world, methane gas usually comprises >99 per cent in gas hydrates with structure I. Only in the Gulf of Mexico and in the Caspian Sea, gas hydrates are found with methane along with significant amounts of ethane and propone having structure II. Although gas hydrates containing 90 per cent methane and 10 per cent H_2S have been recovered offshore from Oregon on ODP Leg 146, they exhibit only structure I, because of similar molecular size of both the gases.

Methane in gas hydrates has essentially two origins: microbial formed during diagenesis and thermal formed during catagenesis of kerogen. Information on the molecular composition of hydrocarbon gases (wetness) in gas hydrate sample, coupled with measurements of carbon isotopic composition of methane, provides a basis for interpreting the origin of methane (7.8.0 and 6.4.0) in gas hydrate. In most of the gas hydrate samples of structure I, methane has a carbon isotopic composition ($\delta^{13}C$) –55 to –110‰ suggesting mainly of microbial origin. This microbial methane is believed to result from methanogenic process taking place in shallow sediment, in which CO_2 ultimately derived from organic matter, is reduced to methane. In contrast, the gas hydrate samples with possible structure II from

Gulf of Mexico and Caspian Sea contain methane with carbon isotopic composition –39 to –45‰. This methane and accompanying heavier hydrocarbon gases are considered to be of thermogenic origin, resulting from the thermal decomposition of organic matter at great depths. Microbial methane is likely to migrate only short distances to form gas hydrates occurring at or near the surface. Thermogenic methane, on the other hand, is likely to migrate longer distances from deeply burried sediment in order to form gas hydrates at or near the sea floor, as is the case of Gulf of Mexico and the Caspian Sea. More recent carbon isotope analysis of sediments in hydrate rich regions support the observation that there is a complex mixture of thermogenic and biogenic carbon sources that influence the methane hydrate formation and stability.

12.8 Stability of Gas Hydrates

The stability of a gas hydrate depends on pressure (P), temperature (T) and the solubility of gas as a function of P and T of the system. The stability is more susceptible to changes in temperature than pressure. In addition to temperature, other physical properties such as thermal conductivity and thermal diffusivity are also critical which control the transfer of heat through hydrate reservoirs. Further, certain chemical and geological factors promote or inhibit hydrate stability. These include:

☆ Gas properties: presence of ethane, propane, CO_2, H_2S etc.,

☆ Pore fluid properties: dissolved ions (Na^+, K^+, Mg^{2+}, Ca^{2+}),

☆ Sediment composition: fine grained sediments (clays)

☆ Geological processes: erosion, rate of sedimentation, slumping, subsidence, uplift, and earthquakes.

The phase quilibria shown in Figure 12.13 apply to pure methane system and structure I gas hydrate in the presence of fresh or saline water. Presence of even a fraction of percent of ethane or propane that produce structure II hydrate, is stable at higher temperatures than structure I at a given pressure, thus increasing its stability. In contrast, the presence of dissolved ions (Na^+, K^+, Ca^{2+}, Mg^{2+}) in pore fluids inhibit the stability of gas hydrate. The inhibitary effect of ionic compounds on the stability of hydrate has critical implications for the evolution and long term stability of hydrate deposits in areas characterized by salt tectonism, evolution of seafloor

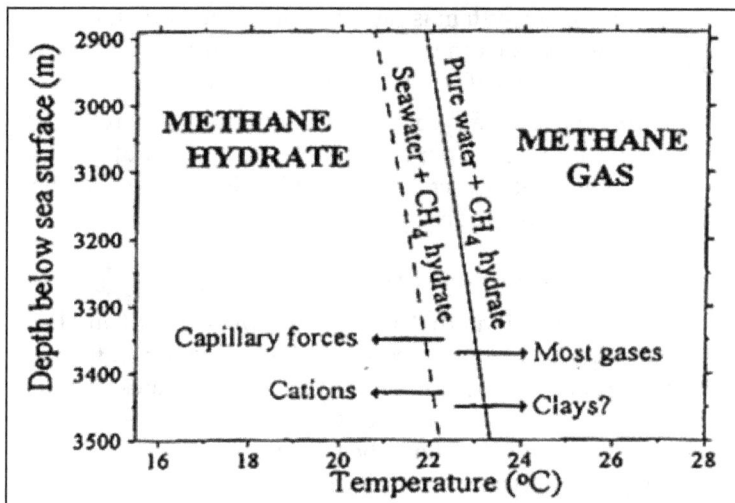

Figure 12.13: Stability of Marine Hydrate for Pure Water (Solid line) and Seawater (Dashed line) Systems. Arrows schematically show various physical and chemical factors that affect the stability of hydrate (*Source*: Max, 2000)

brine basins, and circulation of briny liquids as evident in the Gulf of Mexico (Sassen and McDonald, 1997).

The composition of sediment matrix may exercise a strong influence on the temperature of hydrate dissociation in some settings. Clays appears to promote hydrate formation at higher temperatures. Physical properties of the sediment may play a critical role on the stability of hydrate as well. Large capillary forces that arise in fine grained sediments may inhibit the entry of fluids into the interstices between the grains and thus significantly inhibit hydrate stability, depressing its dissociation temperature.

Methane hydrates often form as disseminated grains and pore fillings in coarse grained sediments whereas in finer silt/clay deposits, they commonly appear as nodules and veins. In general, naturally occurring methane hydrates form in two types of geological settings (Milkov and Sassen, 2001):

(i) Structural accumulations associated with faults and mud volcanoes, where gas leaking from deeper subsurface petroleum system is rapidly transported (via faults, gas chimneys or other fluid conduits) to the gas hydrate stability zone (Cassassucem *et al.*, 2004), and

(*ii*) Stratigraphically bound accumulations in relatively permeable sea floor sediments where bacterial methane has generated *in situ* or has slowly migrated form shallow depths into the gas hydrate stability zone.

12.8.1 Gas Hydrate Stability Zone (GHSZ)

Physical conditions that control the presence of methane hydrate are usually represented by a phase diagram in terms of temperature and depth fields (Figure 12.14). It illustrates the phase quilibrium among gas hydrate, free gas and aqueous solution, and physical parameters controlling the formation of gas hydrate (P, T and salinity). The phase boundary (dashed line) separates colder, higher pressure conditions where methane hydrate is stable to the left of the curve, from conditions to the right where it is not stable (solid line) assuming a typical hydrothermal gradient in water and geothermal gradient in sediment. The figure shows the variation of temperature with depth of water and underlying sediment. The point where the solid line (representing the conditions in sediment) crosses the phase boundary represents the bottom of the zone where methane hydrate is stable (GHSZ). This phase boundary corresponds to a sediment depth of about 300–600 m below the sea floor (bsf) at a water depth of about 3,000 m along the Indian continental margins (Figure 12.14). However, the hydrate zone in the water column is much shallower (about 750 m). When methane in water is of sufficient concentration (near saturation), it forms hydrate in the hydrate stability zone. However, being lighter than water, it floats upward and would dissociate when it crosses the depth where the hydrothermal gradient and the gas hydrate boundary curves intersect each other. On the other hand, if the gas hydrate forms within the sediment, it will be bound in place.

The precise location of base of the GHSZ under known pressure and temperature conditions depends on several factors, the most important being the gas composition. Presence of higher hydrocarbons (ethane, propane etc.) allow the gas hydrates to form at lower pressure and higher temperature (*i.e.*, shallow waters). The presence of salts (NaCl, KCl, $MgCl_2$ etc.), in porewaters shift the gas

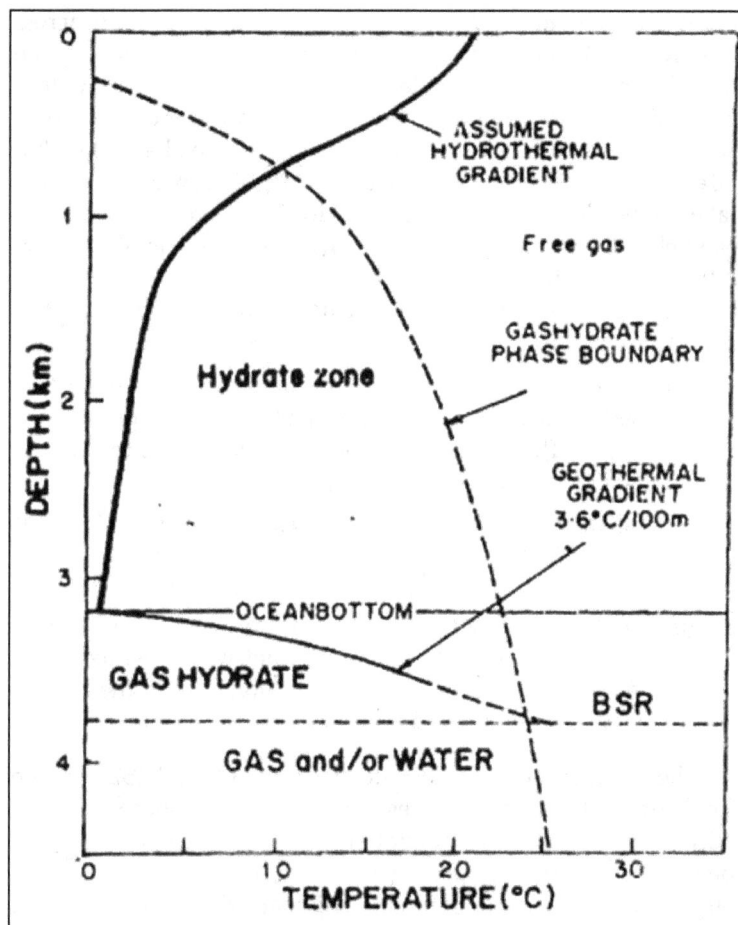

**Figure 12.14: Phase Diagram of Methane Hydrate for Indian
Continental Margins
(*Source*: Hanumantha Rao *et al.*, 1998)**

hydrate phase boundary to the left, low temperatures and high
pressure (*i.e.*, deep waters).

12.8.2 Thickness of GHSZ

As stated above, the vertical extent of the sediment layer from
sea floor to the intersection point of the phase boundary curve with
the geo-thermal gradient curve in the sediment gives the thickness

of GHSZ. In general, the thermal gradient tends to be uniform across broad regions where sediments do not vary. Therefore, for a given water depth, the sub-bottom depth to the base of the GHSZ will be quite constant. However, because a change in water depth causes changes in pressure, it is to be anticipated that the base of GHSZ will extend further below sea floor as water depth increases. It implies that the zone where gas hydrate exists, forms a more or less uniform layer below the sea floor, thickening towards greater depths (Figure 12.15). Though this is commonly true, there are few exceptions because of disturbances of thermal structure in the sediments (Max, 2000). These include:

    (i) *Sea floor land slides*: They remove cooler near surface sediments, leaving warmer-than normal materials near the sea floor which cause local shallowing of the base of GHSZ.

    (ii) *Salt diapers*: They produce warm spots since the salt has greater thermal conductivity than other sediments which forces the base of GHSZ to shallower regions.

    (iii) *Faults*: Circulation of warm fluids upto shallow subbottom regions through faults as channel ways results in the disruption of gas hydrate formation or shift the base of GHSZ to shallow waters.

The thickness of gas hydrate stability zone (GHSZ) can be calculated on the basis of its pressure-temperature phase diagram, bathymetry and sea bottom temperature. Figure 12.16 shows the potential thickness of GHSZ calculated for the Indian continental margins in the form of contour map. This map will help in validating the potential zones of gas hydrates as revealed by BSR (bottom simulating reflection) surveys. The hydrate stability zone lies between 300-400 m thick below sea floor (bsf) in the western offshore, and 400-600 m thick (bsf) in the eastern offshore, varying with the heat flow (Rao, 1999).

12.9 Detection of Gas Hydrates

Seismic reflection and geo-chemical methods are most commonly employed for identifying and locating gas hydrate zones. Seismic attributes such as bottom simulating reflection (BSR) which runs parallel to the sea floor, blanking above BSR, velocity inversion

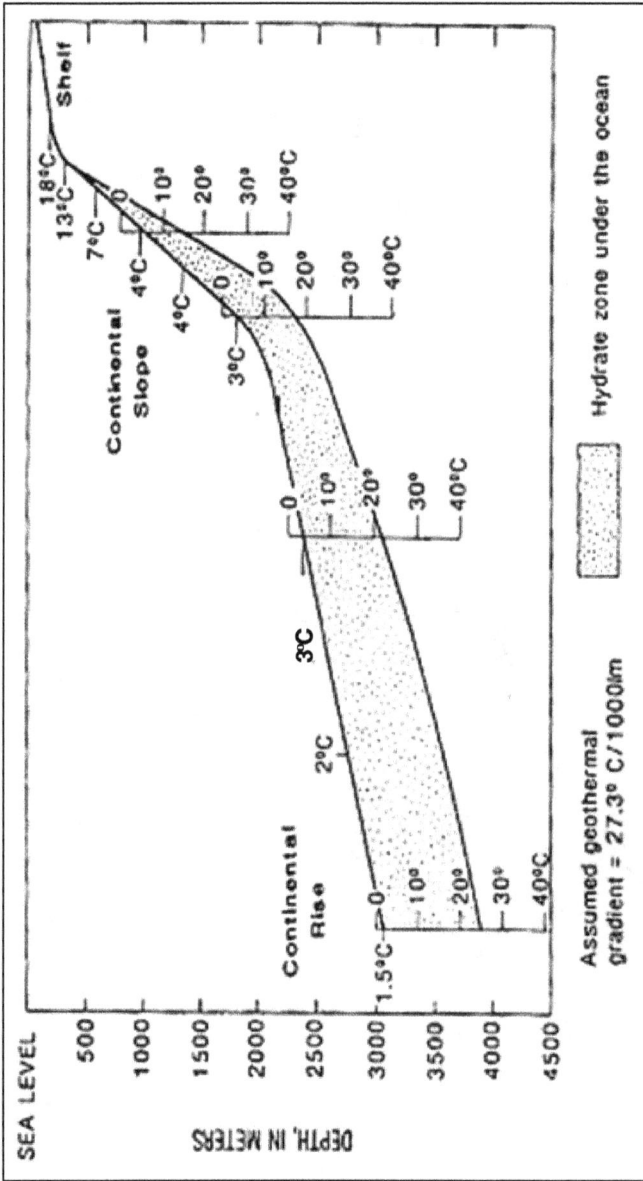

Figure 12.15: Inferred Thickness of the GHSZ (Dot pattern) in Sediments of a Continental Margin Assuming a Typical Geothermal Gradient
(*Source*: Max, 2000)

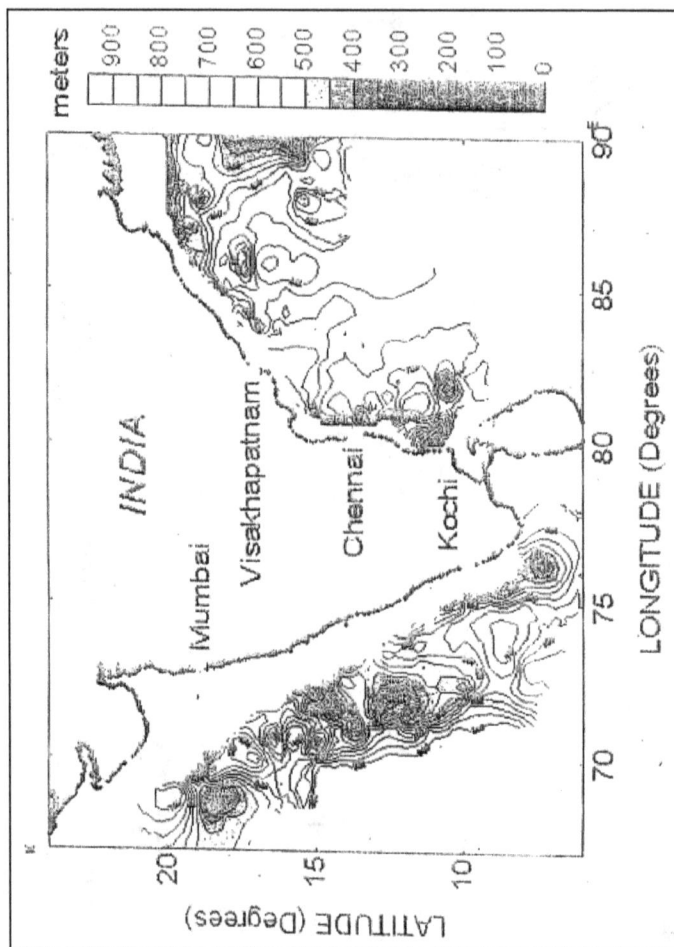

Figure 12.16: Thickness Variation of Gas Hydrate Stability Zone (GHSZ) in Offshore Regions of India (*Source*: Hanumantha Rao *et al.*, 1998)

below BSR and polarity reversal at BSR are frequently used to identify, map and quantify the amount of gas hydrates (Pecher and Holbrook, 2001). However, reconnaissance mapping of BSRs must be calibrated by bore hole data (logging and geo-chemical) because the former may fail to identify and locate gas hydrate zones in some cases.

Measurement of *in situ* properties using logs is primarily useful for (i) identification of hydrate and hydrate bearing sediment and their distribution with depth, (ii) estimation of porosity and methane saturation, and (iii) calibration of drill hole information with surface seismic or other remote sensing techniques. In general, resistivity, sonic (accoustic) velocity show higher values while natural gamma (radio activity) and formation density show lower values in wire line logs in sediments containing gas hydrates relative to water saturated or gas saturated sediments (Bily and Dick, 1974). Thus changes in their values above BSRs conclusively prove the presence of gas hydrates. Geo-chemical proxies like sulphate, chloride and barium in porewaters, methane in sediment cores and microbiological proxies such as sulphate reducing and fermentation bacteria provide indirect evidence of gas hydrate occurrences (Borowski *et al.*, 1999). Depletion of chlorinity and sulphate, and enrichment of barium in porewater of sediment cores are considered as common indirect geochemical proxies. The sulphate methane interface (SMI) which is the boundary between sulphate minimum and threshold of methane generation, in conjunction with enrichment of methane flux and abundance of bacteria namely sulphate reduces (SRB-r) and fermentators (SRB–f) in sediment cores are good indicators for the likely occurrence of gas hydrates below the sea floor (Ramana *et al.*, 2006).

12.9.1 Production Technologies

The volume of methane that contained in a gas hydrate accumulation depends on five primary reservoir parameters. (*i*) areal extent of gas in the gas hydrate accumulation, (*ii*) reservoir thickness, (*iii*) sediment porosity, (*iv*) degree of gas hydrate saturation, and (*v*) hydrate gas yield. For extraction of gas from gas hydrates, three principal methods are being considered : (*i*) thermal stimulation, (*ii*) depressurization, and (*iii*) inhibitor injection.

In the thermal stimulation method, the hydrated sediment is heated enough by pumping hot water or steam to increase the local temperature so that it dissociates in to gas. In the depressurisation method, the pressure of the hydrate bearing formation is reduced to below hydrate stability to cause its dissociation (Sloan, 1998). In the inhibitor injection method, a dissociating agent like methanol is injected in to the hydrate filled sediment. This disturbs the pressure–temperature equilibrium making gas hydrates no longer stable at *in situ* conditions resulting in their dissociation into methane gas. Among the three methods, depressurization combined with hot water injection appears to be the most practical method where free gas lies beneath the gas hydrate (Holder *et al.*, 1984). Circulation of warm surface water into gas hydrate deposits and horizontal drilling technique provide possible future approaches to the exploitation of hydrated methane (Kvenvolden, 1993, b).

12.9.2 Environmental Impacts

Gas hydrates may be of considerable importance in understanding climate changes, both the rapid rise in glacial temperature at the end of the Paleocene (Dickens *et al.*, 1995) and more recent Quaternary climate fluctuations (Nisbet, 1990). During glacial maxima vast quantities of methane (biogenic, thermogenic or mixed origin), become trapped within permafrost and submarine clathrate cemented sediment. As the climate warms, a critical temperature is reached at which point the clathrate destabilize and huge amounts of methane gas will be released. This sudden release of gas may trigger mud volcanoes and pock marks on the seabed, and pingoes in the permafrost. The huge increase of methane gas released into the atmosphere may be responsible for the sudden increase in global temperature (global warming) observed at the end of glacial maxima since methane is a green house gas much more toxic than other green house gases like carbon dioxide.

Decomposition of gas hydrates decrease the shear strength, making the sediment more to failure. It also affects porepressure, dilation (expansion) of sediment volume and development of interstitial gas bubbles-all of which have the potential to weaken the sediment. Major slumps on continental margins are tentatively related to instability associated with the breakdown of gas hydrates. Occurrence of huge submarine slope failures leading to slides and slumping events in the Norwegion Continental Margins in Late Pleistocene are correlated to hydrate decomposition. Further, mass movement of sediment in the sea floor may often be triggered by the gas hydrate dissociation (Kvenvolden, 1993 a; Nisbet and Piper, 1998).

12.9.3 Favourable Conditions for Gas Hydrates in the Indian Offshore Regions

Deep water areas covering the Bay of Bengal, the Andaman Sea and the Arabian Sea from about 600 m (about 6 MPa, and 9°C) to a maximum depth of 3000 m (30 MPa and 1°C) are favourable for the formation of methane hydrates over large areas of sediments (Figure 12.17). The rate of sedimentation of Pliocene to Recent varies between 47 cm/1000 years in the KG (Eastern) Offshore, 13 cm/1000 years in Kerala (Western) Offshore and 4 to 11cm/1000 years in the Andaman sea. The average geothermal gradient of Indian Offshore regions vary between 3-4°C/100 m with maximum values in Kutch (Arabian Sea). It is about 2°C/100 m in the Bay of Bengal and varies between 1.2–1.8°C/100 m in the Andaman Sea. The organic carbon content of recent sediments vary between 0.14 to 6.3 per cent on the west coast, and between 0.5–2.1 per cent on the east coast (Max, 2000). These geological and geochemical conditions are most suitable for the occurrence of methane hydrates in the sea subsurface sediments. Broad estimates on the basis of likelihood of prospectivity in the Indian offshore (Chandra *et al.*, 1998) indicate about 80,000 km² drillable sub-surface area (Table 12.7) and a prognosticated gas resource of 1894 TCM (DGH, 2006 b). Seismic attributes of BSR such as blanking, velocity inversion and polarity reversals within the estimated hydrate stability zones also broadly concur with this estimate.

12.10 Distribution of Gas Hydrates

12.10.1 World Scenario

World wide, about 60 locations of gas hydrates have been identified in the onshore (permafrost) and offshore regions (Figure 12.18). The most important among permafrost areas include: Siberian Basins (Russia), Prudhoe Bay (Alaska, USA), Mackenzie Delta (Canada) and Arctic islands. Some typical Offshore continental margins include: Gulf of Mexico, Oregon, Blake Ridge (USA), Nankai Trench (Japan), Chile triple junction (Chile), Makran Accretionary Wedge (Pakistan) and Offshore regions of Indonesia, Korea, Peru,

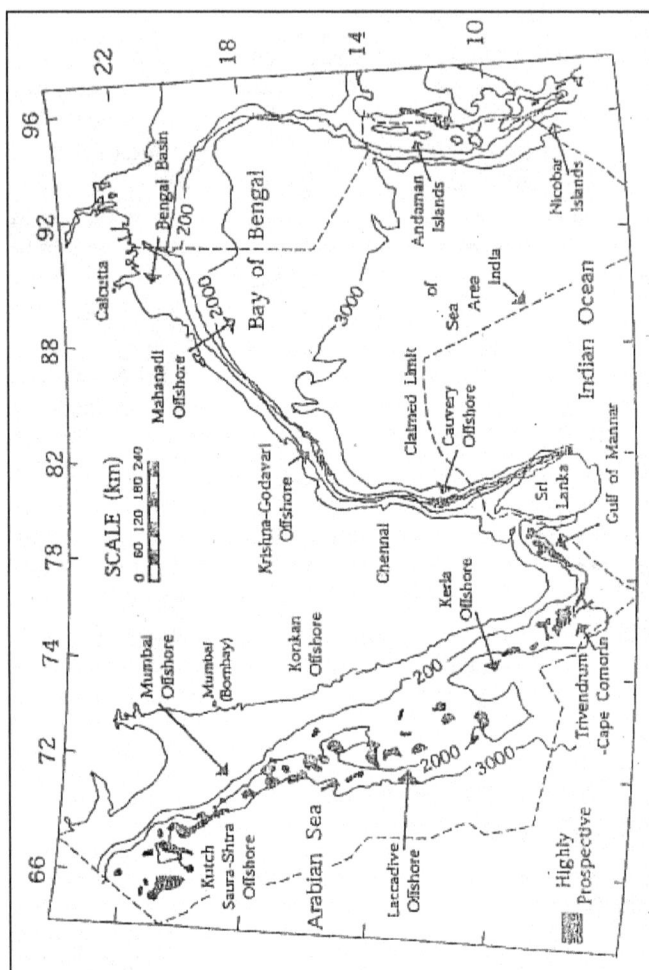

Figure 12.17: Bathymetric Countours, Delineation of the Indian Sea Areas (Dashed) and Prospective Hydrate Deposit Areas (Black) Identified from Seismic Interpretation
(Source: Max, 2000)

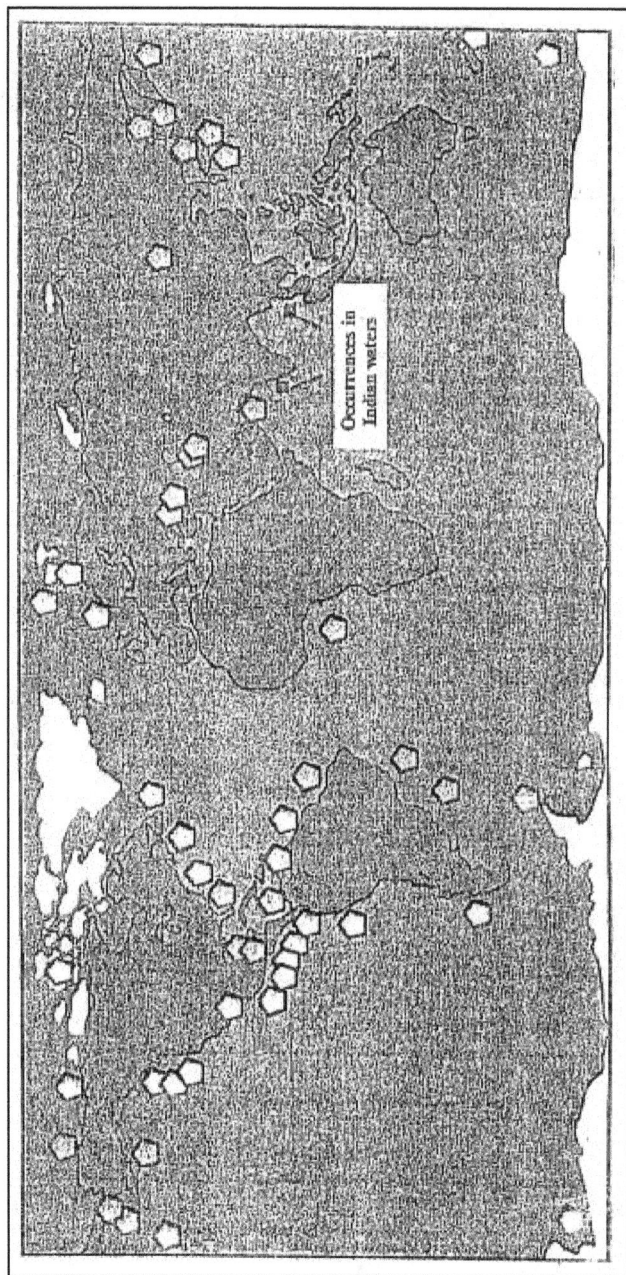

Figure 12.18: Global Occurrences both in Permafrost and Outer Continental Margin Sediments
(*Source:* Subrahmanyam *et al.,* 1998)

India and Nigeria. Global estimates of natural gas (methane) in hydrates are astounding. A "consensus value" of about 7,00,000 x 10^{12} Scft (Kvenvolden, 1993b; Collett, 2002) suggests that methane hydrate contains two orders of magnitude more carbon than all other fossil fuels on the earth.

Table 12.7: Areas of Gas Hydrate Prospects.
Categories–I: Highly prospective; II: Moderately
prospective, and III: Low Prospective

Area Name	Area in Km²			
	I	*II*	*III*	*Total*
Western Offshore				
Kutch	2,900	4,150	3,250	10,300
DCS	4,105	3,550	600	8,275
Bombay	3,075	800	350	4,225
Konkan	3,800	5,476	–	9,276
Padua-Bank	1,100	3,600	6,700	11,400
Lakshadweep	1,850	3,275	–	5,125
Kerala	275	2,800	1,325	4,400
Cape Comorin	3,500	7,059	–	10,559
	20,625	**30,710**	**12,225**	**63,560**
Eastern Offshore				
Gulf of Mannar	1000	725	–	1,725
Cauvery	1,900	3,100	–	5,000
Krishna	2,125	750	–	2,875
Godavari	325	150	–	475
	5,350	**4,725**	**–**	**10,075**
Andaman-Nicobar Offshore				
Nicobar	775	475	250	1,500
Andaman	3,125	1,025	925	5,075
	3,900	**1,500**	**1,175**	**6,575**
Grand Total	**29,875**	**36,935**	**13,400**	**80,210**

Source: Chandra *et al.*, 1998.

12.10.2 Indian Scenario

The first gas hydrate work based on identification of BSR's was reported in the Andaman offshore (Chopra, 1985). Subsequently a national gas hydrate programme (NGHP) was formulated in 1997 by the government of India for exploration and development of gas hydrate resources (DGH, 2006, b). This study enabled firming up of locations for drilling and coring of gas hydrate bearing sediments in four areas namely, Krishna Godavari (K-G) and Mahanadi basins in the east cost, Kerala–Konkan basin in the west coast and the Andaman Offshore. A scientific expedition (NGHP-01) first of its kind in India was carried out on a research drill ship JODES Resolution in the above four prospective offshore areas. Two sites drilled, cored and logged in the K-G basin showed the presence of gas hydrate zones. At one of these sites, a high quality massive gas hydrates zone was established in the depth range of 40-160 m below sea floor (bsf) suggesting the most favourable site for future exploration. Following are the salient features of the Indian gas hydrate exploration:

 (*i*) The Bay of Bengal, unique in its geological, geochemical, thermal regime, and possessing large volume of gas hydrates, is a potential area for exploration.

 (*ii*) The Krishna Godavari (K-G) Offshore is the most favourable area in the Indian passive margins with thick sediment deposited from Miocene to Recent in a growth fault environment.

 (*iii*) The areas lying between 600-2000 m water depth off the coasts of Mahanadi, Godavari and Krishna River deltas are likely to emerge as a huge storehouses of massive hydrate deposits.

 (*iv*) Andaman–Nicobar arc basin displaying favorable geological, geochemical and geophysical attributes, also offer locations for huge hydrate deposits.

In conclusion, it may be stated that methane hydrate trapped in the continental slopes offer the greatest potential, although the resources may be a double edged sword with equally potential environmental hazards. If they proved to be economical, new technological methods of exploring them have to be developed (Odedra *et al.*, 2005). Recent researches on one hand, indicate the possibility of converting large volumes of natural gas occupying in deep ocean areas into gas hydrates and towed them to shallow waters. On the other hand, there is a concern about the possible clogging of pipe lines and explosions from gas hydrate formation/decomposition. Nevertheless, the majority opinion is that methane obtained from hydrate is a clean form of energy and should be considered as a favourable future resource (Subrahmanyam *et al.*, 1998). However, the most important question posing the scientific community is will the gas hydrate become a significant energy resource in the near future ? The worldwide opinion on this essential question is that it is unlikely before 2030. But some countries like India, Japan and South Korea are of the opinion that it may become a critical sustainable source possibly by 2015.

Appendices

Appendix 1: The Geological Time Scale

PHANEROZOIC	PRECAMBRIAN

The Geological Time Scale chart showing the Phanerozoic eon divided into Cenozoic, Mesozoic, and Paleozoic eras, and the Precambrian divided into Proterozoic and Archean eons.

CENOZOIC — Neogene (Pliocene, Miocene), Paleogene (Oligocene, Eocene, Paleocene)

MESOZOIC — Cretaceous (Late, Early), Jurassic (Late, Middle, Early), Triassic (Late, Middle, Early)

PALEOZOIC — Permian, Carboniferous, Devonian, Silurian, Ordovician, Cambrian

PRECAMBRIAN — Proterozoic (Neo-proterozoic, Meso-proterozoic, Paleo-proterozoic), Archean (Neo-archean, Meso-archean, Paleo-archean, Eoarchean)

(Source : Gradstein et al., 2004)

Appendix 2: Abbreviations

AAPG	:	American Association of Petroleum Geologists
amu	:	Atomic Mass Units
API	:	American Petroleum Institute
BTU	:	British Thermal Unit
CBM	:	Coal Bed Methane
CF-IRMS	:	Continuous Flow Isotope Ratio Mass Spectrometer
CPI	:	Carbon Preference Index
daf	:	Dry ash free
DGH	:	Directorate General of Hydrocarbons
DTA	:	Differential Thermal Analysis
Eh	:	Redox Potential
EOM	:	Extractable Organic Matter
EOR	:	Enhanced Oil Recovery
ESR	:	Electron Spin Resonance
FID	:	Flame Ionisation Detector
GC	:	Gas-Liquid Chromatography
GC-MS	:	Gas Chromatography with Mass Spectrometry
GHSZ	:	Gas Hydrate Stability Zone
GOR	:	Gas Oil Ratio
HGRT	:	High Solution Geochemistry Technology
HI	:	Hydrogen Index
HPLC	:	High Pressure Liquid Chromatography

IR	:	Infra Red
IRMS	:	Isotope Ratio Mass Spectrometer
LOM	:	Level of Organic Metamorphism
Mb	:	Million barrels
Md	:	Millidarcy
MMT	:	Million Metric Tonnes
MMT/d	:	Million Metric Tonnes per day
MMT/Y	:	Million Metric Tonnes per year
MMSCM/d	:	Million Metric Standard Cubic Metres per day
MRM-GC-MS	:	Metastable Reaction Monitoring Gas Chromatography Mass Spectrometry
MS	:	Mass Spectrometry
NMR	:	Nuclear Magnetic Resonance
OEG	:	Oil Equivalent of Gas
OM	:	Organic Matter
PDB	:	Pee-Dee formation of South Carolina Belemnite
ppb	:	Parts per billion (10^9)
ppm	:	Parts per million (10^6)
permil	:	Parts per thousand (‰)
ppt	:	Parts per trillion (10^{12})
PI	:	Production Index
Ro	:	Vitrinite reflectance measured in oil immersion
RT	:	Retention Time
S c ft	:	Standard Cubic feet
S c ft/d	:	Standard Cubic feet per day
STP	:	Standard Temperature (70° F or 15.6° C) and Pressure (1 atmosphere or 760 mm of Hg)
SMOW	:	Standard Mean Ocean Water
TAI	:	Thermal Alteration Index
TB	:	Trillion Barrels

TC	:	Total Carbon
TCM (Tm^3)	:	Trillion Cubic Metres
TCM/d	:	Trillion Cubic Metres per day.
TDS	:	Total Dissolved Salts
TG	:	Thermogravimetry
TOC	:	Total Organic Carbon
TTI	:	Time–Temperature Index
USGC	:	United States Geological Survey
UV	:	Ultra-Violet
VM	:	Volatile Matter

Appendix 3: Standard Units and Conversion Factors

1U S Barrel	=	158.984 litres (L)
	=	5.6146 cubic feet (c.ft)
	=	0.15899 cubic metre (cm or m³)
1 Litre (L)	=	0.035316 cubic feet (c.ft)
	=	0.00629 U.S barrel (bbl)
	=	0.001 cubic metre (cm or m³)
1 cubic foot	=	28.136 litres (L)
(c.f.t)	=	0.17811 US barrels (bbl)
	=	0.028317 cubic meters (cm or m³)
	=	28317.01 cubic centimeters (c.c)
1 cubic metre	=	35.315 cubic feet (c ft)
(cm or m³)	=	999.97 litres (L)
	=	6.2898 US barrels (bbl)
	=	1,000,000 cubic centimetres (c.c)
1 Mile (mi)	=	1.609 kilometres (km)
1 BCM (billion cubic metres)	=	35.3 billion c. ft of natural gas
	=	0.89 million metric tones (MMT) of crude oil
	=	1.35 million metric tones (MMT) of coal
	=	36 trillion British thermal units (BTU)
Degrees (°C)	=	5/9 (degrees °F–32)
Degrees (°F)	=	9/5 (degrees °C) + 32)

1 K W H (kwh)	=	3412 BTU =	36,00,000 Joules
1 BTU	=	252 calories	
1 g. calorie	=	4.186 joules	
1 md (milli darcy)	=	9.869×10^{-6} cm^2	
1 atmosphere	=	76 cm of mercury (cm-Hg)	
(Pressure)	=	760 mm of mercury (mm-Hg)	
	=	101.325 kilo pascal (kP$_a$)	
	=	101,325 Pascal (P$_a$)	
	=	14.70 pound per sq. inch (psi)	
1 kilo pascal	=	10,000 dynes/cm^2	
(kP$_a$)	=	1000 Newton per sq. metre (N/m^2)	
	=	0.1451 Pounds per sq. inch.(Psi)	

References

Abelson, P.P (1963). Organic geochemistry and the formation of petroleum. 6th World Pet. cong. Proc, 1,397-407

Ahmed, A.E; Murthy, R.V.S; and Bharktya, D.K. (1993). Depositional environments, structural style and hydrocarbon habitat of Upper Assam Basin. Proc. 2nd Semi. petroliferous basins of India. Vol. I. Biswas, S.K. et al. (Eds). Indian Petroleum Pub; Dehra Dun, p. 438-458.

Alexander, R; Gray, M.D; Kogi, R.I. and Woodhouse, G.W. (1980). Proton magnetic resonance spectroscopy as a technique for measuring the maturity of petroleum. Chem. Geol., 30, 1-14.

Alpern, B; Durand, B; Espitalie,J; and Tissot, B. (1972). Localisation characterization et classification petrographique des substances organiques sedimentaries fossils. In: Advances in organic geochemistry. Pergamon Press, London.

Antia, D.D.J. (2007). Oil polymerisation and fluid expulsion from low temperature, low maturity, over pressured sediments. J. Petrol. Geol; 31, 263-282.

Antia, D.D.J. (2008). Polymerisation theory–Formation of hydrocarbons in sedimentary strata (Hydrates, Clays, Sandstones, Carbonates, Evaporites Volcanoclastics) from CH_4

and CO_2 : Part II : Formation and interpretation of Stage 1 to Stage 5 oils. Indian J. Petrol. Geol; 17, 11-70.

Antia, D.D.J. (2009). Low temperature hydrocarbon polymerisation and hydrocarbon expulsion from continental shelf and continental slope sediments. Indian J. Perol. Geol; 16, 1-30.

Ayers,W.B. (2002). Coal bed gas systems, resources and production, and a review on the contrasting cases from the Sanjuan and Powder River Basin. AAPG Bull., 86, 1853-1890.

Bailey, N.J.L; Evans, C.R; and Milner, C.W.D. (1974). Applying petroleum geochemistry to search for oil: Examples from western Canada Basin. AAPG Bull., 58, 2284-2294.

Bailey, N.J.L; Krouse, H.R; Evans, C.R; and Rogers, M.A. (1973). Alternation of crude oil by water and bacteria-evidence for geochemical and isotopic studies. AAPG Bull., 57, 1276-1290.

Bally, A.W; and Snelson, S. (1980). Realms of subsidence. In: Facts and Principles of World Petroleum Occurrences. Mail, A.D. (Ed). Can. Soc Petrol. Geol. Memoir; 6, 9-75.

Banerjee, A; Jha, M; Mittal, A.K; Thomas, N.J; and Misra, K.N. (2000). The effective source rocks in the north Cambay basin, India. Mar. Pet. Geol; 17, 1111–1129.

Banerjee, A; Pahari, S; Jha, M; Sinha, A.K; Jain, A.K; Kumar, N; Thomas, N.J; Misra, K.N; and Chandra, K. (2002). Effective source rocks in the Cambay basin, India. AAPG Bull; 86, 433-456.

Barker, C (1985). Origin and properties of petroleum, In: Enhanced oil recovery–I. Fundamentals and analysis. Donaldson EC (Ed.). Elsevier, Amsterdam, UK. p. 11-45.

Barker, C and Dickey, P.A. (1984). Hydrocarbon habitat in main producing areas, Saudi Arabia: discussion. Bull. AAPG., 68, 108-109

Bastia, R; Naik, G.C; and Mohapatra, P. (1993). Hydrocarbon prospects of Schuppen Belt, Assam–Arakan Basin. Proc. 2nd Semi. petroliferous basins of India. Vol.1. Biswas, S.K. *et al.* (Eds). Indian Petroleum Pub; Dehra Dun, p. 493-506.

Bastia, R; and Sankaran, V; and Srinivasan, S. (1995). Coal seam methane, its potential in India. In: Proc. PETROTECH–95, Kuldeep Chandra *et al.* (Eds). B.R. Pub. Corp; New Delhi, p. 435-443.

Basu, D.N; Benerjee, A; and Tamhane, D.M. (1980). Source area migration trends of oil and gas in Bombay Offshore Basin, India. AAPG Bull; 64, 209-220.

Basu, D.N; Benerjee, A; and Tomhane, D.M.(1982). Facies distribution and the petroleum geology of the Bombay Offshore Basin, India. Jour. Petrol. Geol; 5, 57-75.

Berard, B.B; Brooks, J; and Sackett; W.M. (1976). Natural gas seepages in the Gulf of Mexico. Earth Planet. Sci. Lett; 31, 48-54.

Benerjie, V; Mittal, A.K; Agarwal, K; Uniyal, A.K; and Chandra, K. (1994). Carbon isotope geochemistry of petroleum associated gases in Krishna-Godavari Baisn, India. Org. Geochem; 21, 373-382.

Bhandari, L.L; Fuloria, R.C; and Sastri, V.V.(1973). Stratigraphy of Assam Valley, India. AAPG Bull; 57, 642-650.

Bhandari, L.L; and Chowdhury, L.R. (1975). Stratigraphic analysis of Kalol and Kadi Formations. AAPG Bull; 59, 856-871.

Bhandari, L.L; Venkatachala, B.S; Kumar, R; Nanjuada Swami, S; Garg, G; and Srivastva, D.C. (Eds) (1983). Petroliferous basins of India. Vol. I. Petrol. Asia Jour; 6, 189 pp.

Bhowmick, P.K.(2009). Petroleum Geochemistry in hydrocarbon exploration : Past, present and Future. J. Appl. Geochem; 11, 276-289.

Biederman; E.W. (1965) Crude oil composition–a clue to migration. World Oil. 161, 78-82.

Bily, C; and Dick, J.W.L. (1974). Naturally occurring gas hydrates in the Mackenzie Delta, N.W.T. Bull. Can. Pet. Geol; 22, 405-412.

Bishop, R.P; Gehman, H.M; and Young, Y.A. (1983). Concepts for estimating hydrocarbon accumulations and dispersion. AAPG Bull; 67, 337-348.

Biswas, S.K. (1982). Rift basins in the western margin of India and their hydrocarbon prospects. AAPG Bull; 66, 1497-1513.

Biswas, S.K. (1987). Regional tetonic framework, structure and evolution of the western marginal basins of India. Tectonophysics; 135, 307-324.

Biswas, S.K. (1989). Hydrocarbon exploration in Western Offshore Basins of India. In: Recent geoscientific studies in Arabian Sea off India. Geol. Surv. India Spl. Pub, No.24, 185-194.

Biswas, S.K. (1991). Exploration in Bombay Offshore Basin: an overview. In : Proc. Conf. Integrated Exploration Research: Achievements and Prospectives. Pandey, J; and Benerjie, V (Eds). KDMIPE, Dehra Dun, p. 15-25.

Biswas, S.K. (1992). Tectonic frame work and evolution of graben basins of India. Indian J. Petrol. Geol; 1, 276-292.

Biswas, S.K. (1995). Prospect of coal bed methane in India. Indian J.Petrol. Geol; 4, 1-23.

Biswas, S.K. (1998). Overview of sedimentary basins of India and their hydrocarbon resource potential. In: Proc. Nat. Symp. Recent researches in sedimentary basins. Tiwari, N.D (Ed). Indian Petroleum Pub; Dehra Dun, p.1-25.

Biswas, S.K; Bhasin, A.L ; and Jokhan Ram. (1993). Classification of Indian sedimentary basins in the frame work of plate tectonics. In: Proc. 2nd semi. petroliferous basins of India. Vol. I. Biswas, S.K et al. (Eds). Indian Petroleum Pub; Dehra Dun, p. 1-46.

Biswas, S.K; Ranga Raju, M.K; Thomas, J; and Bhattacharya, S.K. (1994). Cambay–Hazad petroleum system in South Cambay, India. AAPG Memoir; 60, 615-624.

Bois, C.P; Bouchche, P; and Pelet, R. (1982). Global geological history and distribution of hydrocarbon reserves. Bull. AAPG; 66, 1248-1270

Bolin, B. (1970). The carbon cycle. Scientific American. 233, 125-132.

Borowski, W.S; Paull, C.K; and Williams III, U. (1999). Global and local variations of interstitial sulphate gradients in deep water continental margin sediments; sensitivity to underlying methane and gas hydrates. Mar. Geol; 159, 131-154.

Brahmaji Rao, G; Phillip, P.C; Padmanabhan, N; Siddiqvi, M.A; and Sexena, R.P. (1991). Source rock development in Kommagudem, Mandapeta and Drakshrama. In: Proc. Conf. Integrated Exploration Research: Achievements and Prospectives. Pandey, J; and Benerjie, V (Eds). KDMIPE; Dehra Dun, p. 395-398.

Braun, R.L; and Burnham, A.K. (1987). Analysis of chemical reaction kinetics using a distribution of activation energies and simpler models. Energy Fuels. 1, 153-161.

Braun, R.L; and Burnham, A.K. (1990). Mathematical model of oil generation, degradation and expulsion. Energy Fuels. 4, 132-146.

Bray, E.E; and Evans, E.D. (1961). Distribution of n-paraffins as a clue to recognition of source beds. Geochim. Cosmochin. Acta; 22, 2-15.

Bray, E.E; and Foster, W.R. (1980). A process for primary migration of petroleum. AAPG Bull; 64, 107-114.

Brooks, J. (1977). Organic geochemistry and petroleum exploration. Bull. ONGC; 14, 15-34.

Brooks, J. (1981). Organic maturation of sedimentary organic matter and petroleum exploration: A review. In: Organic maturation studies and fossil fuel exploration. Brooks, J. (Ed.). Academic Press, London.

Brooks, J. and Bindra, T. (1977). Oil-source rock identification of the Jurassic sediments in the northern North Sea. Chem. Geol., 20, 283-294.

Buckley, S.E; Hocott, C.R; and Taggart Jr, M.S. (1958). Distribution of dissolved hydrocarbons in sub-surface waters. In: Habitat of oil. Weeks, L.G. (Ed). AAPG, Tulsa, Oklahoma. p. 850-852.

Burlingame, A.L; and Schnoes, H.K. (1969). Mass spectrometry in organic geochemistry: Methods and results. Eglinton, G; and Murphy, M.J.T (Eds.). Springer-Verlag, New York, p. 89-149.

Carmat, S.W; and Bell St. John. (1986). Gaint oil and gas fields. AAPG Memoir–40, p. 11-53.

Carroll, J. (2001). Natural gas hydrates. Gulf Publishing, Boston.

Carpenter, A.B. Trout, M.L; and Picket, E.E. (1974). Preliminary report on the origin of chemical evolution of lead and zink rich oil field brines in central Mississippi. Econ. Geology, 69, 1191-2006.

Case, L.C.(1945). Exceptional Silurian brine near Bay city, Michigan. AAPG Bull; 29, 567-570.

Case, L.C. (1956). The contrast in initial and present application of the term "connate water". J. Pet. Technol., 8, 12.

Cassassucem, F; Rector, J; and Hoverstain, M. (2004). Study of Gas hydrates in the deep sea Gulf of Mexico from seismic data. The Leading Edge; 23, 366-372.

Chandra, K; Misra, C.S; Samanta, U; Anita Gupta; and Mehrotra, K.L. (1994). Correlation of different maturity parameters in the Ahmedabad–Mehsena block of the Cambay basin. Org. Geochem; 21, 313-321.

Chandra, K; Misra, C.S; Dwivedi, P; and Thomas, N.J. (1996). Energy issues in India and petroleum geochemistry. In: Proc. 4th AAPG, Tanzania, Mpanju, F (Ed). p. 499-508.

Chandra, K; Singh, R.P; and Julka, A.C. (1998). Gas hydrate potential of Indian Offshore. In: Proc. 2nd conf. and exposition of Petroleum Geophys. SPG–98, p. 357-363.

Chandra, P.K; and Chowdhury, L.R. (1969). Stratigraphy of Cambay Basin. Bull. ONGC; 6, 37-50.

Chandra, P.K; and Venkataraman, S. (1988). Exploration history and status of hydrocarbon exploration of Cauvery Basin with particular reference of occurrence of oil and gas pools and their relationship to depositional systems in Paleogene and Cretaceous. In: Proc. 7th OSEA conf. Singapore; p. 353-363.

Chandra, U; Dhawan, R; Mittal, A.K; Dwivedi, P; and Uniyal, A.K. (1995). Stable isotope geochemistry of associated gases from Lakwa–Lakshmani field, Upper Assam Basin, India. In: Proc. PETROTECH–95, Vol. 2. Kuldeep Chandra et al. (Eds). B.R. Pub. Corp. New Delhi; p. 361-364.

Chilingar, G.V; Buryakovsky, L.A; Eremenko, N.A; and Gorfunkel, M.V. (2005). Geology and geochemistry of oil and gas developments in petroleum. Science No. 52, Elsevier, Amsterdam, UK, 370 pp.

Chopra, N.N. (1985). Gas hydrates–an unconventional trap in forearc regions of Andaman Offshore. ONGC Bull; 22, 41-54.

Clementz, D.M; Demaison, G.J; and Daley, A.R. (1979). Well site geochemistry by programmed pyrolysis. Oil and Gas J., 77, 142-146.

Cohen, C.R. (1981). Time and temperature in petroleum formation. Applications of Lopatin's method to petroleum exploration: Discussion. AAPG Bull., 65, 1647-1648.

Cohen, C.R. (1982). Model for a passive to active continental margin transition: implications for hydrocarbon exploration. Bull. AAPG; 66, 708-718.

Coleman, D.D; Liu, C; and Riley, C.M. (1988). Microbial methane in the shallow Paleozoic sediments and glacial deposits of Illinois, USA. Chem. Geol., 71, 23-40.

Collett, T.S. (2002). Energy resource potential of natural gas hydrate. AAPG Bull; 86, 1971-1992.

Collins, A.G. (1975). Geochemistry of oil field waters. Elsevier, New York, 496 pp.

Collins, A.G. (1980). Oil field brines. In: Developments in petroleum geology–II. Hobson, G.D. (Ed.). Applied Sci. Pub., London, p. 177-185.

Collins, A.G; and Wright, C.C. (1985). Enhanced oil recovery injection waters. In: Enhanced oil recovery I. Fundamentals and Analysis. Donaldson, E.D; Chilingarin, G.V and Yen, F.T. (Eds). Elsevier Pub. Amsterdam, p. 151-221.

Collins, B.I; Tedesco, S.A; and Martin, W.F. (1992). Integrated petroleum prospect evaluation-three examples from the Denver Basin, Colarado. J. Geophys. Expl. 43, 67-89.

Connan, J. (1974). Time temperature relation in oil genesis. Bull. AAPG; 58, 2516-2521.

Connan, J. (1984). Biodegradation of crude oils in reservoirs. In: Advances in Petroleum Geochemistry, Brooks, J; and Welte, D (Eds). Vol.1. Academic, London. p. 299-333.

Connan, J; Letran, K; and Vanderwide, B. (1975). Alteration of petroleum in reservoirs. Proc. 9th World Pet. Congr., 2, 171-178.

Curtis, J.B. (2002). Fractured shale gas systems. AAPG Bull; 86, 1921-1938.

Dahl, J.E; Moldowan, J.M; Teerman, S.C; McCaffery, M.A; Sundararaman, P; Pena, M; and Stetting, E.C. (1994). Source rock quality determination from oil biomarkers I–An example from the Aspen shale, Scully's Cap, Wyoming. AAPG Bull., 78, 1057-1526.

Dandekar, A.Y. (2006). Petroleum reservoir rock and fluid properties. Taylor & Fancies Group, ACRC Press, 460 pp.

Das Gupta, A.B. (1977). Geology of Assam–Arakan Region Quart. Jour. Geol. Min. Met. Soc. India, 49.

Das Gupta, S.K. (1974). The stratigraphy of the west Rajasthan Shelf. In: Proc. 4th Collog. Ind. Micropalaental. Strat; p. 219-233.

Datta, A.K. (1983). Geological evolution and hydrocarbon prospects of Rajasthan Basin. Petrol. Asia Jour; 6, 93-100.

Davidson, C.F. (1965). A possible mode of strata–bound copper ores. Econ. Geol; 60, 942-954.

Davis, J.B. (1952). Studies on soil samples from a paraffin dirt. AAPG Bull., 36, 2186-2188.

Davis, J.W; and Collins, A.G. (1971). Solubility of barium and strontium sulphate in strong electrolyte solutions Environ. Sci. Technol; 5, 1039-1043.

Davy, H. (1811). On Combination of oxymuriatic gas and oxygen gas. Philosophical Transaction of Royal Society; 101, 155.

Demaison, G. (1984). The generative basin concept in : Petroleum geochemistry and basin evaluation. Demaison, G and Murris, J (Eds). AAPG Memoir; 35, 1-14.

Demaison, G.J. (1977). Tar sand and super gaint oil fields. AAPG Bull; 61, 1950-1961.

Demaison, G; and Huizinga, B.J. (1991). Genetic Classification of Petroleum Systems. AAPG Bull; 75, 1626–1643.

D.G.H. (2006). India–Petroleum exploration and production Activities–2005-2006. Directorate General of Hydrocarbons, New Delhi; 148 pp.

D.G.H. (2006,a). Coal bed methane (CBM) exploration in India. In: India–Petroleum exploration and production activities–2005-06. Directorate General of Hydrocarbons. New Delhi; p. 133-138.

D.G.H. (2006, b). Gas hydrate R&D activities in India. In: India–Petroleum exploration and Production activities–2005-2006. Directorate General of Hydrocarbons. New Delhi; p. 139-143.

Dhannawat, B.S; and Mukherjee, M.K. (1997). Source rock studies in Jaisalmer Basin, India. Indian J. Petrol. Geol; 6, 25-42.

Dickens, G.R; O'Neil, J.R; Rea, D.K; and Owen, R.M.(1995). Dissociation of oceanic methane hydrate as a cause of the carbon isotope excursion at the end of the Paleocene. Paleo-Oceanog; 10, 965-971.

Dickey, P.A; and Hunt, J.M. (1972). Geochemical and hydrogeological methods of prospecting for stratigraphic traps. In: Stratigraphic oil and gas fields. King, R.E (Ed). AAPG Memoir 16, Tulsa, Oklahoma. p. 136-137.

Dickinson, W.R. (1976). Plate tectonics and hydrocarbon accumulation. AAPG Continuing Education Course Note series, No. 1.

Dhar, P.C; and Bhattacharya, S.K. (1993). Status of Exploration in the Cambay Basin. In: Proc. 2nd semi. Petroliferous basins of India. Vol.2. Biswas, S.K et al. (Eds). Indian Petroleum Pub; Dehra Dun. p. 1-32.

Dore, A.G; Augustson, J.H; Hermanrud, C; Stewart, D.J; and sylta, O. (Eds) (1990). Basin modeling: advances and applications. Elsevier, Amsterdam.

Douglas, A.G. (1969). Gas chromatography. In: Organic geochemistry methods and results. Eglinton, G. and Murphy, M.J.J. (Eds.). Spinger-Verlag, New York, p. 161-180.

Dow, W.G.(1977). Kerogen studies and geological interpretations. J.Geochem. Explor.,72,77-79.

Dunsmore, H.E. (1973). Diagenetic process of lead–zinc emplacement in carbonates. Trans. Inst. Min. Metall; Sec. B; 82, 168-173.

Durand, B. (1980). Kerogen–Insoluble organic matter from sedimentary rocks. Grahman & Trotman, London.

Durrance, E.M. (1986). Radio activity in geology. Ellis Horwood, Chichester.

Dwivedi, P; Prabhakar, V; and Thaplial, J.P. (1991). Geochemical evidence suggesting coal origin for oils of Upper Assam Basin. In: Proc. Conf. Integrated Exploration Research: Achievements and Prospectives. Pandey, J; and Benerjie, V (Eds). KDMIPE; Dehra Dun, p. 407-414.

Eagler, T; and Perry, K, (2002). Creating a road map for unconventional gas R&D. Gas-Tips; 8, 16-20.

Earlich, R. (2001). Nine crazy ideas in science. Oriental Longman, Hyderabad.

Eglinton, G; Hajidrahim, S.K; Maxwell, J.R; Quirke, J.M.E. (1980). Petroporphyrins. Structural elucidation and the applications of HPLC finger printing to geochemical problems In: Advances in organic chemistry. Douglas, A.G; and Maxwell, J (Eds). Pergamon press, oxford, p. 193-203.

Espitalie, J; Laporte, J.L; Madec, M; Marquis, F; Laplat, P; Poulet, J; and Bouteu, A. (1977). Rapid method for source rock characterization and evaluating their petroleum potential and degree of maturity. Revve del' Institute frncis du petrole, 32, 23-42 (In French).

Evans, P. (1932). Tertiary succession in Assam. Trans Min Geol. Inst. India, New Delhi; 85.

Ferry, J.H. (1992). Biochemistry of Methanogenesis. Critical reviews in Biochem. and Molecular Biology; 27, 473-503.

Fuex, A.N. (1977). The use of stable carbon isotopes in hydrocarbon exploration. J. Geochem. Expl., 7, 155-158.

Gallagher, R.M; (1984). Iodine-a path finder for petroleum deposits. In: Unconventional methods in exploration for petroleum and Natural gas III. Davidson, M.J; Gottlieb, B.M; and Prince, E (Eds). Dallas, SMU press, p 162-173.

Gallegoes, E.J; Green, J.W; Lindeman, L.P; LeTourneau, R.L; and Teeter, M. (1967). Petroleum group type analysis by high resolution mass spectrometry. Anal. Chem., 39, 1883-1889.

Gold, T. (1979). Terrestrial sources of carbon and earthquake out gasing. J. Pet. Geol., 1, 3-19.

Gold, T. (1999). The deep hot biosphere. Springer-Verlag, New York.

Gold, T; and Soter, S. (1982). Abiogenic methane and the origin of petroleum. Energy explo. Exploit; 1, 89-104.

Grandstein, F; Ogg, J; and Smith, A. (2004). A Geological Time Scale –2004. Cambridge University Press, UK, p. 5.

Gransch, J.A; and Eisma, E. (1970). Characterization of the insoluble organic matter of sediments by pyrolysis. In: Advances in organic geochemistry, Hobson, G.D; and Speers, G.C (Eds). Pergamon, New York, p. 407-426.

Green, A.R. (1985). Integrated sedimentary basin analysis for Petroleum exploration and Production. In: Proc. 17th Annual OTC paper 4842, Vol. I, p. 9-20.

Gupta, H.K (Ed). (2000). Proc. Indo-Russian joint workshop on gas hydrates. DOD Pub; New Delhi; 178 pp.

Gupta, S.K; Prasad, G.K; Majumdar, S.K; and Siva Sankar, J. (1998). Tectonics and sedimentation in Krishna–Godavari Basin, India. In: Proc. Nat. Symp. Recent researches in sedimentary basins. Tiwari, N.D (Ed). Indian Petroleum Pub; Dehra Dun, pp. 69-77.

Gupta, V; Sarma, M.C; Datta G.C; and Rao, S. (1984). Geochemical evidence for terrestrial source input for the oils of Cambay Basin. Petrol. Asia Jour; 7, 89-91.

Hanumantha Rao, Y; Reddy, S.L; Ramesh Khanna ; Rao, T.G; Thakur, N.K ; and Subrahmanyam, C. (1998). Potential distribution of methane hydrates along the Indian continental margins, Curr. Sci; 74, 466-468.

Harding, T.P. (1984). Graben hydrocarbon occurrences and structural style. Bull. AAPG; 68, 333-362.

Haseldonckx, P. (1979). Relation of palynomorph colour and sedimentary organic matter to thermal maturation and hydrocarbon generation potential: UN. ESCAP/CCOP Tech. Pub. No. 6, p. 41-53.

Hill, D.G; and Nelson, C.R. (2000). Gas productive fractured shales-an overview. Gas-Tips; 6, 4-13.

Hirst, J.P.P; Davis, N; Palmer, A.F; Achache, D; and Reddiford, F.A. (2001). The tight gas challenge: Appraisal results from the Devonian of Algeria Petroleum. Geoscience 7, 13-21.

Hitchon, B.(1963). Geochemical studies of Natural gas. Part III. Inert gases in western Canada natural gases. J. Can. Pet. Technol; 2, 165-174.

Hitchon, B (1974). Application of geochemistry to the search for crude oil and natural gas. In: Introduction to exploration geochemistry, Levinson, A.A (Ed). Applied Pub; Calgary, Canada, p. 509-545.

Hitchon, B; Billings, G.K; and Klovan, J.E. (1971). Geochemistry and origin of formation waters in the western Canada sedimentary basins–III. Factors controlling chemical composition. Geochim. Cosmochim. Acta. 35, 567-98.

Hobson, D.G; and Tiratsoo, F.N. (1975). Introduction to petroleum Geology, Scientific press. Beaconsfield.

Holder, G.D ; Kamath, V.A ; and Godbole, B.P. (1984). The potential of natural gas hydrate as an energy resource. Ann. Rev. Energy; 9, 427-445.

Hood, A; Gutjahr, C.C.M; and Heacock, R.L. (1975). Organic metamorphism and the generation of petroleum. AAPG Bull., 59, 986-999.

Horsefield, B. (1984). Pyrolysis studies and petroleum exploration. In: Advances in petroleum geochemistry. Brooks J. and Welte, D.H. (Eds.). Academic Press, London, p 247-298.

Hoyle, F. (1955). Frontiers of astronomy. Harper and Row, New York.

Huff, K.F. (1978). Frontiers of World oil exploration. Oil and Gas Jour; 76, 214–220.

Hughes, W.B; Holba, A.G and Dzou, I.P. (1995). The ratio of dibenzothiophene to phenanthrene, and pristane to phytane as indicators of depositional environments and lithology of petroleum source rocks. Geochim.Coschim.Acta. 59, 3581-3598.

Hunt, J.M. (1976). Origin of Athabaska oil. AAPG Bull., 60, 1112.

Hunt, J.M. (1996). Petroleum geochemistry and geology, 2nd Edn. W.H. Freeman Co. San Francisco, USA, 617 pp.

Hunt, J.M; and Whelan, J.K. (1978). Light hydrocarbons in sediments of DSDP leg 44 hole. In: Initial reports on the deep sea drilling project. Benson, E.D. and Sheridan, R.E. (Eds.). Vol. 44, US Govt. Printing Office, Washington D.C.

Illing, V.C. (1933). Migration of oil and natural gas. J. Inst. Petrol; 19, 229-274.

India's hydrocarbon vision –2025 Statement (2001). Drilling and Exploration World; July-Sept. 2001, 14-17.

Jackson, J.A. (1997). Glossary of geology, 4ᵗʰ Eds; American Geological Institute, Alexandria, Virginia, USA.

Jamal, S and Peters, J. (1997). CBM Potential and Prospects in India: A case of diversification. In: Proc. PEROTECH-97; New Delhi; p. 155-162.

Johns, W.D; and Shimoyama, A. (1972). Clay mineral and petroleum forming reactions during burial and diagenesis. AAPG Bull., 56, 2160-2167.

Jokhan Ram. (2008). Application of Geochemistry in Petroleum exploration in India–Current and future trends. J. Appl. Geochem; 10, 1-16.

Jokhan Ram; Mohapatra, P; and Ganesan, G.P. (1998). Western Offshore Petroliferous province: Tectono–Sedimentary evolution and petroleum system. In: Proc. Nat. Symp. Recent researches in sedimentary basins. Tiwari, N.D (Ed). Indian Petroleum Pub; Dehra Dun, p. 26-44.

Jones, V.T; Drozd, R.J. (1983). Prediction of oil and gas potential by near surface geochemistry. Bull. AAPG., 67, 932-952

Juntgen, H; and Klein, J. (1975). Formation of natural gas from coaly sediments. Erdol und Kohle-Erdgas-petrochemie; 28, 64-73 (In German).

Juranek, J. (1958). A contribution to the problem of origin of C_1-C_5 hydrocarbons in samples of soil-air in gas survey work. Czechoslovakian Institute of petroleum Research Transactions. Vol. 9, p 57-79.

Kaila, K.L; Krishna, G.C; and Mali, D.M. (1980). Crustal Structure along Mehmadabad–Billimora profile in the Cambay Basin, India from deep seismic soundings. Tectonophysics; 76, 99-130.

Kantsler, A.J; Cook, A.C; and Smith, G.C. (1978). Rank variation, calculated paleothermometer in understanding oil and gas occurrence. Oil and Gas J., 20, 196-205.

Kenney, J.F; Kutchervov, V.A; Bendeliane, N.A; and Alekseev, V.A. (2002). Proc. Natl. Sci. USA. 99, 10976-10981.

Khan, M.S.R; Pande, A; Garg, A.K; and Awasthi, A.K. (1995). Hydrocarbon generation and migration in the southern flank of Karaikal Ridge, Cauvery Basin, India. In: Proc. PETROTECH-95. Kuldeep Chandra et al. (Eds). B.R Pub. Crop. New Delhi; p. 327-331.

Khaveri; K.G. (1984). Free hydrocarbons in Unita basin, Utah. Bull. AAPG; 68, 1193-1197.

Killops, S.D; Woolhouse, A.D; Westen, R.J; and Cook, R.A. (1994). A geochemical appraisal of oil generation in the Taranaki Basin, Newzealand. AAPG Bull; 78, 1560-1588.

Kingston, D.R; Dishroom, C.P and Williams, P.A. (1983, a). Global basin classification systems. AAPG Bull; 67, 2175-2193.

Kingston, D.R; Dishroom, C.P; and Williams, P.A. (1983, b). Hydrocarbon plays and global basin classification. AAPG Bull; 67, 2194-2198.

Klemme, H.D. (1975). Geothermal gradients, heat flow and hydrocarbon recovery. In: Petroleum and Global tectonics. Fischer, A.G and Judson, S (Eds). Prinston univ. Press.

Klemme, H.D. (1980). Petroleum basins–classification and characteristics. Jour. Petrol. Geol; 3, 187-207.

Klusman, R.W; and Voohees, K.J. (1983). A new development in petroleum exploration technology. Mines magazine. March, p 6-10.

Koons, C.B; Bond, T.G; and Pierce, F.L. (1974). Effect of depositional environment and post depositional history on chemical composition of lower Tuscaloosa oils. Bull. AAPG; 58, 1272-1280.

Koshal, V.N.(1993). Organic matter maturation studies in Cambay Basin based on TAI Values. In: Proc. 2nd semi. petroliferous basins of India. Vol. 2. Biswas, S.K. et al. (Eds). Indian Petroleum Pub; Dehra Dun, p. 173-189.

Kudryavtsev, N.A. (1959). Geological proof of the deep origin of petroleum. cited in: the origin of methane in the crust of the earth, Gold, T. US Geological Survey Proc. Paper. 1570 (1993).

Kuldeep Chandra. (1997). Non-Conventional hydrocarbon resources like coal bed methane and gas hydrates : Exploration imperatives for India. Indian J. Geol; 69, 261-267.

Kuldeep Chandra; and Samanta, U. (1984). Organic matter maturation in the sedimentary sequence of Cauvery Basin. Petrol. Asia. Jour; 7, 129-131.

Kuldeep Chandra; Misra, K.N; Misra, C.S. (1994). Petroleum Potential of sedimentary basins of India–Geochemical perspectives In: Proc. 2ⁿᵈ Semi Petroliferous basins of India. Vol. 3. Biswas, S.K *et al.* (Eds). Indian Petroleum Pub; Dehra Dun, p. 201-210.

Kuldeep Chandra; Sahai, S; Awadhesh Rai; Debashis Das; Gupta, U.K; and Singh, A.K. (1999). Coal bed methane resources of India. A case study of Jharia basin. In: Proc. PETROTECH-99, New Delhi; p. 435-443.

Kuldeep Chandra; Raju, D.S.N; Bhandari, A; Misra, C.S. (2001). Petroleum system in the Indian sedimentary basins: Stratigraphic and geochemical perspectives. Bull. ONGC; 38, 1-45.

Kumar, R.K; Prakash, C; Mali, M.R; and Dwivedi, P.(1984). Geological control on the occurrence of crude oil types in Cambay Basin. Petrol. Asia Jour; 7, 87-88.

Kumar, R.K; Singh, S.P; Saxena, P.K; and Das, S.P. (1985). The oil-type in Krishna-Godavari, and results of oil- oil correlations. Petrol. Asia jour; 6, 128-136.

Kumar, R.K; Ghosh, L.M; Dalal, S.R; and Sukla, S.K. (1998). Significance of Geochemical patterns of fluids from Ghotaru, B; Rajasthan Basin. Bull. ONGC; 25, 103-116.

Kumar, S.P. (1983). Geology and hydrocarbon prospects of Krishna-Godavari and Cauvery basins. Petrol. Asia Jour; 6, 57-65.

Kumar, S.P. (1993). Hydrocarbon exploration in Assam Basin– Retrospect and prospect. In: Proc. 2ⁿᵈ Semi. petroliferous basins of India. Vol. 1. Biswas, S.K. *et al.* (Eds). Indian Petroleum Pub; Dehra Dun, p. 545-551..

Kuuskraa, V.A; Boyer, C.M; and Kelafant, J.A. (1992). Coal bed gas. The hunt for quality basin goes abroad. Oil and gas journal; 90, 49-54.

Kvenvolden, K.A. (1993 a). Gas hydrates : Geological perspectives and global change. Rev. Geophys; 31, 173-187.

386 *Petroleum Geochemistry*

Kvenvolden, K.A. (1993 b). Gas hydrates as potential energy resources-A review of their methane content. In: The future of the energy gases. Howell (Ed). U.S. Geological Survey professional paper, Vol. 1570, 551-561.

Lakshman Singh (2000). Oil and gas fields in India. Indian Petroleum Pub. Dehra Dun; 382 pp.

Landes, K.K. (1973). Mother Nature as an oil pollutor. AAPG Bull; 57, 637-641.

Landis, E.R; and Weaver, J.N. (1993). Global coal occurrences. In: Hydrocarbons from coal. Law, B; and Rice, D (Eds). AAPG studies in Geology; 38, p. 1-12.

Lane, E.C; and Garton, E.C. (1935). Base of crude oil. Rep. Invest.– US. Bur. Mines. RI–3279.

Lang, K.R; and Whitney, C.A. (1991). Wanderers in space, Cambridge University press. UK.

Law, B.E. (2002). Basin centred gas systems. AAPG Bull; 86,1891-1919.

Law, B.E; and Curtis, J.B. (2002). Introduction to Unconventional gas petroleum systems. AAPG Bull; 86, 1851-1852.

Law, B.E; and Spencer, C.W. (1993). Gas in tight Reservoirs–an emerging source of energy. In: The future of energy gases. Howell, G.D (Ed). U.S. Geological Survey professional paper; 1570, p. 233-252.

Leach, W.H. (2002). New Technology for CBM production. Opportunities of coal bed methane; oil and gas Investor. December, 3-10.

Lee, H. (1963). The technical and economic apects of helium production in Saskaichewan.. J. Can. Pet. Technol; 2, 16-27.

Lee, M.L; Novotny, M.S; and Bartle, K.D. (1981). Analytical chemistry of polycyclic aromatic compounds. Academic Press, New York.

Levine, J.R. (1993). Coalification. The evolution of a Source rock and Reservoir rock for oil and gas. In: Hydrocarbons from coal. Low, B.E; and Rice, D.W (Eds). AAPG studies in Geology Series; 38, p. 39-77.

Levorsen, A.I. (1967). Geology of petroleum, 2nd Edn., W.H. Freeman, San Francisco, 724 pp.

Lewin, M.D. (1984). Factors controlling the proportionality of vanadium to nickel in crude oils. Geochim. Cosmochim. Acta. 48, 2231-2238.

Leythaeuser, D; Schaefer, R.G; and Yukler, A. (1982). Role of diffusion in primary migration of hydrocarbons. AAPG Bull; 66, 408-429.

Link, W.K. (1952). Significance of oil and gas seeps in world oil exploration. AAPG Bull; 36, 1505-1540.

Lopatin, N.V. (1971). Temperature and geological time as factors in coalification (in Russian). IZV. Akad. Nauk SSR, Seriya Geologicheskaya, No. 3, 95-100.

Mackenzie, A.S. (1984). Applications of biological markers in petroleum geochemistry. In: Advances in petroleum geochemistry. Brooks, J. and Welte, D.H. (Eds.), Vol. 1, Academic Press, London, p. 115-214.

Magoon, L.B; and Dow, W.G. (1994). The Petroleum System. AAPG Memoir; 60, 3-24.

Mallick, R.K; Raju, S.V; and Mathur, N. (1997). Geochemical characterization of genesis of Eocene crude oils in a part of Upper Assam Basin, India. In: Proc. PETROTECH–97, Vol. 1. Swami S.N; and Dwivedi, P (Eds). B.R. Pub. Corp. New Delhi; p. 391-402.

Mallick, R.K; and Raju, S.V. (1999). Coal bed methane potential of sub-surface Barail coal from the eastern part of Upper Assam. Indian J. Petrol.Geol; 8, 35-43.

Mangotra, S.R; Mamgain, A; Yadav, T; Sharma, V.N; and Pande, M.N. (1993). Source rocks along the flanks of Nizira Low. In: Proc. 2nd Semi. petroliferous basins of India. Vol. 1. Biswas, S.K. et al. (Eds). Indian Petroleum Pub; Dehra Dun, p. 563-577.

Marchand, A; and Conard, J. (1980). Electron Paramagnetic Resonance in kerogen studies. In: Kerogen. Durand, B (Ed). Technip Editions, Paris, p. 2239-2254.

Mathur, L.P; and Evans, P. (1964). Oil in India. In: Proc. 22nd IntNatl. Geol. Cong., New Delhi, India. 85

Mathur, L.P; Rao, K.L.N; and Chaube, A.N. (1968). Tectonic frame work of Cambay Basin, India. Bull, ONGC; 5, 7-28.

Mathur R.B; and Nair, K.M.(1993). Exploration of Bombay Offshore Basin. In: Proc. 2nd semi. Petroliferous basins of India. Vol. 2. Biswas, S.K. *et al.* (Eds). Indian petroleum Pub; Dehra Dun, p. 365-393.

Mathur, S; Sethi, H.S; Goswami, B.G; Pande, S.D; Sapru, R. K; Chowdhury, D.R; and Misra, K.N. (1993). Biomarker geochemistry of oils of Neelam Field and some DCS structures. In: Proc. 2nd Semi. Petroliferous basins of India. Vol. 2. Biswas, S.K *et al.* (Eds). Indian Petroleum Pub; Dehra Dun, p. 443-454.

Mattavelli, I; Ricchiuto, T; Grignani, D; and Schoell, M. (1983). Geochemistry and habitate of natural gases in Po Basin, Northern Italy. Bull. AAPG; 67, 2239-2254.

Max, M.D. (2000). Gas hydrate potential of the Indian Sector of the N.E. Arabian sea and Northern Indian Ocean. In: Natural gas hydrates. Max, M.D (Ed). Kluwer Academic Pub; London. p. 213-224.

Max, M.D (Ed). (2000). Natural gas hydrates. Kluwer Academic pub; London, 414 pp.

McCaffrey, M.A; Dahl, J.E; Sundararaman, P; Moldowan, J.M; and Schoell, M (1994 a). Source rock quality determination from oil biomarkers II–A case study using Tertiary reservoired Beaufort Sea oils. AAPG Bull; 78, 1527-1540.

McCaffrey, M.A; Moldowan; J.M; Lipton; P.A; Summons, R.E; Peters, K.E; Jaganathan, A; and Watt, D.S. (1994 b). Paleoenvironmental implications of noval C_{30} steranes in Precambrian to Cenozoic age petroleum and bitumen. Geochim. Cosmochim. Acta. 58, 529-532.

McNab, J.G; Smith, P.V; and Betts, R.L. (1952). The evolution of petroleum. Ind. Eng. Chem; 44, 2556-2563

Meissner, F.F. (1978). Petroleum geology of the Bakken formation, Williston basin, North Dakota and Montana. In: The economic geology of the Williston basin: Billings, Montuna. Geol. Soc., 207-227.

Mello, R; and Moldowan, J.M.(2006). Diamondoids and compound specific analysis applied to age, source, thermal evolution oil cracking oil mixing and petroleum system assessment. AAPG IntNat. Conf. & Exhibt; Perth, Australia.

Mendele'ev, D. (1877). Entstehungund vorkommen des minerals. Dtsch. Chem. Gas. Ber; 10, 229.

Mendele'ev, D. (1902). The principles of chemistry, 2ⁿᵈ English ed; Vol. 1, (Translated from the 6ᵗʰ Russian ed;) Collier, NewYork.

Milkov, V.A ; and Sassen, R. (2001). Economic gelogy of Offshore gas hydrate accumulations and provinces. Mar. and Petrol. Geol; 19, 1-11.

Milner, C.W.D; Rogers, M.A; and Evans, C.R (1977). Petroleum transformations in reservoirs. J. Geochem. Expl., 7, 101-153.

Mitra, P; Zutshi, P.L; Chowrasia, R. A; Caugh, M.L; Anathanarayanan, S; and Sukla, B. (1983). Exploration in western Offshore Basins. Petrol. Asia Jour; 6, 15-24.

Mitra, P; MukherJee, M.K; Mathur, B.K; Bhandari, S.K; Qureshi, S.M; and Bahukhandi, G.C. (1993). Exploration hydrocarbon prospects in Jaisalmer Basin, Rajasthan. In: Proc. 2ⁿᵈ Semi petroliferous basins of India. Vol. 2. Biswas, S.K. *et al.* (Eds). Indian Petroleum Pub; Dehra Dun, p. 236-284.

Mittal, A.K; Pande, H.C; Agarwal, K; Uniyal, A. K; and Thomas, N.J.(2002). Geochemical characteristics of natural gases from Mehsana and Mandhali pay sands of Sohasam oil field of north Cambay Basin. Bull. ONGC; 39, 47-59.

Mohan, M; Kumar P; Narayanan, V; Govindan, A; Soodan, K. S; and Singh, P. (1982). Bio facies Paleoecology and geological history of Bombay Offshore region. ONGC Bull; 19, 13-27.

Mohan, S; and Sangai, P. (1995). ONGC Thardesert–An exploration Overview. In: Proc. PETROTECH-95, Vol.1. Kuldeep Chandra et al. (Eds). B.R Pub. Corp. New Delhi; p. 319-326.

Moldowan, J.M; Dahl, J; Jacobson, S.R; Huiznnga, B.J; Fago, F.J; Shettey, R; Watt, D.S; and Peters, K.E. (1996). Chemostratigraphic reconstruction of biofacies: Molecular evidence linking cyst-forming dianoflagellates with Pre-Triassic ancestors. Geology. 24, 159-162.

Moldowan, J.M; Fago, F.J; Lee, C.Y; Jocobson, S.R; Watt, D.S; Sougui, N.E; Jaganathan, A; and Young, D.C. (1990). Sedimentary 24-n-propyl cholestanes, molecular fossil diagnostic of marine algae. Science. 309-312.

Moldowan, J.M; Huizinga, B.J; Dahl, J.E; Fago, M.J; Taylor, D.W; and Hickey, L.J. (1994). The molecular fossil record of oleanane and its relationship to angiosperms. Science. 265, 768-771.

Moldowan, J.M; Lee, C.Y; Sundararaman, P; Salvatori, R; Alajbeg, A; Gjukic, B; Demaison, G.J; Slougui, N.E; and Watt, D.S. (1992). Source correlations and maturity assessment of select oils and rocks from the central Adriatic basins (Italy and Yugoslavia). In: Biological markers in sediments and petroleum Moldowan, J.M.; Albrecht, P; and Philip, R.P. (Eds.). Printice Hall, New Jersey, p. 370-401.

Moldowan, J.M; Seifert, W.K; and Gallegos, E.J. (1985). Relationship between petroleum composition and depositional environment of petroleum source rocks. AAPG Bull., 69, 1255-1268.

Momper, J.A. (1978). Oil migration limitations suggested by geological and geochemical considerations. In: Physical and chemical controls on petroleum migration. AAPG Continuing Education Course Notes Series, No. 8, Tulsa, p. B1-B60.

Moshier, S.O and Waples, D.W. (1985). Quantitative evaluation of Lower Cretaceous Mannville Group of source rocks for Alberta's oil sands. Bull. AAPG; 69, 161-172.

Mungan, N. (1965). Permeability reduction through changes in the pH and salinity. J. Pet. Technol; 12, 1449-1453.

Munn, M.J. (1909). Studies in the applications of anticlinal theory of oil and gas accumulations. Econ. Geol., 4, 141.

Murris, R.J. (1984) Introduction. In: Petroleum Geochemistry and Basin evolution. Demaison, G; and Murris, R.J. (Eds). AAPG memoir, 35, Tulsa. p. x-xii.

Murthy, K.N. (1983). Geology and hydrocarbon prospects of Assam Shelf–Recent advances and present status. Petrol. Asia Jour; 6, 1-14.

Murty K.V.S; and Rama Krishna, M. (1980). Structure and tectonics of Godavari-Krishna Coastal sedimentary basins. Bull. ONGC; 17, 147-158.

Myers (Jr), M.E; Stollsteimer J; and Wims, A.M. (1975). Determination of gasoline octane numbers from chemical composition. Anal. Chem., 47, 2301-2304.

Naini, B.R; and Talwani, M. (1982). Structural frame work and the evolutionary history of the continental margin of western India. AAPG Memoir; 34, 167-192.

Nakayama, K; and Vansiclen, D.C. (1981). Simulation model for petroleum exploration. Bull. AAPG; 65, 1230-1255.

Neerja, P; Mathur, P; Maheswari, S.C; and Khanduri M.P. (1997). Characterization of Vadaparru–Pasarlapudi–Pallakollu formations, East Godavari Basin. In: Proc. PETROTECH-97. Vol. 2. Swami, S.N; and Dwivedi, P. (Eds). B.R Pub. Corp; New Delhi; p. 9-18.

Nisbet, E.G. (1990). The end of the ice age. Can. J. Earth Sci; 27, 148-157.

Nisbet, E.G; and Piper, D.J.W. (1998). Gaint submarine land slides. Nature, 392, 329-330.

North, F.K. (1979). Characteristics of oil Provinces–A study for students. Bull. Cand. Petrol. Geol; 19, 601-658.

North, F.K. (1980,a). Episodes of source-sediment deposition-I. J. petrol. Geol., 2, 199-218.

North, F.K.(1980,b). Episodes of source-sediment deposition-II. J. petrol. Geol; 2, 323-338.

North, F.K. (1990). Petroleum geology. 2nd impression, Unwin Hyman, Boston, USA.

O'Connor, J.G; Burrow; F.H; and Norris, M.S. (1962). Determination of normal paraffins in C_{30}-C_{32} in paraffin waxes by molecular sieve adsorption: Molecular weight distribution by gas-liquid chromatography, Anal. Chem., 34, 82-85.

Odedra, A; Burley, S.D; Lewis, A; Hardman, M; and Haynes, P. (2005). The World according to gas. In: Petroleum Geology: North-West Europ and Global perspectives. Dore, A.G; and Vinning, B (Eds). Geological Society, London, p. 1-16.

Oil Tracers, L.L.C. (2006). Using oil biomarkers in petroleum exploration. USA; p. 1-14.

Ostroff, A.G. (1979). Introduction to oil field water Technology. National Association of Corrosion Engineers, Huston, Texas, 394 pp.

Palmer, C. (1911). The geological interpretation of water analysis. Bull. 479, US Geol. Surv., 31 pp.

Panda, P.K. (1985). Geothermal maps of India and their significance in resource assessments. Petrol. Asia Jour; 8, 202-210.

Pande, A; Hajra, P.N; Singh, B.P;Khan, M.S.R; Berry, V.K; Tripathi, G.K; and Kuldeep Chandra. (1993). Origin and evolutionary histories of crude oils of Cambay Basin through biomarker composition. In: Proc. 2nd semi. Petroliferous basins of India Vol. 2. Biswas, S.K et al. (Eds). Indian Petroleum Pub; Dehra Dun. p. 137-169.

Pande, A; Hajra, P.N; Singh, B.P; Chandra, U; Sood, R; Kumar, R.K; and Chandra, K. (1994). Studies on oil–oil and oil-source rock correlations through biomarker and isotopes in Cauvery Basin. Bull. ONGC; 31, 59-97.

Pandey, B.P. (1999). Methane from coal beds. Indian J. Pet. Geol; 8, 69-105.

Parker, J.S; and Southwell, C.A. (1929). Chemical investigations of Trinidad well waters and its geological and economical significance. Jour. Inst. Pet. Technol; 15, 138-173.

Paropkari, A.L. (2008). Abiogenic origin of petroleum hydrocarbons. Need to rethink exploration strategies. Curr. Sci; 95, 1018-1019.

Pecher, I.A; and Holbrook, W.S. (2001). Seismic methods for detecting and quantifying marine methane hydrate/free gas reservoirs. In: Natural gas hydrates. Max, M.D (Ed). Kluwer Academic Pub; London, p. 275-294.

Peters, K.E; and Moldowan, J.M. (1991). Effects of source, thermal maturity and biodegradation on the distribution and isomerisation of homohopanes in petroleum. Org. Geochem., 17, 47-61.

Peters, K.E. and Moldowan; J.M. (1993). The biomarker guide interpreting molecular fossils in petroleum and ancient sediments. Printice-Hall, New York.

Philip, P.C; Chopra, V.S; Sridharan, P; and Bala Krishna, M. (1991, a). Geochemistry and petroleum source potential of the sedimentary sequence of Amalapuram block. In: Proc. Conf. Integrated Exploration Research: Achievements and

Prospectives. Pandey, J; and Benerjie, V (Eds). KDMIPE; Dehra Dun, p. 443-447.

Philip, P.C; Chopra, V.S; Sridharan, P; and Bala Krishna, M. (1991, b). Source rocks and hydrocarbons of Ravva structure and their correlation. In: Proc. Conf. Integrated Exploration Research: Achievements and Prospectives. Panday, J; and Benerjie, V (Eds). KDIMIPE; Dehra Dun, p. 455-460.

Philp, R.P; and Crisp, P.J. (1982). Surface geochemical methods used for oil and gas prospecting- A review. J. Geochem. Expl. 17, 1-34.

Philippi, G.T. (1965). On the depth, time and mechanism of petroleum generation. Geochim. Cosmochim. Acta. 29, 1021-1049.

Phizackerley, P.H; and Scott, L.O. (1967). Major tar sand deposits of the World. In: Proc. 7th World Petroleum conrg; Mexico; 3, p. 551-571.

Pirson, S.J. (1983). Geological Well Log Analysis. 3rd edn. Gulf Publishing Co; Houston, Texas.

Poekhau, H.S; Baker, D.R; Hantschel, Th; Horsfield, B; and Wygrala, B. (1997). Basin simulation and the design of the conceptual Basin Model. In: Petroleum and Basin Evolution. Welte, D.H; Horsfield, B; and Barkers, D. R (Eds). Springer–Verlag, Newyork. p. 3-70.

Porfirev, V.B. (1974). Inorganic origin of petroleum. AAPG Bull; 58, 3-33.

Powell, T.G; Creaney, S; and Snowdon, L.R (1982). Limitations of organic petrographic techniques for identification of petroleum source rocks. Bull. AAPG; 66, 430-435.

Prabhakar, K.N; and Zutshi, P.L. (1993). Evolution of southern part of Indian east coast basins. Jour. Geol. Soc. India. 41, 215-230.

Prasad, B; Jain, A.K; and Mathur, Y.K. (1995). A Standard Palynological Zonation scheme for the Cretaceous and Pre-Cretaceous sub-surface sediments of Krishna-Godavari Basin, India. Geo Sci. Jour; 16, 155-233.

Price, L.C. (1976). Aqueous solubility of petroleum as applied to its origin and primary migration. AAPG Bull., 60, 213-244.

Price; L.C. (1980). Utilization and documentation of vertical oil migration in deep basins. J. Petr.Geol., 2, 353-387.

Price, L.C. (1981). Primary migration by molecular solution: Consideration of new data. J. Petroleum Geol; 3, 91-116.

Price, Leigh C. (1986). A critical review and proposed working model of surface geochemical exploration. In: unconventional methods in exploration IV. Southern Methodist University, Dellas, TX, p 245-304.

Price, L.C. (1997). Geological Controls in deep natural gas resources in the Uninited States. U.S. Geological Survey Bullelin; 2146 L.

Price, R.A. (1973). Large scale gravitational flow of supra crustal rocks, southern candian Rockies. In: Gravity and Tectonics. Dejong, K.A; and Scholten, R (Eds). John Wiley, New York; p. 491-502.

Pusey, W.C. (1973). How to evaluate potential gas and oil source rock?. World oil. 176, 71-75.

Rai, A; Chandrasekharan, P; and Misra V.N. (1998). Indian sedimentary basins and their hydrocarbon potential. In: Proc. Nat. Symp. Recent researches in sedimentary basins. Tiwari, N.D (Ed). Indian Petroleum Pub; Dehra Dun, p. 91-103.

Rai, A; Hedge, V.N; Gupta, U.K ; Gita Singh; and Sikka, S.N. (1994). Hydrocarbon resources of Indian sedimentary basins. In: Proc. 2nd Semi. Petroliferous basins of India. Vol. 3. Biswas, S.K et al. (Eds). Indian Petroleum Pub. Dehra Dun; p. 211-222.

Raju, A.T.R. (1968). Geological evolution of Assam and Cambay Tertiary basins of India. AAPG Bull; 52, 2422-2437.

Raju, A.T.R. (1979). Basin analysis and petroleum exploration with some examples from Indian sedimentary basins. Jour. Geol. Soc. India; 20, 49-60.

Raju, A.T.R; and Srinivasan, S. (1983). More hydrocarbons from well explored Cambay Basin. Petrol. Asia Jour; 6, 25-35.

Raju, A.T.R; and Srinivasan, S. (1993). Cambay Basin–Petroleum Habitat. In: Proc. 2nd Semi. Petroliferous basins of India Vol. 2. Biswas, S.K et al. (Eds). Indian Petroleum Pub; Dehra Dun, p. 35-78.

Ramana, M.V; Ram Prasad, T; Desa, M; Sethe, V.A; and Sethe, A.K. (2006). Gas hydrate related proxies inferred from multidisciplinary investigations in the Indian Offshore areas. Curr. Sci; 91, 183-189.

Ranga Raju, M.K. (1987). Turbidities in Petroleum Exploration: Concepts, methods of recognising them with examples from Krishna-Godavari Basins. Bull, ONGC; 24, 31-57.

Ranga Raju, M.K; Agarwal, A; and Prabhakar, K.N. (1993). Tectono stratigraphy, Structural style, evolutionary model and hydrocarbon habitat of Cauvery and Palar basins. In: Proc. 2nd Semi. petroliferous basins of India. Vol.1. Biswas, S.K. et al. (Eds). Indian Petroleum Pub; Dehra Dun, p. 371-388.

Ranga Rao, A. (1983). Geology and hydrocarbon potential of a part of Assam–Arakan Basin and its adjacent region. Petrol. Asia Jour; 6, 127-154.

Rao, C.N.R. (1961). Ultraviolet and visible spectroscopy: chemical applications. Butterworth pub. London.

Rao, G.N. (1991). Exploration results and leads in Krishna-Godavari Basin. Bull. ONGC; 28, 69-87.

Rao, G.N. (1993). Geology and hydrocarbon prospects of east coast sedimentary basins of India with special reference to Krishna-Godavari Basin. Jour. Geol. Soc. India; 41, 444-454.

Rao, G.N. (1994). Sedimentation and stratistructural control for hydrocarbon accumulation in Krishna-Godavari Basin. Indian J. Earth Sci; 21, 153-163.

Rao, G.N. (2001). Sedimentation, Stratigraphy and petroleum potential of Krishna-Godavari Basin-East Coast of India. AAPG Bull; 85, 1623-1643.

Rao, G.N; and Mani, K.S. (1993). A study on generation of abnormal pressures in Krishna-Godavari basin, India. Indian J. Petrol. Geol; 2, 20-30.

Rao, K.L.N. (1969). Litho Stratigraphy of Paleogene–Succession of southern Cambay Basin. Bull. ONGC; 6, 24.37.

Rao, R.P; and Talukdar, S.N.(1980). Petroleum Geology of Bombay High Field, India. In: Gaint oil and gas fields of the decade 1968-1978. Halbouty, M.T (Ed). AAPG Memoir; 30, 487-506.

Rao, Y.H. (1999). C-Program for the calculation of gas hydrate stability zone thickness. Computers & Geo Sci; 25, 705-707.

Richers, D.M; Reed, R.J; Horstman, K.C; Michaels, G.D; Barker, R.N; Lundel, L; and Marrs, R.W. (1982). Land sat and soil-gas geochemical study of Patric Draw oil field, Sweet Water Country, Wyoming. Bull. AAPG; 66, 903-922.

Robinson, R (1963). Duplex origin of petroleum. Nature. 199, 113-114.

Robinson, R (1966). The origin of petroleum. Nature. 212, 1291-1295

Roy Chowdhury, S.C; Mathur, R.B; and Misra, G.S. (1972). Subsurface stratigraphy of Tharod Serau area, Gujarat. Bull. ONGC; 9, 57-74.

Sachenen, A.N. (1945). Chemical constituents of petroleum. Rheinhold, New York.

Sackett, W.M.(1977). Use of hydrocarbon sniffing in offshore exploration. J.Geochem. Explor; 7, 243-254.

Sahi, B. (1986). Hydrocarbon potential of Bombay Offshore Basin. Oil Asia; 6, 16-25.

Samanta, M.K; Sen Gupta, T.K; Bharktya, D.K; and Saikea, A. (1993). Depositional Environments and Reservoir characteristics of oil bearing Tertiary sands of Upper Assam Plain, India. In: Proc. 2nd Semi. petroliferous basins of India. Vol. 1. Biswas, S.K. *et al.* (Eds). Indian Petroleum Pub; Dehra Dun, p. 685-705.

Samanta, U; Misra, C.S; Gupta Anita; Mamgain, A; and Thomas, M.J. (1996). Laboratory simulation of hydrocarbon generation from the Barail Coal Shale unit of Upper Assam, India. In: Proc. 4th AAPG Tanzania, Mpanja, F (Ed); p. 372-380.

Samar, Abbas (1996). The non-organic theory of the genesis of petroleum Curr. Sci.; 71, 677-684.

Sassen, R; and McDonald, I.R. (1997). Hydrocarbons of experimental and natural gas hydrates, Gulf of Mexico continental slope. Org. Geochem; 26, 289-293.

Sastri, V.V; Raju, A.T.R; Venkatachala, B.S; and Acharya, S.K. (1974). Phanerozoic map of India. Map Div; Geol. Survey India ESCAP ATLAS STRATI IECP, Project No. 32, Sheet No. 2, India.

Saxby, J.D. (1980). Atomic H/C ratios and the generation of oil from coals and kerogen. Fuel. 59, 305-307.

Schenk, H.J; Horsfield, B; Krooss, B; Schaefer, R.J and Schwochau, K. (1997). Kinetics of petroleum formation and cracking. In: Petroleum and basin evolution. Welte, D.H; Horsfield, B; Barker, D. (Eds). Springer-Verlag, New York.

Schiessler, R. W., and Flitter, D. (1952). Urea and Thiourea adduction of C_5-C_{42} hydrocarbons. J. Am. Chem. Soc., 74, 1720-1733.

Schoell, M.(1980). The hydrogen and carbon isotopic composition of methane from natural gases of various origins. Geochim. Cosmochim. Acta; 44, 649-661.

Schoell, M. (1984). Stable isotopes in Petroleum Research, In: Advances in Petroleum Geochemistry, Brooks, J and Welte D.H. (Eds). Academic press, London, p 215-245.

Schreiber, B.C; and Hsu, K.J. (1980). Evaporites. In: Developments in Petroleum Geology–2. Hobson, G.D. (Ed). Applied Science Pub. Ltd., London, p. 87-138.

Schwochow, S.D. (2002). CBM: Comming to a basin near you. Oil and gas investor; 12-16.

Scott, A.R. (1993). Composition and origin of coal bed methane gases from selected basins in the United States. In: Proc. Int. Natl. Symp. Coal bed methane. Thompson, D.A (Ed), Vol I, p. 207-216.

Seifert, W.K; and Moldowan, J.M. (1978). Applications of steranes, tempanes, and mono aromatics to the maturation, migration and source of crude oils. Geochim. Cosmochim. Acta; 42, 77-95.

Seifert, W.K; and Moldowan, J.M. (1979). The effects of biodegradation on steranes and terpanes in crude oils. Geochim. Cosmochim. Acta.43, 111-126.

Seifert, W.K; Moldowan, J.M; and Jones, R.W. (1980). Applications of biological marker chemistry to petroleum exploration. Proc. 10th World Petrol. Cong; 2, 425-438.

Selley, R.C. (1998). Elements of Petroleum Geology, Academic press, London, 470 pp.

Sharma, D.D. (2003). Review of coal resources in India and their exploration strategy. J. Geol. Soc. India; 61, 387-402.

Sikka, D.B.(1959). Radiometric survey of red water field, Alberta, Canada. Symp. petroleum Geochemistry, Pordhan university, Newyork, 8 pp.

Silverman, R.S and Epstein, S. (1958). Carbon isotopic composition of petroleum and other sedimentary organic material. AAPG Bull; 42, 998-1012.

Singh, A; and Singh, B.D. (1999). Methane gas: An unconventional energy resource. Curr. Sci; 76, 1546-1553.

Singh Dhruvendra; Srivastava, D.K; Gupta, V.P; and Singh, N.P (1995). Thermal maturation modeling, hydrocarbon generation and hydrocarbon prospects in Gulf of Cambay Basin, India. In: Proc. PETROTECH-95, Vol. 2. Kuldeep Chandra et al. (Eds). New Delhi; p. 171-182.

Singh Narin; Pangtey, S.K; and Mathur, M. (1993). Geochemical characteristics of oil of Demalgoam field Upper Assam. In: Proc. 2nd Semi. petroliferous basins of India. Vol. 1. Biswas, S.K. et al. (Eds). Indian Petroleum Pub; Dehra Dun, p. 625-633.

Singh, S.P; Manju Mathur; Sexena, J.G; and Thapliyal, J.P (1987). Geochemical Characterstics of oil occurrence in the multilayered reservoirs of the Broach Syncline, Cambay. Bull. ONGC; 24, 191-201.

Sinninghe Damste, J.S., Kenig, F; Koopmans, M.P; Koster, J., Schouten, S; Hayes, J.M., and de Leeuw, J.W. (1995). Evidence for gammacerane as an indicator of water column stratification. Geochim. Cosmochim. Acta. 59, 1895-1900.

Sloan, E.D. (1998). Clathrate hydrates of natural gases. 2nd Edn; Marcel-Dekker, New York, 705 pp.

Sluijk, D; and Nederlof M.H.(1984). World wide geological experience as a systematic basis for prospect apprial. In: Petroleum geochemistry and basin evaluation, Demaison, G and Murris, R.J (Eds). AAPG Memors, 35, Tulsa, pp 15-26.

Smith, N.A.C. (1927). The interpretation of crude oil analysis. Rep. Invest-U.S. Bur. Mines. RI-2806.

Sokoloff, W. (1889). Kosmicher ursprung der Bitumina. Bull. Soc. Imperiale des Naturalistes de Mascou, n. Ser, 3, 720-739.

Sokolov, V.A; Zhuse, T.P; Vassaeyvich, N.B; Angnov, P.L; Grigoryev, G.G; and Kozlov, V.P. (1963). Migration processes of gas and oil, their intensity and directionality. Proc. 6th World Petroleum Congr; Frankfurt. 1, 493-505.

Special report on CBM. (2001). Drilling and Exploration World; July–Sept., p.27-31.

Speight, J.M. (2007). The chemistry and Technology of petroleum. 4th edn., CRC press, USA.

Srivastava, D.C; Singh, B.K; Bechari, K; Nayal, A.K; Bhatnagar, A.K; and Saika, H.C. (1993). Lithofacies distribution and paleo environmental analysis of Cretaceous and Tertiary sediments in Nagapattinam sub-basin. In: Proc. 2nd Semi petroliferous basins of India. Vol.1. Biswas, S.K. et al. (Eds). Indian Petroleum Pub; Dehra Dun, p. 237-244.

Srivastva, D.K; Dhruvendra Singh; and Alat, C.A. (1993). Thermal maturation of Tertiary sediments in Upper Assam Shelf, India. In: Proc. 2nd Semi. petroliferous basins of India. Vol. 1. Biswas, S.K. et al. (Eds). Indian Petroleum Pub; Dehra Dun, p. 719-738.

Stach, E; Mackowsky, M. Th; Teichmuller, R; Taylor, G.H; and Chandra, D. (1982). Coal Petrology; Bortntraeger, Stuttgart, 3rd Edn.

Stahl, W; Wollanke, G; and Boigk, H. (1977). Carbon and nitrogen isotope data for Upper Carboniferous and Rotliegend natural gas from north Germany and their relationship to the maturity of the organic source material. In: Advances in organic geochemistry Campos, R; and Goni, J (Eds). Madrid, p. 539-560.

Staplin, F.L. (1969). Sedimentary organic matter, organic metamorphism, and oil and gas occurrence. Bull. Can. Petr. Geol., 17, 47-66.

Stiff (Jr), H.A. (1951). The interpretation of chemical water analysis by means of patterns. Tec. Note 84, petrol Technol, Trans. Amer. Inst. Min. Met. Engrs., 192, 376-379.

Storch, H.H; Golumbic, N; and Anderson, R.B. (1951). The Fischer Tropsch and Related synthesis. Wiley, New york.

Stout, S.A; and Douglas, G.S. (2004). Diamondoid hydrocarbons–applications in the chemical finger printing of natural gas condensate and gasoline. Environmental Forensocs, 5, 225-235.

Subrahmanyam, C; Reddy, S.L; Thakur, N.K; Gangadhara Rao, T; and Kalachandra Sain. (1998). Gas hydrates–A synoptic view. Jour. Geol. Sci, India; 52, 497-512.

Subroto, E.A; Alexander, R; and Kagi, R.I. (1991). 30-norhopanes: their occurrence in sediments and crude oils. Chem Geol; 93, 179-192

Sudhakar, R. (1991). Scope of Integrated Exploration Research in Krishna-Godavari, Cauvery and Andaman basins. In: Proc. Conf. Integrated Exploration Research Achievements and Prospectives. Pandey, J; and Benerjie, V (Eds). KDMIPE; Dehra Dun, p. 69-80.

Sudhakar, R; and Basu, D.N. (1973). A reappraisal of the Paleogene stratigraphy of southern Cambay Basin. ONGC Bull; 10, 55-76.

Suggate, R.A. (1959). New Zealand coals, their geochemical setting and influence on their properties. NewZealand Dept. Sci. Industry. Bull; 134,111

Sulin, V.A. (1946). Waters of oil reservoirs in the system of natural waters (in Russian). Moscow Gostoptekhizdat, 35-96.

Sunwall, M.T. and Pushkar, P. (1979). The isotopic composition of strontium in brines from petroleum fields of southern Ohio. Chem. Geol., 24, 189-197.

Tang, Y; Jendon, P.D; and Teeman, B.C. (1991). Thermogenic methane formation in low rank Coals-Published models and results from laboratory pyrolysis of lignite. In : Organic Geochemistry-Advances and applications in the natural environment. Manning, D.A.C (Ed). Manchester University press, p. 329-331.

Taylor, G.(1983). CO_2 projects to test recovery theories. AAPG Explorer, June 1, 20-21.

Tedesco, S.A. (1995). Surface geochemistry in petroleum exploration. Chapman & Hall, New York, 206 pp.

Thode, H.G. (1981). Sulphur isotopic ratios in petroleum research and exploration: Willston Basin. AAPG. Bull; 65, 1527-1535.

Thom, W.T. Jr. (1934). Present status of the carbon–ratio theory. In: Problems of petroleum geology. Rather, W.E.W and Lahee, F.H (Eds). AAPG, Tulsa, Oklahoma, 69-95.

Thomas, N.J; Uniyal, A.K; Panda, A; and Samant, A.K. (1991). Geochemistry in petroleum exploration in Krishna–Godavari and Cauvery Basins. In: Proc. Conf. Integrated Exploration Research: Achievements and Prospectives. Pandey, J; and Benerjie, V (Eds). KDMIPE; Dehra Dun, p. 483-494.

Thomas, N.J; and Sharma, V.N. (1993). Thermal evolution of source rocks in Cauvery Basin. In: Proc. 2nd semi. petroliferous basin of India. Vol. 1. Biswas, S.K. *et al*. (Eds). Indian Petroleum Pub; Dehra Dun, p. 245-254.

Thomson, T.L. (1976). Plate tectonics in oil and gas exploration of continental margins. AAPG Bull; 60, 1463-1501.

Tickell, F.G. (1921). Summary of operations. Bull. Calif. Div. Mines Geol., 6,7-16.

Ting, F.I.C. (1975). Fluorescence characteristics of thermally altered exinite (Sportenites). Fuel. 54, 2000-2005.

Tissot, B. (1969). Premieres donnees surles mechanisms *et al*., cinetique de la formation d'un schema reactionnel sur ordinatuer. Rev. Inst. Fr. Pet, 24, 470-501.

Tissot, B. (1977). The application of the results of organic geochemical studies in oil and gas exploration. In: Developments in petroleum Geology I. Hobson, G.D (Ed). Applied Science pub; London, p. 52-83.

Tissot, B. (1984). Characterization of heavy crude oils and petroleum residues. Technip, Paris.

Tissot, B; Demaison, G. J; Masson, P; Delteil, J.R. and Combaz, A (1980). Paleoenvironment and petroleum potential of middle Cretaceous black shales in Atlanta basin. AAPG Bull; 64, 2051-2063.

Tissot, B; Durand, B; Espitalie, J; and Combaz, A. (1974, a). Influence of nature and diagenesis of organic matter in formation of petroleum. Bull. AAPG; 58, 499-506.

Tissot, B; Espitalie, J; Deroo, G; Temperre, C; and Jonathan, D. (1974, b) Origin and migration of hydrocarbons in the Eastern Sahara (Algeria). In: Advances in Organic Geochemistry. Tissot, B; and Bienner, F (Eds). Technip Editions; Paris, p. 315-334.

Tissot, B. and Espitalie, J. (1975). Levolution thermique de la maticore organique des sediments: Applications d'une simulation mathematique. Rev. Inst.Fr. Pet, 30, 743-777.

Tissot, B. Pelet, R; and Ungerer, P. (1987). Thermal history of sedimentary basins, maturation indices and kinetics of oil and gas generation. AAPG Bull., 71, 1445-1466.

Tissot, B.P and Welte, D.H. (1984). Petroleum formation and occurrence. 2nd edn., Springer- Verlag, New York, 699 pp.

Turcotte, D.L; and Ahern, J.L. (1977). On the thermal and subsidence history of sedimentary basins. Jour. Geophy. Res; 82, 3762-3766.

Ungerer, P. (1990). State of the art of research in kinetic modeling of oil formation and expulsion. Org. Geochem; 16, 1-25.

Ungerer, P; Besis, F; Chenet, P.Y; Durand, B; Nagaret, E; Chiarelli, A; Oudin, J.L, and Perrian, J.F. (1984). Geological and geochemical models in oil exploration, principles and practical examples. In: Petroleum geochemistry and basin modeling. Demaison, G; and Muris, R.J (Eds). AAPG Memoir, 35, Tulsa, p 53-78.

Vassoeyvich, N.B., Korchagina, Yu K; Lopatin, N.V; and Chernyshev, V.V. (1970). Principle phases of oil formation. Moskov. Uni. Vestnik, No6, 3-27 (in Russian). English transl; Inter Nat. Geology Rev; 12, 1276-1296.

Vassoeyvich, N.B., Akramkhodzhaev, A.M and Geodekyan, N.A. (1974). Principal zone of oil formation. In: Advances in organic geochemistry, Tissot, B and Bienner, F (Eds). Technip, Paris, 309-314.

Venakatarengan, R; and Ray D. (1993). Geology and petroleum systems, Krishna-Godavari Basin. In: Proc. 2nd semi. petroliferous basins of India. Vol. I. Biswas, S.K. *et al.* (Eds). Indian Petroleum Pub; Dehra Dun, p. 331-353.

Von Gaertner, H.R; and Schmitz, H.H. (1963). Organic matter in Posidonia shale as indicate of residual oil deposits. Proc. 6th World Pet. Conf; Frankfurt, p. 355-363.

Wang, Z; Stout, S.A; and Fingas, M. (2006). Forensic finger printing of biomarkers for oil & spill characterization and source identification. Environmental Forensics, 7, 105-146.

Waples, D. (1981). Organic Geochemistry for Exploration Geologists. Burgess Pub. Co, Minnesota, USA, 151 pp.

Waples, D.W. (1979). Simple method for source rock evaluation. Bull. AAPG; 63, 239-245.

Waples, D.W. (1980). Time and temperature in petroleum exploration. Applications of Lopatin's method to petroleum exploration. AAPG. Bull., 64, 916-926.

Waples, D.W. (1984). Thermal models for oil generation. In: Advances in petroleum Geochemistry. Vol. 1, Brooks, J and Welte, D.H. (Eds). Academic press; London, p. 7-67.

Waples, D.W. (1985). Geochemistry in petroleum Exploration. Reidel Pub. Co. USA, 232 pp.

Watts, A.B; and Rayon, W.B.F. (1976). Flexure of the lithosphere and continental margin basins. Tectonophysics. 36, 25-44.

Webster, R.L. (1984). Petroleum source rocks and stratigraphy of the Bakken Formation in North Dakota. In: Hydrocarbon source rocks of the Greater Rockey Mountain Region, Woodward, J; Meissner, F.F; and Clayton, J.L. (Eds). Denver Rocky Mountain Association of geologists, 57-81.

Weitkamp, A.W. and Gutberlet, L.C. (1968). Application of a micro retort to petroleum in shale pyrolysis. Am. Chem. Soc. Div. Pet. Chem. Preprints 13, 2, F71-F85.

Wellsbury, P; and Parkes, R.J. (2000). Deep biosphere source of methane for Oceanic hydrates. In: Natural gas hydrates. Max, M.D (Ed). Kluwer Academic Pub; Boston. p. 91-104.

Welte, D.H; Hagemann, H.W; Hollerbach, A; Leythaeuser, D; and Stahl, W. (1975). Correlation between petroleum and source rock. Proc. IX World Pet. Cong; 2, 179-191.

Welte, D.H; Horsefield, B; and Baker, D.R. (Eds). (1997). Petroleum and basin evolution. Springer-Verlag, New York, 535 pp.

Welte, D.H; and Yukler, M.A. (1981). Petroleum origin and accumulation in basin evolution-a quantitative model. Bull. AAPG; 65, 1387-1396.

Whelan, J.K. (1979). C_1-C_7 hydrocarbons from IPOD Hole 397/397A. In: Initial reports of deep sea drilling project. Ryan, W.B.F. and

Ulrich von Rod (Eds). Vol.47, U.S. Govt. Printing Office, Washington, D.C.

White, D. (1915). Geology: some relations in origin between coal and petroleum. J. Wash. Acad. Sci; 5, 189-212.

White, D.E. (1957). Magmatic connate and metamorphic waters. Geol. Soc. Am. Bull; 68, 1659-1682.

Whiticar, M.J; Faber, E; and Schoell, M. (1986). Biogenic methane formation in marine and fresh water environments : CO_2 reduction vs. acetate fermentation–Isotope evidence. Geochim. Cosmochim. Acta; 50, 693-709.

Williams, J.A. (1974). Characterization of oil types in the Williston Basin. Bull. AAPG., 58, 1243-1252.

Yarullin, K.S.(1961). Characteristics of the distribution of gas and oil pools in the Cis-Uralian Trough. Dokl Akad Nauk SSSR, 141(1), 1142-1145.

Yen, T.F; and Chilingar, G.V. (Eds). (1976). Oil-shale. Development in Petroleum Science; No.5, Elsevier, Amsterdam.

Youngs, B.C. (1975). The hydrology of the Gidgealpa Formation of the western and central Cooper Basin. Rep. Invest–South Aust; Geol. Surv. 43, 1-35.

Yukler, M.A; and Welte, D.H. (1980). A 3-D Deterministic model to determine geologic history and hydrocarbon generation, migration and accumulation in a sedimentary basin. In: Fossil. Editions Technip, Paris, p 267-285.

Yukler, M.A; and Kokesh. F. (1984). Review of models used in petroleum resource estimation and organic geochemistry. In: Advances in petroleum geochemistry. Brooks, J; and Welte, D.H. (Eds). Academic press, London. p 69-113.

Zieglar, D.L; and Spotts, J.H. (1978). Reservoir and source bed history of Great Valley, California. Bull. AAPG; 62, 813-826.

Zumberge, J.E. (1987). Prediction of source rock characteristics based on terpane biomarkers in crude oils: A multi-variate statistical approach. Geochim. Cosmochim. Acta. 51, 1625-1637.

Bibliography

(a) Books

Allen, P.A; and Allen, J.R. 1990. Basin Analysis : Principles and applications, Blackwell, Oxford. 427 pp.

Beck, R.J. (ed). 2002. Worldwide Out Look Demand. Pennwell Corporation.

Bjoroy, M; (ed). 1983. Advances in organic geochemistry chichester, Wiley, N.Y.

Breger, I. A (ed). 1963. Organic geochemistry. Pergamon Press, N.Y.

Buryakovsky, L.A; and Agamaliyev, R.A. 1990. Text book on application of mathematical methods in geochemistry and hydrochemistry. Azerbaijan Institute of oil and chemistry publishing. 140 pp.

Chapman, R.E. 1973. Petroleum Geology: A concise study. Elsevier, Amsterdam, 304 pp.

Chilingarian,G.V; and Yen, T.F; (eds).1978. Bitumen, Asphalts and Tar Sands. Elsevier, Amsterdam.

Columbo, U; and Hobson, G.D; (eds). 1962. Advances in organic geochemistry. Pergamon, Amsterdam.

Cox, J.L. 1983. Natural gas hydrates : Properties, Occurrence and Recovery. Butterworth Pub, Boston.

Cubitt, J.M; and England, W.A; (ed) 1995. The geochemistry of reservoirs. Spec. Publ-No 86, Geol. Soc. London.

de Haan; (ed) 1995. New Developments in enhanced oil recovery. Spec. Publ; Geol. Soc. London. 84, 286 pp.

Duchscherer, W. 1984. Geochemical Hydrocarbon Prospecting. Pennwell books, Tulsa; Oklahoma.

Eremenkov, N.A; and Chilingarian, G.V; (ed). 1991. Petroleum Geology Hand Book. OSI Publications, Los Angels. 600 pp.

Gayer, R; and Harris, J. 1996. Coal bed methane and coal geology. Spec. Publ. Geol. Soc. London. 109 pp.

Henriet, J.P; and Mienert, J. 1998. Gas hydrates, Relevance to World Margin Stability, and Climate. Geol. Soc. London. Spec. Publ; No. 37.

Hobson, G.D; (ed). 1973. Modern Petroleum Technology. 4th edn. Applied Science Pub; London.

Kaplan, I.R; Laberg, J.S; (eds). 1974. Natural gases in marine sediments. Plenum, New York. 324 pp.

Kortsev, A.A; Tabasaranskii, Z.A; Subota, M.I; and Magilevskii, G.A. 1959. Geochemical methods of Prospecting and Exploration for Petroleum and Natural Gas. Uni. California Press, Berkely, CA.

Mc Caslin, J; (ed). 1975. International Petroleum Encyclopedia. Tulsa, Oklahoma. 480 pp.

Miller, S.L; and Wasserburg, G.J; (eds). 1964. Isotope and cosmic chemistry. Pergamon, Amsterdam.

Moody, G.B; (ed). 1961. Petroleum Exploration Hand Book. Mc Graw-Hill, New York.

Morrison, R.J; and Boyd, R.N. 1974. Organic geochemistry, Allyn and Bacyon, Inc; Rockleigh, New Jersey.

Neuman, I.B; Pazynska-lahme; and Severin, D. 1981. Composition and properties of Petroleum. Wiley, London.

Peters, K.E; Walters, C.C; and Moldowan, J.W.(2005). The biomarker guide (2nd edn) Cambridge University Press, Cambridge, UK.

Valkovic, V. 1978. Trace elements in petroleum. Pennwell, Tulsa. 216 pp.

Vassoyevich, N.B. 1986. Geochemistry of organic matter and origin of petroleum Nauka Press, Moscow. 196 pp.

Verweij, J.M. 1993. Hydrocarbon migration system analysis. Elsevier Pub. Amsterdam.

Von Gaertner, H.R; and Wehner, H. 1971. Advances in organic geochemistry. Pergamon Press.

(b) Memoirs

Ali, S.M.F.1974. Application of *in situ* methods of oil recovery to tar sands. Can. Soc. Petrol. Geol. Mem; 3, 199-211.

Barker, C; 1979. Organic geochemistry in petroleum exploration. AAPG course note series No.10.

Chandra Kuldeep; Raju, D.S.N; and Misra, P.K. (1993). Sea level changes, anoxic conditions, organic matter enrichment and petroleum source rock potential of the Cretaceous sequence of the Cauvery Basin, India. AAPG Studies in Geology # 37.

Cordel, R.J. (1972). Depth of oil generation, a review and critique. AAPG. Geol; 56, 2029-2067.

England, W.A. 1994. Secondary migration and accumulation of hydrocarbons. AAPG Mem; 60, 211-217.

Halbourty, M.J.(ed). 1970. Geology of gaint petroleum fields. AAPG Mem; 14, 8-16.

Hills, L.V.(ed). 1974. Oil sands : Fuel of the future. Can Soc. Pet. Geol. Mem; No.3, Calgary, Alberta, Canada.

Klemme, H.D; and Shabad, T. 1970. World's gaint oil and gas fields, geological factors affecting their formation and basin classification. AAPG Mem; 14, 502–555.

Law, E.E; Ulmischek, G.F; and Salvin, V.I, (eds).1998. Abnormal pressures in hydrocarbon Environments. AAPG Mem; 70, Tulsa, Oklahoma. 264 pp.

Mittal, A.D.(ed). 1980. Facts and Principles of World Petroleum occurrences. Can. Soc. Pet. Geol. Mem; Calgary, Alberta, Canada.

Rice, D.D.(ed). 1986. Oil and gas assessment. AAPG. Studies in Geology.

Walters, E.J. 1974. Review. of the World's major oil and sand deposits. Mem-Can. Soc. Pet. Geol; 3, 240-263.

Waples D.W. 1994. Modeling of sedimentary basins and systems AAPG Mem; 60, 307-322.

Welte, D.H; and Yukler, M.A. 1984. Petroleum origin and accumulation in basin evolution–a quantitative model. AAPG Mem; 35, 27-78.

Winestock, A.G. 1974. Developing a steam recovery technology. Mem. Can. Soc. Pet. Geol; 3, 190-198.

White, D.A; Garrett, R.W; March.G.R. 1975. Assessing regional oil and gas potential : Methods of estimating the volume of undiscovered oil and gas resource. AAPG studies in Geology-I; 143-159

(c) Journals

Allexan, S; Fausnaugh, J; Goudge, D; and Tedesco, S. A.(1986). The use of iodine in geochemical exploration for hydrocarbons. Assoc. Pet. Geochem.. Expl. Bull; 2, 71-93.

Barker, C. (1974). Pyrolysis techniques for source rock evaluation. AAPG Bull; 58, 2349-2361.

Barker, E.G. (1959). Origin and migration of oil. Science 129, 871-874.

Bhandari, A; Prasad, I.V.S.V; Kapoor, P.N; Meena Varshey; Madhavan, A.K.S; Pahari, S; and Singh, R.R. (2008). Depositional environment, distribution of source rock and geochemistry of oils and gases, Krishna-Godavari Basin. J. Appl. Geochem; 10, 17-32.

Datta, G.C; Sivan, P; and Singh, R.R. (2009). Applications of hetero atoms of sulphur and nitrogen in crude oils for hydrocarbon exploration. J. Appl. Geochem; 11, 290-299.

Demaison, G.J; and Moore, G.T. (1980). Anoxic environments and oil source bed genesis. Org. Geochem; 2, 9-31.

Dembicki, H (Jr); Horsfield, B; and Ho, T.T.Y. (1983). Source rock evaluation by pyrolysis gas chromatography. Bull. AAPG; 67, 1094-1103.

Dickey, P.A; and Cox, W.C. (1977). Oil and gas in reservoirs with subnormal pressures. AAPG Bull; 61, 2134-2142.

Erdman, J.G; and Morris. (1974). Geochemical correlation of petroleum. AAPG Bull; 58, 2326-2337.

Gornitz, V; and Fung, I. (1994). Potential distribution of methane hydrates in the World Oceans : Global Biogeochemical cycles; 8, 335-347.

Gupta, A.K. (2004). Marine Gas hydrates their economics and environments importance. Curr. Sci; 86, 1198-1199.

Hajra, P.N; Rudra, M; Das, S.K; and Bhattacharya, M. (2002). Predicting gas contents of Indian Gondwana coals, Bull. ONGC; 39, 61-74.

Halderson, H.H; and Damsleth, E. (1993). Challenges in reservoir characterization. AAPG Bull; 77, 541-551.

Hedberg, H.D.(1968). Significance of high-wax oils with respect to genesis of petroleum. AAPG Bull; 52, 736-750.

Heroux, Y, Chgnou, A; and Bertrant, R.(1979). Compilation and correlation of major thermal maturation indicators, Bull. AAPG, 63, 2128-2144.

Hitchon, B. (ed). 1977. Application of Geochemistry to the search of crude oil and natural gas. J. Geochem. Expl; Special issue, 7 (2), 293 pp.

Hunt, J. M. (1978). Characterization of bitumen and coals. AAPG Bull; 62, 301-303.

Hunt, J.M. (1981). Surface geochemical prospecting. Pro. and Con. AAPG Bull; 65-939.

Hunt, J.M; Philp, R.P; and Kvenvolden, K.A (2002). Early Developments in petroleum geochemistry. Org. Geochem; 33, 1025-1052.

Illich, H.A.(1983). Pristane, phytane and lower molecular weight isoprenoid distributions in oils. Bull. AAPG; 67, 385-393.

Karisiddaih, S.M; Borole, D.V; Ramalingeswara Rao, B; Paropkari, A.L; Joau, H.M; Muralidhar, K; Sarkar, G.P; Biswas G; and Narendrakumar. Studies on the pore water sulphate, chloride, and sedimentary methane to understand the sulphate reduction process in eastern Arabian sea. Curr. Sci; 91, 54-60.

Kelafant, J.R. (1999). A review of Global Coal Bed Methane Resources and Activity, Indian J. Pet. Geol; 8, 1-18.

Klemme, H.D. (1975). Gaint oil fields related to their geological setting: A possible guide to exploration. Bull. Can. Pet. Geol; 23, 30-66.

Klusman, R.W; Saeed, M.A; and Abu-Ali, M.A. (1992). The potential use of biogeochemistry in the detection of petroleum micro seepage. AAPG Bull; 76, 851-863.

Knebel G.M; and Rodiguez-Eraso. (1956). Habitate of some oil. AAPG Bull; 40, 547-561.

Kvenvolden, K.A. (1995). A review of the geochemistry of methane in natural gas hydrate. Org. Geochem; 23, 997-1008.

Leythaeuser, D; Mackenzie, A; Schaefer, R.G; and Bjoroy, M. (1984). A navel approach for recognition and quantification of hydrocarbon migration effects in shale-sand stone sequences. Bull. AAPG; 68, 196-219.

Mac Gregor, D.S.(1993). Relationship between seepage, tectonics and subsurface petroleum reserves. Mar. Pet. Geol; 10, 606-619.

Mehrotratra, N.C; Venkatachala, B.S; and Kapoor, P.N. (2010). Palynology in hydrocarbon exploration: High impact palynological studies in Western Offshore and Krishna-Godavari Basins. J. Geol. Soc. India; 75, 364-379.

Miknis, F.P; Smith, J.W; Maughan, E.K. and Maciel, G.F. (1982). Nuclear magnetic resonance: technique for direct non-destructive evaluation of source-rock potential. Bull. AAPG; 66, 1396-1401.

Miodrag, S. (1975). Should we consider geochemistry an important exploratory technique? Oil and gas Journal, 106-110.

Misra, C.S, Samanta, U; Gupta, A.K; Thomas, N.J. (1996). Hydrous pyrolysis of type III kerogen: source fractionation effects during primary migration in natural and artificially matured samples. Org. Geochem; 25, 489-505.

Murchison, D.G. (1987). Recent developments in organic petrology and organic geochemistry–review with reference to oil from coal. In : Coal and coal bearing strata, recent advances. Scott, A.C. (Ed). Geol.. Soc. London. Spec. Pub. No. 32, p. 257-302.

Narasimhachari, M.V. (1994). The determination of optical kinetic parameters from pyrolysis data. Bull. ONGC; 37, 41-58.

Ostroff, A.G. (1967). Comparison of some formation water classification systems. AAPG Bull; 51, 404.

Pepper, A.S; and Dodd, T.A. (1995). Simple kinetic models of petroleum formation. Part II. Oil and gas generation from kerogen. Mar. Pet. Geol; 12, 291-320.

Peters, J; Singh, S.K. and Punjrath, N.K. (2002). Classification of coal fields of India with respect to their coal bed methane potential. Bull ONGC; 39, 1-11.

Peters, K.E; and Fowler, M.G. (2002). Application of Petroleum Geochemistry to exploration and reservoir management. Org. Geochem; 33, 5-36.

Pophare, A.M; and Varade, A.M. (2004). Coal bed methane potential of coal seams in Sawang Colliery, East Bokaro coal fields, Jarkhand,. India J. Pet. Geol; 13, 41-53.

Prasad, I.V.S.V; Madhavan, A.K.S; Sinha, A.K; Pahari, S; and Singh, R.R. (2008). Kinetics based hydrocarbon generation and expulsion modeling. Some examples and a discussion on early mature oil window in Upper Assam. J. Appl. Geochem; 10, 41-53.

Rai S.A; Gupta, U.K; Saini, S.S; and Das, D. (1999). Coal bed methane sources of east Bokaro Basin. Indian J. Pet. Geol; 8, 54-68.

Raju, S.V; and Mathur, N. (1995). Petroleum geochemistry of a part of Upper Assam Basin, India : A brief review. Org. Geochem; 23, 55-70.

Rice, D.D; and Claypool, G.E. (1981). Generation, accumulation and resource potential of biogenic gas. Bull AAPG; 65, 5-25.

Samanta, V; Mishra, C.S; and Misra, K.N. (1994). Indian high wax crude oils and depositional environments of their source rocks, Mar. Pet. Geol; 11, 756-759.

Sehgal, S; Banerjee, V; and Chandra, K. (1991). Gas hydrates- a review. ONGC Bull; 28, 179-202.

Singh, R.R; Sinha, A.K; Bhatnagar, A.K; and Prasad, I.V.S.V.(2008). Application of reservoir geochemistry in exploration and production of oil and gas. J. Appl. Geochem; 10, 33-40.

Singh, S.K; Punjrath, N.K; and Peters, J. (1999). Coal bed methane potential of Gondwana Sediments–an Indian experience. Indian J. Pet. Geol; 8, 19-34.

Sloan, E.D. (1998). Gas hydrates: review of physical/chemical properties. Energy & Fuels; 12, 191-196.

Smith, P.V; jr. (1954). Studies on origin of petroleum : Occurrence of hydrocarbons in recent sediments. AAPG Bull; 38, 377-404.

Sofer, Z.(1984). Stable carbon isotopic compositions of crude oils : application to source depositional environments and petroleum alterations. Bull. AAPG; 68, 31-49.

Tegellar, E.W; and Noble, R.A. (1994). Kinetics of hydrocarbon generation as a function of the molecular structure of kerogen as revealed by Pyrolysis-gas chromatography. Org. Geochem; 22, 543-574.

Thompson, K.F.M. (1983). Classification and thermal history of petroleum based on light hydrocarbons. Geochim. Cosmochim. Acta; 47, 303-316.

Tissot, B. (1979). Effect on prolific petroleum source rocks and major coal deposits caused by sea level changes. Nature (London). 277, 462-465.

Waples, D.W. (2000). The kinetics of the reservoir oil destruction and gas formation : Constraints from experimental data, and from thermodynamics. Org. Geochem; 31, 553-575.

Waples, D.W; Kamat, H; and Saizu, M (1992, a). The art of maturity modeling : Part 1. Finding a satisfactory geologic model. AAPG Bull; 73, 31-46.

Waples, D.W; Kamat, H; and Saizu, M. (1992,b). Art of maturity modeling : Part 2. Alternative models and sensitivity analysis. AAPG Bull; 76, 47-66.

Welte, D.H. (1965). Relation between oil and source rock. Bull AAPG, 49, 2246-2267.

Welte, D.H. (1972). Petroleum exploration and organic geochemistry. J. Geochem. Explor; 1, 117-136.

Wenneckess, J.H.N.(1981). Tar sands. AAPG Bull; 65, 2290-2293.

While, D.A; and Gehman, H M. (1979). Methods of estimating oil and gas resources. AAPG Bull; 63, 2183-2203.

Young, A; Monaghan, P.H; and Schweisberger. (1977, a). Calculation of ages of hydro carbons in oils : Physical chemistry applied to petroleum geochemistry. Part 1. AAPG Bull; 61, 573-600.

Young, A; Monagham, P.H; and Schweisbegar, (1977, b). Distribution of hydrocarbons between oils and associated fine-grained sediments : Application of physical chemistry to petroleum geochemistry. Part 2. AAPG Bull; 61, 1407-1436.

(d) Proceedings of World Congress and International Seminars/Symposia

Abrams, M.A; and Segall, M.P.(2001). Best Practices for Detecting, Identifying and Characterising near surface migration of hydrocarbons with marine sediments. OTC 13039, Offshore Technology Conference, Huston, Texas.

Bally, A.W. (1975). A geodynamic scenario for hydrocarbon occurrences. 9th World Pet. Cong. Proc; 2, 33-44.

Datta, G.C; Dwivedi, S.P; Gupta, V; Banerjie, V; and Kumar, R.K. (1987). Stable carbon isotopic studies in crude oils of Cambay Basin. Proc. 1st Intnat. Conf. Petroleum Geochemistry and Exploration Afro-Asian Region. Kumar, R.K. (ed). Bulkema, Rotterdam, p. 483-486.

Evans, C.R; and Staplin, F.L.(1971). Regional facies of organic metamorphism in geochemical exploration. Proc. 3rd. Intnat. Symp. Geochem. Expl. Special vol. 2, 517-520. Canadian Inst. Min. Mellurgy.

Hajra, P.N; Choudhury, A.T; Biswas, D; Sen, P; and Prasad, A. (2003). Critical assessment of CBM Potential of India. Proc. PETROTECH, New Delhi, p. 277-282.

Halbourty, M.T; and Moody J.D. (1979). World ultimate reserves of crude oil. 10th World Pet. Cong. Proc; 2, 291-301.

Johri, A; Hajra, P.N and Sah, A.K. (2003). Coal bed methane development in India–challenges and opportunities. Proc. PETROTECH, New Delhi; p. 345-352.

Jokhan Ram. (2008). Neoproterozoic successions in peninsular India and their hydrocarbon prospectivity. Proc. Intnat. Conf. Geology and hydrocarbon potential of the Neoproterozoic –Cambrian

Basins in India, Pakistan and Middle East. Univ. Jammu, India. p.36-56.

Kuldeep Chanda; Srivastava, D.C; Sharma, R; and Pal, M. (1994). Hydrocarbon Prospects and Status of exploration in Gondwana Basins of India. 9th Intnet. Symp. Hyderabad, India, p. 1177-1198.

Kuldeep Chandra; Singh, R.P; and Julka, A.C. (1998). Gas hydrate Potential of Indian Offshore area. Proc. 2nd conf. Exp. Pet. Geophys. SPG –98; Chennai, p. 357-368.

Lijmbach, G.M.G. (1975). On the origin of petroleum. Proc. 9th World Pet. Cong. Vol. 2, Applied Science Pub; London, p. 357-369.

Momper, J.A.(1975). Time and temperature relations affecting the origin, expulsion and preservation of oil and gas. Proc. 9th World Pet. Cong. Geology; 2, Applied Sci. Pub. London.

*Prasad, I.V.S.V; Sinha, A.K; Thomas, N.J; Misra, K.N; and Kuldeep Chandra.(2000). Correlation between maceral composition and kinetic. Parameters of Barail Coals from Upper Assam Basin, India. Proc. 5th Intnat. Conf. & Exhibitor Petroleum geochemistry and exploration the Afro –Asian Region Garg, A.K. *et al.* (eds). B R. Pub Corp. Delhi, Inida, 433-439.

Prasad, I.V.S.V; Sinha, A.K; Pahari, S; Singh, R.R; and Bhandari, A. (2006). Geochemical feed back : a novel style for calibrated burrial and thermal reconstruction in Mahim Grabin area– Mumbai Offshore Basin. Proc. 3rd Intnat. Conf & Exhibition. APG, Goa, India.

Philippi, G. T. (1957). Identification of oil source beds by chemical means. Proc. 20th Int. Nat. Geol. Cong; Sec 3, p. 25-38.

Rao, K.L.N. (1997). Resource assessment of coal beds in the Northern Cambay Basin, Gujarat, India. Intnat. CBM Symp; Paper No. 9774, Tuscaloosa, Alabama.

Sethi, A.K; Sethi, A.V; and Ramana, M.V. (2004). Potential natural gas hydrate resource in Indian Offshore area. AAPG. Hedberg Conf; Vancouver, Canada. p. 1-6.

Singh, R.R; Saxena, J.G; Sahota, S.K; and Chandra, K. (1987). On the use of iodine as an indicator of petroleum in Indian basins. Proc. 1st Intnat. Conf. Afra-Asian Region on Petroleum Geochemistry and Exploration. p. 105-107.

Sitapati Rao, P.V.S.K; Garg, V; Kumar, A; and Sah, A.K. (2003). Natural gas hydrates–An Indian perspective. Proc. PETROTECH, New Delhi, p. 389-396.

Smith, J.E.(1991). Migration accumulation and Retention of petroleum in the earth. 8th World Cong. Proc; 2, 13-26.

Tissot, B; Deroo, G; and Espitalie, J. (1975). Comparative study of the duration of formation and expulsion of oil in several geological provinces. Proc. 9th World Pet. Cong. Vol.2 Applied Science Pub; London, p.159-169.

Weisman, R. (1980). Development of geochemistry and its contribution to hydrocarbon exploration. Proc. 10th World Pet. Cong; Heydon Publishers, London. 369–385.

Author Index

Subject Index